Topics in Applied Physics   Volume 48

Topics in Applied Physics Founded by Helmut K. V. Lotsch

Optical Information Processing

Fundamentals

Edited by S. H. Lee

With Contributions by
D. P. Casasent   J. W. Goodman   G. R. Knight
S. H. Lee   W. T. Rhodes   A. A. Sawchuk

With 197 Figures

Springer-Verlag Berlin Heidelberg New York 1981

Professor Dr. *Sing H. Lee*

Department of Applied Physics and Information Sciences
University of California, San Diego
La Jolla, CA 92093, USA

ISBN 3-540-10522-0 Springer-Verlag Berlin Heidelberg New York
ISBN 0-387-10522-0 Springer-Verlag New York Heidelberg Berlin

Library of Congress Cataloging in Publication Data. Main entry under title: Optical information processing. (Topics in applied physics; v. 48) Bibliography: p. Includes index. 1. Optical data processing. I. Lee, S.H.(Sing H.), 1939–. II. Casasent, David Paul. III. Series. TA1630.O642   621.36'7   81-8988   AACR2

© by Springer-Verlag Berlin Heidelberg 1981
Printed in Germany

Monophoto typesetting, offset printing and bookbinding: Brühlsche Universitätsdruckerei, Giessen
2153/3130-543210

Preface

Optical information processing is an advancing field which has received much attention since the nineteen sixties. It involves processing a two-dimensional array of information using light. The attractive feature of optical processing is the parallel processing capability, which offers great potentials in processing capacity and speed. Optical processing is especially useful if the information to be processed is in optical form. When good electrical to optical interface devices are available, optical processing will also be useful for electrical information.

To enter and then participate in new developments in the advancing field of optical processing, it will be necessary to be familiar with its fundamental principles. Basic knowledge about realtime interface devices and hybrid electronic/optical system will prove to be very useful to comprehend much of the current developmental efforts in the field. This volume covers these important basics, although attempts are made to keep these discussions as concise as possible so that sufficient space remains available to discuss subjects of current research interest such as space variant and nonlinear processing. Therefore, this volume can serve as a good text for new graduate students or undergraduates with advanced standing, and as a good reference for those researchers who are already working in certain sub-areas of the field but wish to be more familiar with other sub-areas. Applications are included in such a manner as to illustrate the usefulness of certain processing principles. For more extensive coverage, the companion Topics in Applied Physics Vol. 23 is highly recommended.

June 1981   *Sing H. Lee*

Contents

Contributors

Casasent, David P.
> Department of Electrical Engineering, Carnegie-Mellon University, Pittsburgh, PA 15213, USA

Goodman, Joseph W.
> Department of Electrical Engineering, Stanford University, Stanford, CA 94305, USA

Knight, Gordon R.
> Optimem, 150 Charcot Avenue, San Jose, CA 95131, USA

Lee, Sing H.
> Department of Electrical Engineering and Computer Sciences, University of California, San Diego, La Jolla, CA 92093, USA

Rhodes, William T.
> School of Electrical Engineering, Georgia Institute of Technology, Atlanta, GA 30332, USA

Sawchuk, Alexander A.
> Image Processing Institute, Department of Electrical Engineering, University of Southern California, Los Angeles, CA 90007, USA

1. Basic Principles

S. H. Lee

With 28 Figures

1.1 Historical Overview

To comprehend the development in the advancing field of optical information processing, we shall begin this chapter with a brief summary of its history. Main discussions in the remaining chapters are the simple, scalar diffraction theory of light, the important Fourier transform and imaging properties of lenses. Since the Fourier transform properties require coherent light and the imaging properties are affected by the degree of coherency of the light used, the subject of coherence is also considered. This is followed by discussions on the usefulness of the transfer function concept in further understanding the basic properties of diffraction and lens, and by discussions on several topics of practical interest concerning these basic properties.

Historically, optical processing dates from 1859, when *Foucault* first described the knife-edge test in which the direct image light was removed and the scattered or diffracted light was kept [1.1]. In 1873, *Abbe* advanced a theory in which diffraction plays an important role in coherent image formation [1.2]. In 1906, *Porter* demonstrated *Abbe*'s theory experimentally [1.3]. *Zernike* developed the Nobel prize-winning concepts of phase contrast microscopy in 1935 [1.4]. In 1946 *Duffieux* published his important study on the use of the Fourier integral in optical problems [1.5]. In the fifties, *Elias* and co-workers provided the initial exchange between the disciplines of optics and communication theory [1.6, 7]. Later *O'Neill* contributed a great deal to reconciling the two viewpoints by presenting a unified theory [1.8]. *Maréchal* motivated future expansion of interest in the optical processing field by successfully applying coherent spatial-filtering techniques to improve the quality of photographs [1.9].

In the 1960's, optical processing activities reached a new height with its successful application to synthetic-aperture radar [1.10, 11]. The inventions of the holographic spatial filter by *Vander Lugt* [1.12] and of the computer generated spatial filter by *Lohmann* and *Brown* [1.13] also form the important cornerstones of applying optical processing to the lucrative field of pattern recognition. Much research has also been carried out in developing realtime interface devices to be used between electronic or incoherent optic and coherent optic systems (Chap. 4).

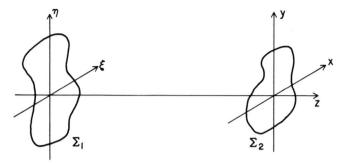

Fig. 1.1. Diffraction

In the seventies, the importance of combining electronic digital computers with optical analog processors to form hybrid processors was established (Chap. 5). Much attention has also been given to extending the flexibility of optical processors beyond the linear, space invariant regime (Chaps. 6, 7).

1.2 Diffraction Phenomena

In optical processing, the information to be processed is often obtained by illuminating a photographic transparency with a laser beam or modulating a coherent wave front by an optical interface device, or from a self-illuminating object. In any case, to design optical systems for processing the information it is important to understand the basic physical phenomena of diffraction.

The term *diffraction* has been conveniently defined by *Sommerfeld* [1.14] as "any deviation of light rays from rectilinear paths which cannot be interpreted as reflection or refraction". The diffraction of monochromatic light (of wavelength λ) by a finite aperture Σ_1 in an infinite opaque screen, as indicated in Fig. 1.1, is described mathematically by the Huygens-Fresnel superposition integral

$$u_2(x, y) = \iint_{\Sigma_1} h(x, y; \xi, \eta) u_1(\xi, \eta) d\xi d\eta, \tag{1.1}$$

where $u_1(\xi, \eta)$ and $u_2(x, y)$ are the field amplitudes at points (ξ, η) and (x, y), respectively, and $h(x, y; \xi, \eta)$ is the impulse response

$$h(x, y; \xi, \eta) = (1/j\lambda r) \exp(jkr) \cos(\mathbf{n}, \mathbf{r}). \tag{1.2}$$

In (1.2), k is $(2\pi/\lambda)$ and the angle $(\mathbf{n}, \mathbf{r})$ is that between vectors $\mathbf{n}$ and $\mathbf{r}$ [$\mathbf{n}$ being normal to the plane containing aperture Σ_1]. The limit associated with the superposition integral can be extended to infinity, if it is understood that $u_1(\xi, \eta)$ is identical to zero outside the aperture Σ_1.

Since the common situations of interest in optical processing problems involves the distance z between aperture and observation planes greater than the maximum linear dimensions of the aperture Σ_1 and the observation region Σ_2, the obliquity factor $\cos(\boldsymbol{n}, \boldsymbol{r})$ can readily be approximated by

$$\cos(\boldsymbol{n}, \boldsymbol{r}) \simeq 1, \tag{1.3a}$$

where the accuracy is to within 5 % if the angle $(\boldsymbol{n}, \boldsymbol{r})$ does not exceed 18°. Under similar conditions, r in the denominator of (1.2) will not differ significantly from z, allowing the impulse response function $h(x, y; \xi, \eta)$ to be approximated as

$$h(x, y; \xi, \eta) \simeq (1/j\lambda z) \exp(jkr). \tag{1.3b}$$

a) The Fresnel Diffraction

Further simplification suggested by Fresnel was to apply a binomial expansion to the square root associated with r in the exponent of (1.3b):

$$r = [z^2 + (x - \xi)^2 + (y - \eta)^2]^{1/2}$$
$$\simeq z[1 + (x - \xi)^2/2z^2 + (y - \eta)^2/2z^2]. \tag{1.3c}$$

Substituting the approximations expressed in (1.3a–c) into the superposition integral of (1.1), the following convolution relationship is obtained:

$$u_2(x, y) = (1/j\lambda z) \exp(jkz) \iint u_1(\xi, \eta) \exp\{j(k/2z)[(x - \xi)^2 + (y - \eta)^2]\} d\xi d\eta$$
$$= u_1(x, y) * h(x, y), \tag{1.4a}$$

where

$$h(x, y) = (1/j\lambda z) \exp(jkz) \exp[(jk/2z)(x^2 + y^2)]$$

and the symbol $*$ stands for convolution.

Alternatively, the quadratic terms in the exponent may be expanded to yield

$$u_2(x, y) = (1/j\lambda z) \exp[jkz + jk(x^2 + y^2)/2z] \iint u_1(\xi, \eta) \exp[jk(\xi^2 + \eta^2)/2z]$$
$$\cdot \exp[-j2\pi(x\xi + y\eta)/\lambda z] d\xi d\eta. \tag{1.4b}$$

Thus, aside from multiplicative amplitude and phase factors of $(1/j\lambda z) \exp[jkz + jk(x^2 + y^2)/2z]$ which are independent of (ξ, η), the function $u_2(x, y)$ may be found from a Fourier transform of $u_1(\xi, \eta) \exp[(jk/2z)(\xi^2 + \eta^2)]$, where the transform must be evaluated at spatial frequencies $(v_x = x/\lambda z, v_y = y/\lambda z)$ to assure the correct space scaling in the observation plane. Figures

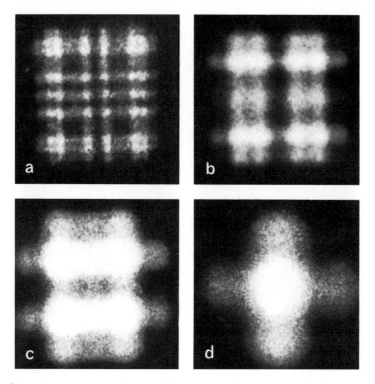

Fig. 1.2a–d. Some typical Fresnel diffraction patterns associated with a rectangular aperture, Σ_1. In (a) Σ_2 is closest to the aperture and the remaining photographs were taken at increasing distances from the aperture [Ref. 1.22, p. 30]

1.2, 3 show some typical Fresnel diffraction patterns associated with rectangular and circular apertures respectively.

The convolution nature of (1.4a) suggests that perhaps some additional insight can be gained by examining the Fresnel diffraction in the spatial frequency domain. Hence, the Fourier transform of (1.4a) is taken to give

$$\hat{u}_2(x/\lambda z, y/\lambda z) = \hat{u}_1(x/\lambda z, y/\lambda z)\hat{h}(x/\lambda z, y/\lambda z), \qquad (1.4c)$$

where

$$\hat{h}(x/\lambda z, y/\lambda z) = \exp(jkz)\exp\{-j\pi\lambda z[(x/\lambda z)^2 + (y/\lambda z)^2]\}.$$

The effect of propagating a distance z in the Fresnel diffraction region therefore consists of two parts: the first exponential factor represents an overall phase retardation experienced by any component of the angular spectrum, and the second exponential factor represents a phase dispersion with quadratic spatial frequency dependence.

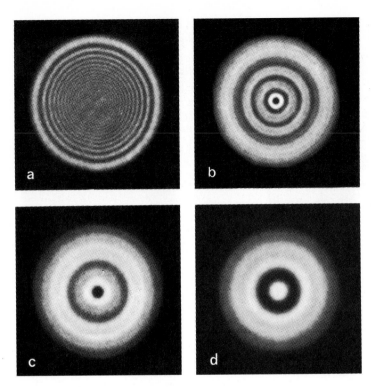

Fig. 1.3a–d. Some typical Fresnel diffraction patterns associated with a circular aperture, Σ_1. In (a) Σ_2 is closest to the aperture and the remaining photographs were taken at increasing distances from the aperture [Ref. 1.22, p. 30]

b) The Fraunhofer Diffraction

Diffraction pattern calculations can be further simplified if restrictions more stringent than those used in the Fresnel approximation are adopted. If the stronger (Fraunhofer) assumption

$$z \gg (k/2)(\xi^2 + \eta^2)_{max}$$

is adopted, the quadratic phase factor $\exp[jk(\xi^2 + \eta^2)/2z]$ inside the convolution integral of (1.4b) is approximately unity over the entire aperture, and the observed field distribution can be found directly from a Fourier transform of the aperture distribution itself. Thus in the region of Fraunhofer diffraction,

$$u_2(x, y) = (1/j\lambda z) \exp\{jkz[1 + (x^2 + y^2)/2z^2]\}$$
$$\cdot \iint u_1(\xi, \eta) \exp[-j2\pi(x\xi + y\eta)/\lambda z] \, d\xi \, d\eta . \tag{1.5}$$

Figure 1.4 shows photographs of the Fraunhofer diffraction patterns of rectangular and circular apertures.

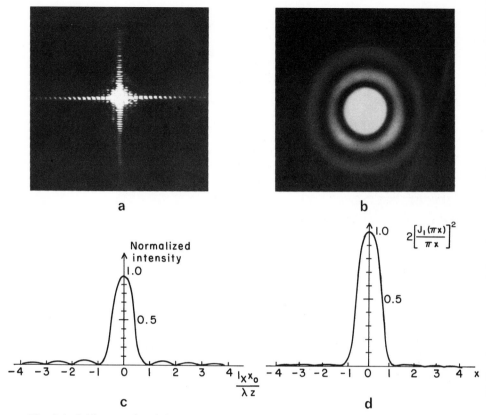

Fig. 1.4a–d. Photographs of the Fraunhofer diffraction patterns of uniformly illuminated (**a**) rectangular aperture, (**b**) circular aperture, (**c**) cross section of (**a**), and (**d**) cross section of (**b**) [Ref. 1.15, Figs. 4-2–5]

1.3 Fourier Transform Properties of Ideal Thin Lens

One of the most important components of optical information processing and optical imaging systems are lenses. A lens is composed of optically dense material, such as glass or fused quartz, in which the propagation velocity of an optical field is less than the velocity in air. A lens is said to be a *thin* lens if a ray entering at coordinates (x, y) on one face emerges at approximately the same coordinates on the opposite face, i.e., if there is negligible translation of the ray within the lens. Thus, a thin lens simply delays an incident wavefront by an amount proportional to the thickness of the lens at each point.

It can be shown that a plane wave passing through a thin lens experiences a phase delay factor of $t_l(x, y)$:

$$t_l(x, y) = \exp[jkn\Delta_0] \exp[-jk(x^2 + y^2)/2F], \tag{1.6a}$$

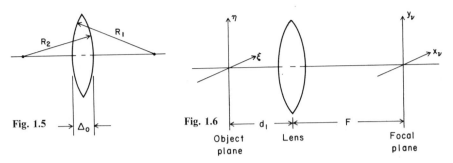

Fig. 1.5 Δ_0

Fig. 1.6   Object
plane   Lens   Focal
plane

d_1 F

Fig. 1.5. Thin lens. As rays travel from left to right, each convex surface encountered is taken to have a *positive* radius of curvature R_1, while each concave surface is taken to have a negative radius of curvature (i.e., R_2 is negative)

Fig. 1.6. Fourier transforming with object placed in front of lens

where n is the index of refraction of the lens material, F is the focal length defined by

$$1/F = (n-1)(1/R_1 - 1/R_2), \tag{1.6b}$$

and Δ_0, R_1 and R_2 are shown in Fig. 1.5. Strictly speaking (1.6a) is valid only under the paraxial approximation [1.15, Sect. 5.1; 1.16, Sect. 5.1], i.e., only the portions of the wavefront that lie near the lens axis experience the phase delay of (1.6a). The first term $\exp(jkn\Delta_0)$ is simply a constant phase delay, while the second term $\exp[-jk(x^2 + y^2)/2F]$ may be recognized as a quadratic phase factor associated with a spherical wave.

Next, let us see how a simple converging lens can be used to perform complex analog Fourier transformations in two dimensions. (The Fourier transform operation generally requires either complex and expensive electronic spectrum analyzers or a long computing time with a digital electronic computer; yet it can be performed with extreme simplicity in a coherent optical system.)

1.3.1 Object Placed Against the Lens

Recalling the two results that (a) the quadratic phase factor inside the convolution integral of (1.4b) is approximately unity in the Fraunhofer diffraction region providing the Fourier transform of the aperture distribution as the observed field distribution (see Sect. 1.2), and that (b) a thin lens introduces the phase delay factor of (1.6a) (see the earlier part of this section), we would like to show that the observed field distribution $u_v(x_v, y_v)$ at the back focal plane of a lens is Fourier transform related to the transmittance of an object $t_0(\xi, \eta)$ placed against the lens [Fig. 1.6, $d_1 = 0$].

Assuming a monochromatic plane wave which illuminates the object transmittance $t_o(\xi, \eta)$ has an amplitude c_1, the light amplitude behind the lens will be

$$u(\xi, \eta) = c_1 \exp(jkn\Delta_0) t_o(\xi, \eta) \exp[-jk(\xi^2 + \eta^2)/2F].$$ (1.7)

The observed field $u_v(x_v, y_v)$ at the back focal plane can then be obtained by substituting (1.7) into (1.4b) and setting z equal to F:

$$u_v(x_v, y_v) = (1/j\lambda F) \exp(jkF) \exp[jk(x_v^2 + y_v^2)/2F] \int\int c_1 \exp(jkn\Delta_0)$$
$$\cdot t_o(\xi, \eta) \exp[-j2\pi(x_v\xi + y_v\eta)/\lambda F] d\xi d\eta$$
$$= c_2 \exp[jk(x_v^2 + y_v^2)/2F] \hat{t}_o(x_v, y_v),$$ (1.8)

where $\hat{t}_o(x_v, y_v)$ is the Fourier transform of $t_o(\xi, \eta)$,

$$\hat{t}_o(x_v, y_v) = \int\int t_o(\xi, \eta) \exp[-j2\pi(x_v\xi + y_v\eta)/\lambda F] d\xi d\eta,$$

and c_2 is a constant independent of (x_v, y_v),

$$c_2 = c_1(1/j\lambda F) \exp[jk(F + n\Delta_0)].$$

Hence, $u_v(x_v, y_v)$ and $t_o(\xi, \eta)$ are Fourier transform related, although the relation is not an exact one due to the presence of the quadratic phase factor $\exp[jk(x_v^2 + y_v^2)/2F]$ that preceeds the integral. The phase factor contributes a simple phase curvature to the object spectrum. In most cases it is the intensity across the back focal plane that is of real interest. Measurement of the intensity distribution, $u_v(x_v, y_v) u_v^*(x_v, y_v)$, yields knowledge of the power spectrum of the object; the phase distribution has no importance in such a measurement.

1.3.2 Object Placed in Front of the Lens

To obtain an exact Fourier transform relationship between $t_o(\xi, \eta)$ and $u_v(x_v, y_v)$, let us study the effect of moving the object away from the lens to some distance d_1 in front of it (Fig. 1.6). This effect can be readily analyzed by first representing $t_o(\xi, \eta)$ as a superposition of plane waves,

$$t_o(\xi, \eta) = \int\int \hat{t}_o(v_x, v_y) \exp[j2\pi(v_x\xi + v_y\eta)] dv_x dv_y,$$ (1.9)

where $v_x = x_v/\lambda F$ and $v_y = y_v/\lambda F$, then considering the effect of propagation through a distance d_1 on each of the plane waves $\exp[j2\pi(v_x\xi + v_y\eta)]$. From (1.4c) we have

$$\hat{h}(v_x, v_y) = \exp(jkd_1) \exp[-j\pi\lambda d_1(v_x^2 + v_y^2)].$$ (1.10)

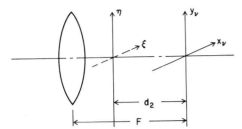

Fig. 1.7. Fourier transforming with object placed behind the lens

The angular spectrum of the field in front of the lens is, therefore,

$$\hat{t}_1(v_x, v_y) = \hat{t}_0(v_x, v_y) \exp(jkd_1) \exp[-j\pi\lambda d_1(v_x^2 + v_y^2)]. \tag{1.11}$$

Substituting $\hat{t}_1(v_x, v_y)$ from (1.11) into $\hat{t}_0(x_v, y_v)$ of (1.8), we obtain the field at the back focal plane for the configuration of Fig. 1.6 as

$$
\begin{aligned}
u_v(x_v, y_v) &= c_2 \exp(jkd_1) \exp[jk(x_v^2 + y_v^2)/2F] \\
&\quad \cdot \exp[-j\pi\lambda d_1(v_x^2 + v_y^2)] \hat{t}_0(v_x, v_y) \\
&= c_3 \exp[jk(1 - d_1/F)(x_v^2 + y_v^2)/2F] \hat{t}_0(v_x, v_y),
\end{aligned} \tag{1.12}
$$

where $c_3 = c_2 \exp(jkd_1)$. Thus the amplitude and phase of light at (x_v, y_v) are again related to the amplitude and phase of the object spectrum at frequencies $(x_v/\lambda F, y_v/\lambda F)$, except for the phase factor $\exp[jk(1 - d_1/F)(x_v^2 + y_v^2)/2F]$ which preceeds $\hat{t}_0(v_x, v_y)$. This phase factor vanishes for the special case $d_1 = F$. In other words, *when the object is placed in the front focal plane of the lens, we can expect to find its exact Fourier transform in the back focal plane.*

1.3.3 Object Placed Behind the Lens

In optical pattern recognition the need for correlating input patterns of unknown size with a reference pattern of a fixed size is frequently encountered. One common method for performing this kind of correlation is to carry out a search for a match between the sizes of Fourier transforms of the input and the reference patterns because the scale of Fourier transform of the input pattern can be varied by placing it behind the lens at various distances away from the back focal plane (Fig. 1.7).

To verify the scaling properties on the input Fourier transform by the optical configuration shown in Fig. 1.7, let us utilize the geometrical-optics approximation to obtain an expression for the field illuminating the object $t_0(\xi, \eta)$ as

$$u_1(\xi, \eta) = (c_1 F/d_2) \exp[-jk(\xi^2 + \eta^2)/2d_2]. \tag{1.13}$$

The amplitude of $u_1(\xi, \eta)$ is equal to c_1 when the input pattern is placed right behind the lens, but it increases because of the lens focusing effect as the input is moved towards the back focal plane (i.e., d_2 is reduced). The phase factor of $u_1(\xi, \eta)$ is that of a spherical wave whose radius of curvature is F behind the lens and reduces to d_2 between the lens and the back focal plane.

If (1.13) is accepted to be the field illuminating the object, the field behind the object is then $u_1(\xi, \eta)t_o(\xi, \eta)$. This field will diffract according to the Fresnel diffraction equation (1.4b) over the distance d_2. Hence, (1.4b) can be used to show that the field in the back focal plane is

$$u_v(x_v, y_v) = (1/j\lambda d_2)\exp(jkd_2)\exp[jk(x_v^2 + y_v^2)/2d_2]\int\int (c_1 F/d_2)t_o(\xi, \eta)$$
$$\cdot \exp[-j2\pi(x_v\xi + y_v\eta)/\lambda d_2]d\xi d\eta$$
$$= c_4 \exp[jk(x_v^2 + y_v^2)/2d_2]$$
$$\cdot \int\int t_o(\xi, \eta)\exp[-j2\pi(x_v\xi + y_v\eta)/\lambda d_2]d\xi d\eta, \tag{1.14}$$

where $c_4 = (1/j\lambda d_2)(c_1 F/d_2)\exp(jkd_2)$. Aside from the quadratic phase factor $\exp[jk(x_v^2 + y_v^2)/2d_2]$, the field on the back focal plane $u_v(x_v, y_v)$ is the Fourier transform of that portion of the object $t_o(\xi, \eta)$ illuminated by $u_1(\xi, \eta)$. The size of the Fourier transform is variable by moving the input between the lens and the back focal plane. By decreasing d_2, the size of the transform is made smaller because $v_x = x_v/\lambda d_2$, $v_y = y_v/\lambda d_2$. The phase factor $\exp[jk(x_v^2 + y_v^2)/2d_2]$ has no importance in the intensity measurement of $u_v(x_v, y_v)u_v^*(x_v, y_v)$.

1.3.4 Some Properties of the Fourier Transform

Since lenses can be used to perform Fourier transform operations, it will naturally be useful to become more familiar with important properties of the Fourier transform, which include at least the following few (more detailed discussions can be found in [1.17, 18]):

1) If $g(x)$ and $\hat{g}(v_x)$ are a Fourier transform pair, then it can be easily proved that the following three pairs of functions are also Fourier transform related: $g(-x)$ and $\hat{g}(-v_x)$, $g^*(x)$ and $\hat{g}^*(-v_x)$, $g^*(-x)$ and $\hat{g}^*(v_x)$.

2) The transform of a sum of two functions is simply the sum of their individual transforms:

$$\mathscr{F}\{ag_1(x) + bg_2(x)\} = a\hat{g}_1(v_x) + b\hat{g}_2(v_x).$$

This is called the linearity theorem.

3) Translation of an object pattern in the space domain introduces a *linear* phase shift in the frequency domain:

$$\mathscr{F}\{g(x-a, y-b)\} = \hat{g}(v_x, v_y)\exp[-j2\pi(v_x a + v_y b)].$$

This is called the shift theorem. The linear phase shift factor disappears when the intensity of the Fourier spectrum is detected. Hence, the modulus of the Fourier transform of an object function is invariant with respect to a shift of its origin.

4) An enlarged object pattern causes a reduced, but brighter Fourier spectrum:

$$\mathscr{F}\{g(ax, by)\} = \frac{1}{|ab|} \hat{g}(v_x/a, v_y/b).$$

This is called the similarity theorem. (To avoid this scaling effect in optical pattern recognition applications, the Mellin transform can be employed because the modulus of the Mellin transform of an object function is invariant with respect to a magnification of the object function in the same way as the modulus of the Fourier transform of an object function is invariant with respect to a shift of origin [Ref. 1.19, pp. 23–25].)

5) The total amount of energy measurable in the Fourier plane must be the same as that passing through the object plane (i.e., energy must be conserved)

$$\iint |g(x, y)|^2 dx dy = \iint |\hat{g}(v_x, v_y)|^2 dv_x dv_y.$$

6) The convolution of two functions in the space domain is entirely equivalent to the more simple operation of multiplying their individual transforms

$$\mathscr{F}\{\iint g(\xi, \eta) h(x-\xi, y-\eta) d\xi d\eta\} = \mathscr{F}\{g(x, y) * h(x, y)\} = \hat{g}(v_x, v_y) \hat{h}(v_x, v_y).$$

The convolution operation frequently arises in the theory of linear systems.

7) The Fourier transform of the autocorrelation function of $g(x, y)$ is given by its power spectrum:

$$\mathscr{F}\{\iint g(\xi, \eta) g^*(\xi-x, \eta-y) d\xi d\eta\} = |\hat{g}(v_x, v_y)|^2.$$

This is called the autocorrelation theorem.

1.3.5 Some Useful Fourier Transform Pairs and Their Physical Significance

Table 1.1 lists some useful Fourier transform pairs. Comments on their significances follow below.

Table 1.1. Some useful Fourier transform pairs and their physical significances

Functions	Transforms	Comments (see below)
$\delta(x, y)$	1	1
$\delta(x-a, y-b)$	$\exp[-j2\pi(v_x a + v_y b)]$	2
$\exp[-\pi(x^2 + y^2)]$	$\exp[-\pi(v_x^2 + v_y^2)]$	3
$\cos(2\pi v_1 x)$	$\frac{1}{2}[\delta(v_x - v_1) + \delta(v_x + v_1)]$	4
$\text{rect}(x)\text{rect}(y)$, where $\text{rect}(x) = \begin{cases} 1 & \text{for } \|x\| \le \frac{1}{2} \\ 0 & \text{otherwise} \end{cases}$	$\text{sinc}(v_x)\text{sinc}(v_y)$ where $\text{sinc}(v_x) = \dfrac{\sin(\pi v_x)}{\pi v_x}$	5
$\Lambda(x)\Lambda(y)$, where $\Lambda(x) = \begin{cases} (1 - \|x\|) & \text{for } \|x\| \le 1 \\ 0 & \text{otherwise} \end{cases}$	$\text{sinc}^2(v_x)\text{sinc}^2(v_y)$	6
$\text{circ}(r) = \begin{cases} 1 & \text{for } r \le 1 \\ 0 & \text{otherwise} \end{cases}$	$\dfrac{J_1(2\pi\varrho)}{\varrho}$	7
$\text{sgn}(x)\text{sgn}(y)$, where $\text{sgn}(x) = \begin{cases} 1 & x > 0 \\ 0 & x = 0 \\ -1 & x < 0 \end{cases}$	$\left(\dfrac{1}{j\pi v_x}\right)\left(\dfrac{1}{j\pi v_y}\right)$	8
$\text{comb}\left(\dfrac{x}{a}\right)\text{comb}\left(\dfrac{y}{b}\right)$, where $\text{comb}\left(\dfrac{x}{a}\right) = \sum\limits_{n=\infty}^{\infty} \delta\left(\dfrac{x-n}{a}\right)$	$ab\,\text{comb}(av_x)\text{comb}(bv_y)$	9
$\partial[g(x, y)]/\partial x$	$j2\pi v_x \hat{g}(v_x, v_y)$	10

Comment 1: A point light source at the optical axis is transformed into an on-axis plane wave.

Comment 2: An off-axis point source of light is transformed into a plane wave propagating off-axis.

Comment 3: A beam of gaussian amplitude profile is transformed into another gaussian beam.

Comment 4: A sinusoidal grating has a spectrum of two frequency components at $\pm v_1$. This transform pair relationship can be readily applied to determine the Fourier spectrum of any object pattern, if the object pattern is considered to be composed of many sinusoidal gratings of various frequencies and amplitudes.

Comment 5: A rectangular aperture uniformly illuminated is transformed into a sinc function (Fig. 1.4a).

Comment 6: An aperture with a triangularly tappered transmittance function is transformed into a sinc² function. This relationship can be used to introduce apodization effects by reducing the lens transmissivity near the edges [1.16, pp. 252–253]. Compared with the transform of a rect function, that of a triangular function will have reduced intensity in the outer rings or side-lobes. One application is to enable a dim object to be imaged even though it may be near a bright one.

Comment 7: A circular aperture uniformly illuminated provides the $J_1(2\pi\varrho)/\varrho$ Fourier-Bessel transform (Fig. 1.4b).

Comment 8: A π-phase step function is transformed into a function whose modulus is inversely proportional to the spectrum coordinate. This is an important transform relationship for designing an optical system to perform a Hilbert transform [Ref. 1.20, pp. 151–153].

Comment 9: The comb function consists of an array of delta functions spaced at intervals of width a in the x direction and width b in the y direction. It is an important function in the sampling theory. The transform of a comb function is another comb function, except that the spacings between the delta functions in the transform plane are inversely related to those in the object plane.

Comment 10: This is the derivative theorem which is frequently found to be useful in image enhancement work. The derivative operation helps to eliminate or greatly reduce the background level, thereby the contrast of image signal is increased.

1.4 Imaging Properties of Lenses

1.4.1 Imaging with Coherent Light

It was seen in Sect. 1.3.2 that a single lens can produce a Fourier transform of a light field in its front focal plane. This property can be used to perform two Fourier transforms with two lenses to return to the original field distributions as shown in Fig. 1.8. The operation of the two-lens imaging system is as follows: the input function $t_o(\xi, \eta)$ is illuminated by a plane wave to give the field $u_v(x_v, y_v)$ in the back focal plane of L_1 according to (1.12):

$$u_v(x_v, y_v) = (c_3'/F_1)\hat{t}_o(x_v/\lambda F_1, y_v/\lambda F_1),\tag{1.15}$$

where $c_3'/F_1 = c_3$. The next lens, L_2, picks up $u_v(x_v, y_v)$ and transforms it to yield $u(x', y')$ in the image plane

$$u(x', y') = \frac{F_1}{F_2} t_o(-F_1 x'/F_2, -F_1 y'/F_2)$$

$$= \frac{1}{M} t_o(-x'/M, -y'/M),\tag{1.16}$$

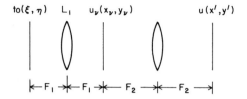

Fig. 1.8. A two-lens imaging system

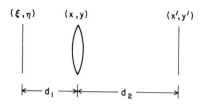

Fig. 1.9. A single-lens imaging system

where M is the magnification factor (F_2/F_1). The negative signs in (1.16) can be interpreted as being due to the inability of a lens to perform an inverse transformation – only direct Fourier transforms.

To carry out simple imaging operations, it is not necessary, however, to have two lenses. The following discussion is intended to establish the fact that single-lens can also be used for imaging (Fig. 1.9).

For an optical system to produce high-quality images, the impulse response should closely approximate

$$h(\xi, \eta; x', y') \simeq c_5 \delta(x' \pm M\xi, y' \pm M\eta), \tag{1.17}$$

where c_5 is a complex constant, and the plus or minus signs are included to allow for possible image inversion. To prove that (1.17) holds, let us consider an object point $\delta(\xi, \eta)$. The spherical wave diverging from $\delta(\xi, \eta)$ can be expressed as

$$h(\xi, \eta; x, y) \simeq (1/j\lambda d_1) \exp\{j(k/2d_1)[(x - \xi)^2 + (y - \eta)^2]\} \tag{1.18a}$$

under the paraxial approximation. After passage through the lens, the field becomes

$$h(\xi, \eta; x, y) \simeq (1/j\lambda d_1) \exp\{j(k/2d_1)$$
$$\cdot [(x - \xi)^2 + (y - \eta)^2]\} \exp[-j(k/2F)(x^2 + y^2)]. \tag{1.18b}$$

Finally, using the Fresnel diffraction equation (1.4a) to account for propagation over the distance d_2, we have

$$h(\xi, \eta; x', y') = (1/j\lambda d_2) \exp(jkd_2) \iint (1/j\lambda d_1)$$
$$\cdot \exp\{(jk/2d_1)[(x-\xi)^2 + (y-\eta)^2]\}$$
$$\cdot \exp[-j(k/2F)(x^2 + y^2)]$$
$$\cdot \exp\{(jk/2d_2)[(x'-x)^2 + (y'-y)^2]\}dxdy$$
$$= c_5 \iint \exp\left[j(k/2)\left(\frac{1}{d_1} + \frac{1}{d_2} - \frac{1}{F}\right)(x^2 + y^2)\right]$$
$$\cdot \exp\left\{-jk\left[\left(\frac{\xi}{d_1} + \frac{x'}{d_2}\right)x + \left(\frac{\eta}{d_1} + \frac{y'}{d_2}\right)y\right]\right\}dxdy, \qquad (1.19)$$

where

$$c_5 = -(1/\lambda^2 d_1 d_2) \exp(jkd_2)$$
$$\cdot \exp[j(k/2d_1)(\xi^2 + \eta^2)] \exp[j(k/2d_2)(x'^2 + y'^2)].$$

When the lens law from elementary geometrical optics is satisfied, the object and image distances are related by

$$(1/d_1) + (1/d_2) - (1/F) = 0, \qquad (1.20)$$

and (1.19) is simplified to

$$h(\xi, \eta; x', y') = c_5 \iint \exp\{-j(2\pi/\lambda d_2)[(x' + M\xi)x + (y' + M\eta)y]\}dxdy$$
$$= c_6 \delta(x' + M\xi, y' + M\eta), \qquad (1.21)$$

where $c_6 = \lambda^2 d_2^2 c_5$ and $M = d_2/d_1$. M is a positive number when real images are formed. Thus, the impulse response of an imaging system due to an object point at (ξ, η) is an image point at $(x' = -M\xi, y' = -M\eta)$. In view of the linearity of wave propagation phenomenon, we can in all cases express the image field $u(x', y')$ by the following superposition integral:

$$u(x', y') = \iint h(\xi, \eta; x', y')t_o(\xi, \eta)d\xi d\eta$$
$$= (c_7/M)t_o(-x'/M, -y'/M), \qquad (1.22)$$

where

$$c_7 = \exp(jkd_2) \exp[j(k/2d_1)(\xi^2 + \eta^2)] \exp[j(k/2d_2)(x'^2 + y'^2)].$$

However, the image formed by a two-lens system differs from that formed by a single-lens system. The differences can be found in the complex constant c_7, which contains the quadratic phase factors $\exp[j(k/2d_1)(\xi^2 + \eta^2)]$ and $\exp[j(k/2d_2)(x'^2 + y'^2)]$ dependent on coordinates (ξ, η) and (x', y'), respectively.

These quadratic factors are simply indicative of the phase curvature over the (input) object and (output) image planes. To eliminate the factor $\exp[j(k/2d_2)(x'^2+y'^2)]$, note that in the vast majority of cases of interest, the light distribution behind the lens will be the end product of the imaging operation to be detected directly by eyes or photographic films which respond only to light intensity. To eliminate the factor $\exp[j(k/2d_1)(\xi^2+\eta^2)]$, we further need to adopt the reasonable assumption that the light distribution at image point (x', y') is primarily contributed from a small region of the object space centered on its ideal geometrical object point. Within that small region, the argument of $\exp[j(k/2d_1)(\xi^2+\eta^2)]$ changes no more than a fraction of a radian; we may then use the approximation

$$\exp[j(k/2d_1)(\xi^2+\eta^2)] \simeq \exp[j(k/2d_1)(x'^2+y'^2)/M^2].\tag{1.23}$$

Since the (ξ, η) dependence can be removed by relating it to (x', y') dependence, the phase factor $\exp[j(k/2d_1)(\xi^2+\eta^2)]$ may now be dropped by again noting that it will not affect the intensity measurements in the (x', y') plane.

1.4.2 Imaging with Partial Coherent Light

In our discussion on imaging up to here, monochromatic plane wave illumination of the object has been assumed. This assumption is overly restrictive, for the illumination generated by most sources (including multimode lasers) is never very monochromatic and the illumination is seldom plane (unless it comes from an infinitely small source). In fact, the amplitude and phase of the illumination fluctuate randomly with time if the source is nonmonochromatic and extended in size. The mutual coherence function is a convenient means of describing such illumination. The mutual coherence function is defined in terms of the cross correlation of the complex fields at two points in space at different times [1.16, Sect. 8.5]:

$$\Gamma(x_1', x_2'; \tau) = \Gamma_{12}(\tau) = \langle u(x_1', t+\tau)u^*(x_2', t)\rangle,\tag{1.24}$$

where u is the complex scalar light field, x_1' and x_2' are the points at which u is measured, the subscript in Γ_{12} is used in place of the arguments x_1' and x_2', τ is the time difference between measurements and the angular brackets denote the infinite time average. For simplicity, we show only the x' coordinate, but in general the y' coordinate would also appear[1].

---

1 To measure the mutual coherence function, an interference experiment between the two light fields $u(x_1', t+\tau)$ and $u(x_2', t)$ needs to be performed. When the interference fringes have sharp contrast, the mutual coherence is high. When no interference fringes occur, the two fields are mutually incoherent. Partial coherence corresponds to the intermediate situations where the fringe contrast takes on intermediate values. The mutual coherence function contains two major components: temporal coherence $\langle u(x_1', t+\tau)u(x_1', t)\rangle$ is measured by conducting the Michelson interference experiment, the spatial coherence $\langle u(x_1', t)u(x_2', t)\rangle$, by a double slit experiment [1.22].

It can be shown [Ref. 1.21, Chap. 3] that the mutual coherence function will satisfy the wave equation

$$\nabla_s^2 \Gamma_{12}(\tau) = \frac{1}{C^2} \frac{\delta^2 \Gamma_{12}(\tau)}{\delta \tau^2}, \tag{1.25}$$

where $s = 1, 2$. That is, there are two equations, one for each value of s representing the two coordinates x_1' and x_2'. To show that $\Gamma_{12}(\tau)$ satisfies (1.25), (1.24) is used and differentiation with respect to x_1' and x_2' performed. The order of the differential and the time integral is then reversed.

If $\Gamma_{12}(\tau)$ satisfies the wave equation, it should also satisfy the imaging equation that was derived from the diffraction formula. The imaging equation for a single-lens system (1.22) was derived in terms of a single wavelength. To fit $\Gamma_{12}(\tau)$ into the same form, we consider the Fourier components of the temporal function. Let $\Gamma_{12}(\tau)$ and $\hat{\Gamma}_{12}(\nu)$ be the temporal transform pair. We can follow each temporal frequency component of the mutual coherence function through the imaging system and then recombine the components. If we use $\hat{\Gamma}_{12}(\nu)$ in the diffraction formulas, we find [Ref. 1.21, Chap. 7]

$$\hat{\Gamma}_{12}(\nu) = \hat{\Gamma}(x_1', x_2'; \nu) = \iint h(\xi_1, x_1'; \nu) h^*(\xi_2, x_2'; \nu) \hat{\Gamma}(\xi_1, \xi_2, \nu) d\xi_1 d\xi_2, \tag{1.26}$$

where ξ is the coordinate in the object plane, x' is that in the image plane, h is the impulse response of the imaging system, and the subscripts refer to different points along the same coordinate, not the two coordinates of a plane. To follow the derivation of (1.26), (1.24) must be used. The expression in τ is obtained from (1.26) by taking an inverse transform

$$\Gamma_{12}(\tau) = \Gamma(x_1', x_2'; \tau) = \iiint h(x_1' - \xi_1, \nu) h^*(x_2' - \xi_2, \nu) \hat{\Gamma}(\xi_1, \xi_2, \nu)$$
$$\cdot \exp(j2\pi\nu\tau) d\nu d\xi_1 d\xi_2, \tag{1.27}$$

where spatial stationarity has been assumed. This expression can be simplified if the quasimonochromatic approximation of $\Delta\nu \ll \bar{\nu}$ is made, where $\Delta\nu$ is the spectral width and $\bar{\nu}$ is the mean temporal frequency. Under the quasi-monochromatic approximation,

$$\Gamma_{12}(\tau) \simeq \exp(j2\pi\bar{\nu}\tau) \int \hat{\Gamma}_{12}(\nu) d\nu = \exp(j2\pi\bar{\nu}\tau) \Gamma_{12}(0), \tag{1.28}$$

$$\hat{\Gamma}_{12}(\nu) = \Gamma_{12}(0)\delta(\nu - \bar{\nu}). \tag{1.29}$$

The use of (1.29) in (1.27) gives

$$\Gamma_{12}(\tau) = \exp(j2\pi\bar{\nu}\tau) \iint h(x_1' - \xi_1, \bar{\nu}) h^*(x_2' - \xi_2, \bar{\nu}) \Gamma(\xi_1, \xi_2, 0) d\xi_1 d\xi_2. \tag{1.30}$$

Consequently, *we can find the mutual coherence in the image plane from the mutual coherence function in the object plane.*

It would be interesting to see how (1.30) reduces to the previously obtained expressions for the special case of coherent imaging. In the case of coherent illumination, the mutual coherence function is

$$\Gamma(\xi_1,\xi_2,0)=t_o(\xi_1)t_o^*(\xi_2).\tag{1.31}$$

Equation (1.30) then becomes

$$\Gamma_{12}(0)=\iint h(x_1'-\xi_1)h^*(x_2'-\xi_2)t_o(\xi_1)t_o^*(\xi_2)d\xi_1 d\xi_2,\tag{1.32}$$

where the explicit variable $\bar{v}$ has been dropped. The intensity in the image plane is equal to the mutual intensity function for the image, so that $I_i(x_1')=\Gamma_{11}(0)$ and

$$I_i(x_1')=\iint h(x_1'-\xi_1)h^*(x_1'-\xi_2)t_o(\xi_1)t_o^*(\xi_2)d\xi_1 d\xi_2.\tag{1.33}$$

Because the integrals are separable,

$$I_i(x_1')=|\int h(x_1'-\xi_1)t_o(\xi_1)d\xi_1|^2=|u(x_1')|^2,\tag{1.34}$$

which is to be compared with (1.22). *Thus, the coherent imaging process is linear in amplitudes.*

1.4.3 Imaging with Incoherent Light

Another important special case of partial coherent imaging (1.30) is incoherent imaging. In the case of (spatially) incoherent illumination, the field amplitudes across the object vary in statistically independent fashions

$$\Gamma(\xi_1,\xi_2,0)=\langle t_o(\xi_1,t+\tau)t_o^*(\xi_2,t)\rangle=I_o(\xi_1)\delta(\xi_1-\xi_2),\tag{1.35}$$

since the correlation of the complex wave amplitudes is zero for every case except autocorrelation [Ref. 1.15, p. 109] and [Ref. 1.16, p. 252]. The use of (1.35) in (1.30) yields

$$\Gamma_{12}(0)=\iint h(x_1'-\xi_1)h^*(x_2'-\xi_2)I_o(\xi_1)\delta(\xi_1-\xi_2)d\xi_1 d\xi_2$$

$$=\int h(x_1'-\xi_1)h^*(x_2'-\xi_1)I_o(\xi_1)d\xi_1.\tag{1.36}$$

The image intensity $I_i(x')$ is then

$$I_i(x_1')=\Gamma_{11}(0)=\int|h(x_1'-\xi_1)|^2 I_o(\xi_1)d\xi_1,\tag{1.37a}$$

or in its two-dimensional form

$$I_i(x', y') = \iint |h(x' - \xi, y' - \eta)|^2 I_o(\xi, \eta) \, d\xi \, d\eta . \qquad (1.37b)$$

Hence, *an incoherently illuminated imaging system is linear in intensity.* The image intensity is found as a convolution of $I_o(\xi, \eta)$ with an impulse response $|h|^2$, which is the square modulus of the impulse response obtained with coherent illumination.

1.5 Effects of Lens Pupil Function

To this point of our lens operation discussion we have entirely neglected, for simplicity, the lens pupil function $P(x, y)$, which takes into account the finite extent of the lens aperture and any phase errors introduced by the lens. The finite extent of the lens will give rise to pupil functions which are zero outside the lens and unity within the lens aperture:

$$P(x, y) = \text{rect}(x/l) \, \text{rect}(y/l) \qquad \text{for a square aperture} \qquad (1.38a)$$

$$P(x, y) = \text{circ}[(x^2 + y^2)^{1/2}/(D/2)] \qquad \text{for a circular aperture} . \qquad (1.38b)$$

Roughness on lens surfaces will introduce an undesirable phase factor

$$P(x, y) = \exp[jkW_r(x, y)] \, \text{rect}(x/l) \, \text{rect}(y/l), \qquad (1.38c)$$

where $W_r(x, y)$ is a random function with the statistical properties of $\bar{W}_r = 0$ and $\bar{W}_r^2 = \sigma^2$ (the bars indicate a spatial average). A defocussed lens or a lens with aberrations will also introduce undesirable phase factors [1.15, 23]

$$P(x, y) = \exp[jW_{d,a}(x, y)] \, \text{rect}(x/l) \, \text{rect}(y/l), \qquad (1.38d)$$

where

$$W_d(x, y) = \left(\frac{1}{d_1} + \frac{1}{d_2} - \frac{1}{F} \right)(x^2 + y^2)/2 \qquad \text{for defocussing}$$

and

$$W_a(x, y) = W_{40}(x^2 + y^2)^2 + W_{60}(x^2 + y^2)^3 \qquad \text{for spherical aberration;} \quad (1.38e)$$

W_{40} and W_{60} are proportionality constants.

In this section we shall first discuss the effects of $P(x, y)$ on the Fourier transform operation, then its effects on the imaging operations.

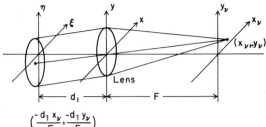

Fig. 1.10. Vignetting of the object (effect of the lens aperture on Fourier transform) [Ref. 1.15, p. 87] and [Ref. 1.16, p. 114]

1.5.1 Effects of $P(x, y)$ on Fourier Transform Operations

With reference to Fig. 1.10, the finite size of the lens restricts the region of $t_o(\xi, \eta)$ which can contribute to the transform at (x_v, y_v). The geometrical-optics approximation may be used to determine the extent of this restriction, and is reasonably accurate if the object distance d_1 is sufficiently short so that diffraction effects occurring between the object plane and the lens entrance pupil are small. With this approximation the restricted region can be found to be the size of the lens aperture on the (ξ, η) plane centered at $(-d_1 x_v/F, -d_1 y_v/F)$. The location of the center can be found at the intercept of the line joining the coordinates (x_v, y_v) with the center of the lens and the object plane [Ref. 1.15, pp. 87–88; 1.16, pp. 112–114]. Thus,

$$u_v(x_v, y_v) = c_3 \exp[(jk/2F)(1 - d_1/F)(x_v^2 + y_v^2)]$$
$$\cdot \iint t_o(\xi, \eta) P(\xi + d_1 x_v/F, \eta + d_1 y_v/F)$$
$$\cdot \exp[-j2\pi(x_v\xi + y_v\eta)/\lambda F] d\xi d\eta . \tag{1.39}$$

The limitation of the effective object by the finite lens aperture is known as a *vignetting* effect. Note that this vignetting effect is minimized when the object is placed close to the lens and when the lens aperture is larger than the object. In practice, in order to eliminate vignetting it is often preferred to place the object directly against the lens, although in analysis it is generally convenient to place the object in the front focal plane where the Fourier transform relation is exact.

1.5.2 Effects of $P(x, y)$ on Coherent Imaging

To include the effects of $P(x, y)$, one can follow through the analysis of Sect. 1.4.1 to show that (1.21) should be modified to

$$h(\xi, \eta; x', y') = c_5 \iint P(x, y) \exp\{-j(2\pi/\lambda d_2)$$
$$\cdot [(x' - M\xi)x + (y' - M\eta)y]\} dx dy , \tag{1.40}$$

when the coordinates of the image plane is redefined such that the image of an object point at (ξ, η) centers at $(x' = M\xi, y' = M\eta)$. With the changes of variables

$$\tilde{x} = x/\lambda d_2, \; \tilde{y} = y/\lambda d_2, \; \xi' = M\xi, \; \eta' = M\eta, \tag{1.41}$$

we obtain

$$h(\xi, \eta; x', y') = h(x' - \xi', y' - \eta')$$
$$= c_6 \iint P(\lambda d_2 \tilde{x}, \lambda d_2 \tilde{y})$$
$$\cdot \exp\{-j2\pi[(x' - \xi')\tilde{x} + (y' - \eta')\tilde{y}]\} d\tilde{x} d\tilde{y}, \tag{1.42}$$

where $c_6 = \lambda^2 d_2^2 c_5$. Thus, the impulse response of a coherent imaging system (due to an object point) will no longer be an image point, but the diffraction pattern of the pupil function [Ref. 1.15, Sect. 6.1] and [Ref. 1.16, pp. 119–120].

With the effect of $P(x, y)$ included, the superposition integral of (1.22) in terms of the new variables of (1.41) is

$$u(x', y') = \iint h(x' - \xi, y' - \eta)t_o(\xi, \eta)d\xi d\eta$$
$$= \iint h'(x' - \xi', y' - \eta')t'_o(\xi', \eta')d\xi' d\eta', \tag{1.43}$$

where $h' = (1/c_6)h$ and $t'_o(\xi', \eta') = (c_6/M^2)t_o(\xi, \eta)$. t'_o is a replica of the object t_o, except in brightness (because of the $1/M^2$ factor) and scale (because $\xi' = M\xi$); i.e., t'_o is the image predicted by geometrical-optics. Thus, the image, $u(x', y')$ of (1.43), may be regarded as a convolution of the image predicted by geometrical-optics with an impulse response that is determined by the pupil function of the system. Since the impulse response h' has a nonzero width, the convolution operation yields an actual image smoother than the object. This smoothing operation can strongly attenuate the fine detail of the object, with a corresponding loss of the image fidelity resulting.

1.5.3 Effects of $P(x, y)$ on Incoherent Imaging

Incoherent imaging has been shown in Sect. 1.4.3 to produce an intensity which is the convolution of the object intensity with an impulse response $|h|^2$;

$$I_i(x', y') = \iint |h(x' - \xi, y' - \eta)|^2 I_o(\xi, \eta)d\xi d\eta \tag{1.37b}$$
$$= c_8 \iint |h'(x' - \xi', y' - \eta')|^2 |t'_o(\xi', \eta')|^2 d\xi' d\eta'. \tag{1.37c}$$

Since $P(x, y)$ has broadening effects on the impulse response $h(x' - \xi', y' - \eta')$ as described by (1.42), it will have similar (but not identical) effects on the impulse response $|h(x' - \xi', y' - \eta')|^2$. In the following sections we shall further examine the difference between coherent and incoherent imaging by studying the spatial frequency responses of the optical imaging systems.

1.6 Frequency Responses of Optical Imaging Systems

1.6.1 Frequency Response of a Coherent Imaging System

The discussions in Sects. 1.4.1, 1.5.2 have yielded a space-invariant form of amplitude mapping:

$$u(x', y') = \iint h'(x' - \xi', y' - \eta') t_0'(\xi', \eta') d\xi' d\eta' . \tag{1.43}$$

It is space invariant because of the $(x' - \xi')$ and $(y' - \eta')$ dependencies of h'. It is also linear. Therefore, the transfer function concepts widely used in studying linear electrical systems can be directly applied to studying coherent imaging. To do so, let us define the following frequency spectra of the system input and output, respectively:

$$\hat{t}_0'(v_x, v_y) = \iint t_0'(\xi', \eta') \exp[-j2\pi(v_x\xi' + v_y\eta')] \, d\xi' d\eta'$$
$$\hat{u}(v_x, v_y) = \iint u(x', y') \exp[-j2\pi(v_x x' + v_y y')] dx' dy' , \tag{1.44}$$

where $v_x = x/\lambda d_2$ when the image spatial frequency spectrum is studied. In addition, we define the transfer function as the Fourier transform of the space-invariant impulse response,

$$\hat{h}(v_x, v_y) = \iint h'(x', y') \exp[-j2\pi(v_x x' + v_y y')] dx' dy' . \tag{1.45}$$

Now, applying the convolution theorem to (1.43) we have

$$\hat{u}(v_x, v_y) = \hat{h}(v_x, v_y) \hat{t}_0'(v_x, v_y) . \tag{1.46}$$

Thus, the effects of the coherent imaging system have been expressed in the frequency domain. $\hat{h}(v_x, v_y)$ in (1.46) will be referred to as the *coherent transfer function* (CTF) and is related to the pupil function through the Fourier transform relations (1.45) and (1.42).

$$\hat{h}(v_x, v_y) = \mathscr{F}\{\mathscr{F}[P(\lambda d_2 \tilde{x}, \lambda d\tilde{y})]\} = P(-\lambda d_2 v_x, -\lambda d_2 v_y) . \tag{1.47}$$

Equation (1.47) indicates that the coherent transfer function is proportional to the reflected pupil function, as implied by the minus signs in the argument of P. In the simple cases where P is either unity or zero, a finite passband in the coherent transfer function exists within which the coherent imaging system passes all frequency components without amplitude and phase distortion [Ref. 1.15, Sect. 6.2] and [Ref. 1.16, Sect. 8.1]. At the boundary of this passband, the frequency response drops to zero, implying that frequency components outside the passband are completely attenuated (Fig. 1.11). Mathematically, for the

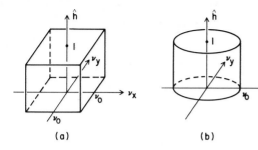

Fig. 1.11a, b. The coherent transfer functions of perfect lenses with (**a**) square (**b**) circular apertures [Ref. 1.15, p. 112]

simple square (width l) and circular (diameter D) apertures, $P(x, y)$ are, respectively,

$$P(x, y) = \text{rect}(x/l)\,\text{rect}(y/l)$$
$$P(x, y) = \text{circ}[(x^2 + y^2)^{1/2}/(D/2)], \tag{1.48}$$

and the corresponding $\hat{h}(v_x, v_y)$ are

$$\hat{h}(v_x, v_y) = \text{rect}\left(\frac{\lambda d_2 v_x}{l}\right)\text{rect}\left(\frac{\lambda d_2 v_y}{l}\right),$$
$$\hat{h}(v_x, v_y) = \text{circ}[(v_x^2 + v_y^2)^{1/2}/(D/2\lambda d_2)]. \tag{1.49}$$

The cutoff frequencies in terms of the image spatial frequency are, respectively,

$$v_0 = l/2\lambda d_2 \quad \text{and} \quad \varrho_0 = D/2\lambda d_2, \tag{1.50a}$$

and in terms of the corresponding object spatial frequency [Ref. 1.16, p. 221],

$$v_0 = l/2\lambda d_1 \quad \text{and} \quad \varrho_0 = D/2\lambda d_1. \tag{1.50b}$$

Equation (1.50a) and (1.50b) are related by the magnification factor $M(=d_2/d_1)$ as expected. (Physically the spatial frequency cutoff of a lens can be described as occurring when the plane wave pair representing a certain spatial frequency no longer overlaps to form fringes in the image plane [Ref. 1.16, Fig. 8.3].)

1.6.2 Frequency Response of an Incoherent Imaging System

The discussions in Sects. 1.4.3, 1.5.3 have yielded the linear mapping relationship in intensity

$$I_i(x', y') = c_7 \iint |h'(x' - \xi', y' - \eta')|^2 I_o(\xi', \eta')\,d\xi'd\eta', \tag{1.37c}$$

where $|h'|^2$ is the square modulus of the impulse response obtained with coherent illumination. To frequency-analyze an incoherent imaging system, let us define the normalized frequency spectra of $\hat{I}_o$ and $\hat{I}_i$ as

$$\hat{I}_o(v_x, v_y) = \frac{\iint I_o(\xi', \eta') \exp[-j2\pi(v_x \xi' + v_y \eta')] d\xi' d\eta'}{\iint I_o(\xi', \eta') d\xi' d\eta'}, \tag{1.51a}$$

$$\hat{I}_i(v_x, v_y) = \frac{\iint I_i(x', y') \exp[-j2\pi(v_x x' + v_y y')] dx' dy'}{\iint I_i(x', y') dx' dy'}. \tag{1.51b}$$

The spectra are normalized by their "d–c components" or constant background values partly for mathematical convenience and partly for the fact that the visual quality of an image is to a large extent dependent on the "contrast", or the relative intensities of the information bearing portions of the image and the ever-present background. If the transfer function of the incoherent imaging is similarly normalized,

$$\mathscr{H}(v_x, v_y) = \frac{\iint |h'(x', y')|^2 \exp[-j2\pi(v_x x' + v_y y')] dx' dy'}{\iint |h'(x', y')|^2 dx' dy'}. \tag{1.52}$$

Use of the convolution theorem to (1.37c) then yields

$$\hat{I}_i(v_x, v_y) = \mathscr{H}(v_x, v_y) \hat{I}_o(v_x, v_y). \tag{1.53}$$

Equation (1.53) indicates that the spatial frequency spectrum of the image intensity can be obtained by applying the complex weighting factor $\mathscr{H}(v_x, v_y)$ to the spatial frequency spectrum of the object intensity. The function $\mathscr{H}(v_x, v_y)$ is commonly known as the *optical transfer function* (OTF) and the modulus $|\mathscr{H}|$ as the modulation transfer function (MTF) [Ref. 1.15, Sect. 6.3] and [Ref. 1.16, pp. 240–244]. In measurements of the performance of the optical systems it is the MTF that is usually measured. This avoids the problems of determining the phase of OTF. (We shall see that if the lens has aberrations, OTF can have a negative sign in some regions.)

OTF is related to CTF (the coherent transfer function) through (1.52). Application of the convolution theorem to (1.52) gives

$$\mathscr{H}(v_x, v_y) = \frac{\hat{h}(v_x, v_y) * \hat{h}^*(v_x, v_y)}{\hat{h}(v_x, v_y) * \hat{h}^*(v_x, v_y)\Big|_{\substack{v_x = 0 \\ v_y = 0}}}, \tag{1.54}$$

where $\hat{h}(v_x, v_y)$ is the CTF of (1.45) and is related to the pupil function through (1.47).

Generally the OTF posseses the following properties:

(a) $\mathscr{H}(0, 0) = 1$, \hfill (1.55a)

(b) $\mathscr{H}(-v_x, -v_y) = \mathscr{H}^*(v_x, v_y)$, \hfill (1.55b)

(c) $|\mathscr{H}(v_x, v_y)| \leq |\mathscr{H}(0, 0)|$. \hfill (1.55c)

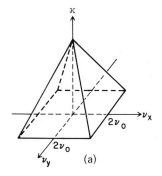

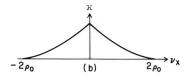

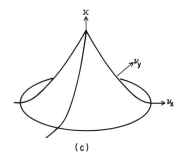

Fig. 1.12. (a) Incoherent transfer function for a rectangular lens [Ref. 1.16, p. 242]. **(b, c)** Incoherent transfer function of a circular lens. **(b)** Cross section in the $\mathscr{H} - v_x$ plane. **(c)** Three-dimensional sketch of $\mathscr{H}$ as a function of v_x and v_y [Ref. 1.16, p. 245]

Property (a) follows directly by substitution of $(v_x = 0, v_y = 0)$ in (1.54). Property (b) can be proven by noting that the Fourier transform of a real function is Hermitian. The proof that the MTF at any spatial frequency is always less than its zero frequency value of unity requires the use of Schwarz' inequality. These proofs will be left to the reader as exercises or one may look up the solution [Ref. 1.15, pp. 115–116]. To overcome the restrictive MTF property of (c), a considerable amount of research effort has been devoted lately. Advances made are summarily discussed in Sect. 3.3. We proceed here to provide several simple examples of calculating the OTF.

The OTFs for the aberration free-lens systems of simple rectangular or circular apertures can be found from combining (1.54) and (1.49) to be, respectively,

$$\mathscr{H}(v_x, v_y) = \Lambda(v_x/2v_0)\Lambda(v_y/2v_0),\tag{1.56a}$$

$$\mathscr{H}(\varrho) = \begin{cases} \frac{2}{\pi}\cos^{-1}(\varrho/2\varrho_0) - \frac{\varrho}{\pi\varrho_0}\left[1 - \left(\frac{\varrho}{2\varrho_0}\right)^2\right]^{1/2}, & \varrho \leq 2\varrho_0 \\ 0, & \text{otherwise} \end{cases}\tag{1.56b}$$

where v_0 and ϱ_0 are given by (1.50a). These OTFs are shown in Fig. 1.12.

The OTF for a lens system either defocused or with aberration can be found from combining (1.54), (1.47), and (1.38d, e). Specifically, the OTF for a defocused lens system of a rectangular aperture (width l) is

$$\mathscr{H}(v_x, v_y) = \Lambda\left(\frac{v_x}{2v_0}\right)\Lambda\left(\frac{v_y}{2v_0}\right)\text{sinc}\left[\frac{8\varepsilon}{\lambda}\left(\frac{v_x}{2v_0}\right)\left(1 - \frac{|v_x|}{2v_0}\right)\right]$$

$$\cdot\text{sinc}\left[\frac{8\varepsilon}{\lambda}\left(\frac{v_y}{2v_0}\right)\left(1 - \frac{|v_y|}{2v_0}\right)\right],\tag{1.57}$$

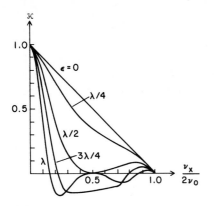

Fig. 1.13. Cross section of the OTF for a focusing error and a square pupil [Ref. 1.15, p. 125]

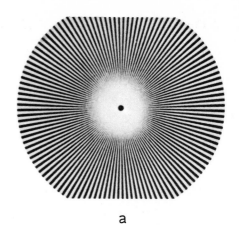

a

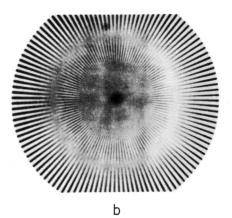

b

Fig. 1.14. (a) Focused and **(b)** misfocused ▶ images of a spoke target [Ref. 1.15, p. 126]

where

$$\varepsilon = \left(\frac{1}{d_1} + \frac{1}{d_2} - \frac{1}{F} \right) \frac{l^2}{8}.$$

A cross section of this OTF is shown in Fig. 1.13 and photographs illustrating the effects of defocusing on a spoke target image are shown in Fig. 1.14.

1.6.3 Comparison of Coherent and Incoherent Imaging

There are several ways to compare imaging systems. The selection of the best imaging system is usually a function of the criterion used (e. g., image frequency content, two point resolution) and the object (intensity and phase) being imaged [Ref. 1.15, Sect. 6.5] and [Ref. 1.16, Sect. 8.4]. In addition, subjective factors such as the reaction of a human viewer may be required.

a) Comparison Based on Image Frequency Contents

A good example that illustrates the intrinsic differences between images formed with coherent and incoherent illuminations can be found in considering two test objects with the same intensity transmittance but different phase distributions. Let the intensity transmittance of the test objects be, in both cases,

$$\tau(\xi, \eta) = \cos^2(2\pi\tilde{v}\xi), \tag{1.58}$$

whereas their amplitude transmittances are, respectively,

$$t_a(\xi, \eta) = \cos 2\pi\tilde{v}\xi, \quad \tilde{v} = 1000 \text{ cycles/cm}, \tag{1.59a}$$

$$t_b(\xi, \eta) = |\cos 2\pi\tilde{v}\xi|. \tag{1.59b}$$

The test objects are illuminated by light of wavelength 10^{-4} cm and at 20 cm away from the 5 cm diameter imaging lens of focal length 10 cm. Thus, the CTF has a cut-off frequency of ϱ_0,

$$\varrho_0 = D/2\lambda d_1 = 1250 \text{ cycles/cm}, $$

and the OTF cut-off frequency is 2500 cycles/cm. From Sects. 1.6.1, 1.6.2 we can also see that the spectrum of the image intensity obtained with coherent illumination is

$$\mathcal{F}\{I_c\} = \mathcal{F}\{(h' * t_o')(h' * t_o')^*\} = (\hat{h}\hat{t}_o) * (\hat{h}\hat{t}_o)^* \tag{1.60}$$

and the spectrum of the image intensity with incoherent illumination is

$$\mathcal{F}\{I_i\} = \mathcal{F}\{h'h'^* * t_o't_o'^*\} = (\hat{h} * \hat{h}^*)(\hat{t}_o * \hat{t}_o^*). \tag{1.61}$$

The use of (1.60) and (1.61) in analyzing the imaging of the object $t_a(\xi, \eta)$ is illustrated in Fig. 1.15a, b. Note that the contrast of the image intensity distribution is poorer for the incoherent case than for the coherent case. Thus coherent imaging must be said to be better for this particular object.

The corresponding comparison for object t_b of (1.59b) and τ of (1.58) is shown in Fig. 1.15b, c. The object amplitude is now periodic with fundamental frequency $2\tilde{v}$. But, since $2\tilde{v} > \varrho_0$, no variations of image intensity will be present for the coherent case while the incoherent system forms the same image it did for object t_a. Thus, the incoherent illumination must be considered better in this second case.

In summary, from the point of view of image spectral contents whether coherent imaging is better than incoherent imaging, depends much on the detailed structure of the object, and in particular on its phase distribution. It is not possible to conclude that one type of illumination is to be preferred in all cases.

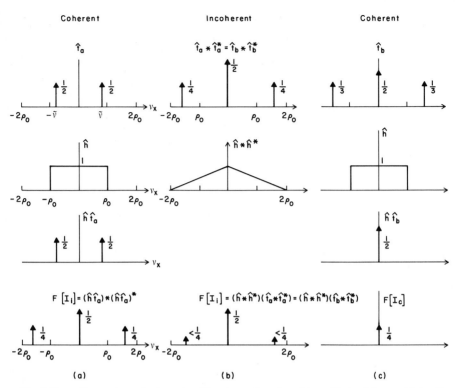

Fig. 1.15a–c. Comparison of coherent and incoherent imaging of two objects of the same intensity transmittance but different phase distribution [Ref. 1.15, p. 129] and [Ref. 1.16, p. 246]

b) Comparison Based on Two Point Resolution

A second comparison criterion may be based on the relative abilities of the coherent and incoherent imaging systems to resolve two closely-spaced object points. In the case of incoherent imaging, the illumination of the two object points is mutually incoherent, no matter how closely the objects are spaced. Each object point gives rise to an Airy pattern caused by the finite lens aperture of diameter D. Consequently, the incoherent image of the two object points is an incoherent superposition of the two Airy patterns

$$I(x') = \left| 2\frac{J_1\left[\frac{\pi D}{\lambda d_2}(x' - \Delta x')\right]}{\frac{\pi D}{\lambda d_2}(x' - \Delta x')} \right|^2 + \left| 2\frac{J_1\left[\frac{\pi D}{\lambda d_2}(x' + \Delta x')\right]}{\frac{\pi D}{\lambda d_2}(x' + \Delta x')} \right|^2. \tag{1.62}$$

According to the so-called Rayleigh criterion of resolution, two incoherent point sources are barely resolved when the center of the Airy pattern generated

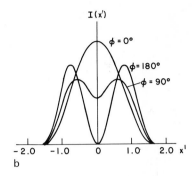

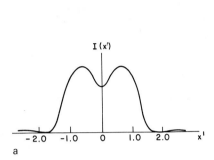

Fig. 1.16a, b. Image intensity for two (**a**) incoherent (**b**) coherent point sources separated by the Rayleigh distance [Ref. 1.15, pp. 130–131]

by one source falls on the first zero of the Airy pattern generated by the second. The minimum resolvable separation of the images is therefore

$$2(\Delta x') = 1.22 \left(\frac{\lambda d_2}{D} \right). \tag{1.63}$$

With this separation, $I(x')$ has a central dip about 19% of the maximum intensity (Fig. 1.16a).

For comparison, we now consider whether the two point objects, separated also by the distance given in (1.63), would be easier or harder to resolve with coherent illumination than with incoherent illumination. Since the illumination of the two points is mutually coherent, the amplitudes of the two Airy patterns can interfere constructively or destructively, depending on the relative phase ϕ between the two sources

$$I(x') = \left| 2 \frac{J_1 \left[\frac{\pi D}{\lambda d_2} (x' - \Delta x') \right]}{\frac{\pi D}{\lambda d_2} (x' - \Delta x')} + 2e^{j\phi} \frac{J_1 \left[\frac{\pi D}{\lambda d_2} (x' + \Delta x') \right]}{\frac{\pi D}{\lambda d_2} (x' + \Delta x')} \right|^2. \tag{1.64}$$

If the two points are 90° out of phase, the image with coherent illumination is the same as with incoherent illumination (Fig. 1.16b). If the two points are in phase, the amplitudes add at $x' = 0$; consequently the dip in the image intensity is absent and the two points are not as well resolved as for incoherent illumination. If the two points are in phase opposition, the dip is always present and deeper than 19%; the two points are resolved better with coherent illumination. However, the two peaks for $\phi = 180°$ are separated wider than those for $\phi = 90°$. Thus, we must know the phase to properly interpret the image.

Fig. 1.17a, b. Photographs of a coherently illuminated, diffusely reflecting objects using different F-numbers (focal length/lens diameter). (a) F# = 8, (b) F# = 64 [Ref. 1.16, p. 237]

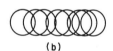

(a)

(b)

Fig. 1.18. (a) Points on object. (b) Overlapping resolution cells of the image. The waves from each object point have different phases and will constructively or destructively interfere in the image plane [Ref. 1.16, p. 236]

c) Comparisons Based on Some Subjective Factors

In comparisons with incoherent imaging, there are two visual effects often found quite bothersome with coherent imaging: the speckle effect and the effect of tiny dust particles on a lens which lead to pronounced diffraction patterns superimposed on the image. The speckle effect is observed in images of objects with optically rough surfaces illuminated by a highly coherent source. The size of the speckle is related to the resolution of the imaging system (Fig. 1.17). If the resolution is sufficiently high, the individual detail of the rough surface is resolved, and speckles do not appear. However, if the resolution is lower, the waves from a number of object points would interfere constructively or destructively, producing speckles when the image is formed (Fig. 1.18). It is easy to see that as the point spread function or impulse response becomes smaller, fewer waves will overlap at a given image point and any speckle will become smaller [Ref. 1.16, pp. 234–235].

The other visual effect, which is caused by tiny dust particles on a lens and leads to pronounced diffraction patterns on the coherent image, is sometimes called coherent noise. An illustration is given in Fig. 3.1 and will be further

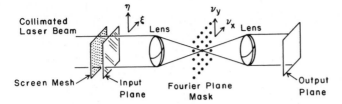

Fig. 1.19. Modified optical processor for spatial frequency diversity filtering. A screen mesh diffracts the light into a number of orders that illuminate the input transparency. Spatial filters (here a high-pass filter is shown) are periodic arrays in the Fourier plane [1.24]

discussed in Sect. 3.1. It is only noted here that to reduce this coherent noise effect there are a few alternative schemes based on the simple idea of illuminating the object from multiple directions (Fig. 1.19). When the object is illuminated from multiple directions, the same image is expected to be formed on the image plane. However, the diffraction patterns due to tiny particles will be moved around, since the plane on which there are dust particles and the image plane do not satisfy the lens law (1.20). Thus, the coherent noise effect is reduced or the signal-to-noise ratio is improved [1.24].

1.6.4 Superresolution

It follows from Sects. 1.6.1 and 1.6.2 that the image resolution is limited fundamentally by the bandwidth of the optical system. The wider the system bandwidth, the better the image resolution that can be expected [see (1.43, 45, 47, 37c, and 52)]. Attempts to improve on the resolutions of imaging systems beyond the fundamental (classical) limit, which is the subject matter of superresolution, have recently been reported in the literature. Since the approaches to achieve superresolution bring out some concepts which are helpful in further comprehending the basic performances of imaging systems, and interesting from information processing points of view, we shall discuss superresolution here at least briefly. Both optical and digital approaches are included in the discussions.

a) Optical Approach

The optical approaches toward superresolution generally rely on the new theorem on the ultimate performance limit of optical systems, which states that not the bandwidth of the transfer function ($\hat{h}$ or $\mathscr{H}$) but only the number of degrees of freedom (N) of the optical message transmitted by a given optical system is invariant:

$$N = 2 \text{ (object area) (optical bandwidth)}$$
$$= 2(\Delta x \Delta y)(\Delta v_x \Delta v_y), \tag{1.65}$$

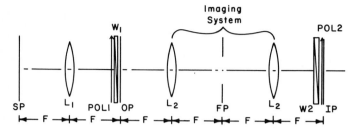

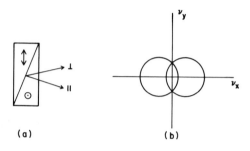

Fig. 1.20. Principal arrangement of Gärtner and Lohmann's experiment on superresolution for nonbirefringent objects. SP, OP, FP, and IP are source plane, object plane, Fourier plane and image plane, respectively. POL1 and POL2 are polarizer and analyzer. W1 and W2 are Wollaston prisms [1.26]

Fig. 1.21. (a) Wollaston prism. A light beam polarized at 45° with respect to the two polarization axes of the Wollaston prism will emerge as two beams propagating in two different directions for the two orthogonally polarized components. **(b)** The effective pupil function for the imaging system in Fig. 1.20 [1.26]

where the factor of 2 comes from the two possible orthogonal states of polarization, and the product of object area and optical bandwidth is frequently called the space-bandwidth product of the optical system [1.25]. This new theorem suggests that it should be possible (a) to double the bandwidth when transmitting information about one state of polarization only, (b) to extend the bandwidth by reducing the object area, and (c) to extend the bandwidth in the v_x direction while reducing it in the v_y direction.

Figure 1.20 shows a possible optical system for doubling the bandwidth by transmitting the information about one state of polarization only [1.26]. This system is the same as the two-lens imaging system of Fig. 1.8 except that there are two polarizers and two Wollaston prisms added to the object and imaging planes as indicated. Polarizer POL 1 is oriented at 45° with respect to the two polarization axes of the Wollaston prism W 1 so that the monochromatic, polarized plane wave passing through POL 1 emerges from the prism in two different directions for the two orthogonally polarized components (Fig. 1.21a). The object $u(\xi, \eta)$ being illuminated by the two plane waves propagating in two different directions will experience the pupil function of Fig. 1.21b. Since the bandwidth $\hat{h}(v_x, v_y)$ of an optical system is directly related to the pupil function $P(-\lambda d_2 v_x, -\lambda d_2 v_y)$ and the $P(-\lambda d_2 v_x, -\lambda d_2 v_y)$ of Fig. 1.21b exceeds the pupil function of the same lens illuminated by only one plane wave, the image resolution is improved as compared to that obtainable from the same optical

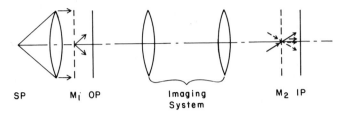

Fig. 1.22. The optical arrangement illustrating increased resolution for a reduced object field. OP, IP are object and image planes; M_1 and M_2 are grating masks [1.25]

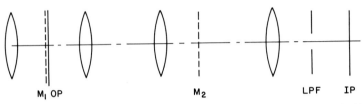

Fig. 1.23. An experimental setup for superresolution image of one-dimensional objects. LFP is the low pass filtering plane [1.27]

system without utilizing POL 1 and W 1. The purpose of the second Wollaston prism W 2 is to compensate the influence of W 1, and the analyzer POL 2 is physically oriented parallel to the orientation of POL 1.

The optical arrangement illustrating increased resolution for a reduced object field is sketched in Fig. 1.22 [1.25]. Masks M_1 and M_2 are identical gratings oriented parallel to each other, each of which produces two diffraction orders. The object $u(\xi, \eta)$ being illuminated by the two diffraction orders from M_1 once again sees a pupil function similar to that shown in Fig. 1.21b and thus the bandwidth of the optical system is extended. The purpose of mask M_2 is to properly recombine the two diffraction orders transmitted through the optical system to yield the final image of improved resolution. Since the object $u(\xi, \eta)$ is masked by a grating, the usable object function is reduced from that without a mask.

To be able to extend the bandwidth of the optical system in the v_x direction while reducing it in the v_y direction is especially helpful for those objects which vary strongly as a function of x but only slowly in y [1.27]. These objects occupy a cigar-shaped area in the spatial frequency domain and are badly matched to the transfer function of a lens which is usually circular. To overcome this problem of mismatch, an optical scheme is suggested in Fig. 1.23, in which sinusoidal grating masks are again found to be very helpful. Figure 1.24 explains the operations of the optical scheme (Fig. 1.23) by showing the spatial frequency contents of the optical information at various planes down the optical axis. The object placed in plane OP has a cigar-shaped spectrum of Fig. 1.24a. This object is modulated by a tilted sinusoidal mask M_1, whose

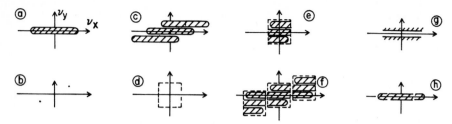

Fig. 1.24a–h. Explanation in spatial frequency domain for the operation of the optical scheme in Fig. 1.23: **(a)** object, **(b)** mask M_1, **(c)** effective object consisting of O and M_1, **(d)** transfer domain of lens L, **(e)** passed through lens, **(f)** after interacting with mask M_2, **(g)** y-smearing by v_y low-pass, **(h)** final image [1.27]

spectrum is shown in Fig. 1.24b. The spectrum of the modulated object (Fig. 1.24c) is the convolution of the two spectra just mentioned (Fig. 1.24a, b). The square aperture of the lens (Fig. 1.24d) let pass only parts (Fig. 1.24e) of the arriving spectrum of Fig. 1.24c. But, notice that the transmitted pattern (Fig. 1.24e) contains all portions of the object spectrum (Fig. 1.24a) although in a distorted arrangement. This transmitted pattern of a distorted spectrum is to be diffracted by a second sinusoidal mask M_2. If the grating constant and tilt angle of mask M_2 is chosen to be identical to those of mask M_1, the diffracted pattern will have the spectrum of Fig. 1.24f. Especially important in the spectrum of Fig. 1.24f is that we find around the v_x axis, all portions of the original object spectrum (Fig. 1.24a). Away from the v_x axis, there are some additional frequency components which obviously do not belong to the object spectrum and should be eliminated by introducing a horizontal slit (Fig. 1.24g) to perform low pass filtering. The result of low pass filtering is the output image whose spectrum (Fig. 1.24h) has a wider v_x extension than the limiting aperture (Fig. 1.24d). Thus, we have achieved superresolution in the v_x direction.

In summary, the optical approaches to superresolution involve the use of either a polarizing prism or a mask to preprocess the object spectrum, avoiding the loss of spatial frequency components through the band-limited optical system. After passage through the system, the optical information will be demodulated by either a polarizing prism or a mask, rearranging the spatial frequency contents properly to yield an output image of improved resolution.

b) Digital Approach

The digital approach to superresolution is generally concerned with processing the image information which is obtained from an object after the passage through an optical system. When the transfer function of the optical system has a bandwidth smaller than that of the object spectrum, certain spatial frequency components are lost in the image spectrum. To regain the lost spectrum, processing can be carried out digitally if the image spectrum is an analytical function [Ref. 1.15, Sect. 6.6; 1.28].

Suppose that the object is bounded by a rectangle of sides L_x and L_y. Then by the sampling theorem, the object spectrum $G_o(v_x, v_y)$ can be written in terms of its sample values at $(n/L_x, m/L_y)$:

$$G_o(v_x, v_y) = \sum_{n=-\infty}^{\infty} \sum_{m=-\infty}^{\infty} G_o\left(\frac{n}{L_x}, \frac{m}{L_y}\right)$$
$$\cdot \text{sinc}\left[2L_x\left(v_x - \frac{n}{2L_x}\right)\right] \text{sinc}\left[2L_y\left(v_y - \frac{m}{2L_y}\right)\right]. \tag{1.66}$$

Now due to the passband limit of the optical system, values of $G_o(n/L_x, m/L_y)$ can be found for only low-integer values of (n, m) up to $(\pm N, \pm M)$. In other words, the image spectrum is

$$G_i(v_x, v_y) = \sum_{n=-N}^{M} \sum_{m=-M}^{M} G_o\left(\frac{n}{L_x}, \frac{m}{L_y}\right)$$
$$\cdot \text{sinc}\left[2L_x\left(v_x - \frac{n}{2L_x}\right)\right] \text{sinc}\left[2L_y\left(v_y - \frac{m}{2L_y}\right)\right]. \tag{1.67}$$

We would like, of course, to extend our knowledge about the object to larger integer values, say $(\pm N', \pm M')$:

$$G_o(v_x, v_y) \simeq \sum_{n=-N'}^{N'} \sum_{m=-M'}^{M'} G_o\left(\frac{n}{L_x}, \frac{m}{L_y}\right)$$
$$\cdot \text{sinc}\left[2L_x\left(v_x - \frac{n}{2L_x}\right)\right] \text{sinc}\left[2L_y\left(v_y - \frac{m}{2L_y}\right)\right]. \tag{1.68}$$

To determine the sample values outside the passband, we measure the values of $G_i(v_x, v_y)$ at $(2N'+1)(2M'+1)$ frequencies within the passband or calculate these $(2N'+1)(2M'+1)$ values of $G_i(v_x, v_y)$ from (1.67). Then these $(2N'+1)(2M'+1)$ values of $G_i(v_x, v_y)$ are substituted into the left of (1.68) to generate $(2N'+1)$ $(2M'+1)$ linear equations for $(2N'+1)(2M'+1)$ unknowns of $G_o(n/L_x, m/L_y)$ on the right of (1.68). The substitution of $(2N'+1)(2M'+1)$ values of $G_i(v_x, v_y)$ into the left of (1.68) is justified because within the passband the coherent transfer function is unity and the image spectrum is identical to the object spectrum. Since the set of $(2N'+1)(2M'+1)$ linear equations can in principle be solved (by digital computers) to yield many values for $G_o(n/L_x, m/L_y)$ outside the passband, an image with improved resolution can now be obtained. The success of this approach, however, depends on the amount of noise being picked up which invariably accompanies any measurements of the object spectrum within the passband. Imprecision of the knowledge within the passband can be greatly amplified when the system of linear equations is solved for sample values outside the passband, for such sample values are expressed as linear combinations of all the measured values, each of which contributes some noise.

1.7 Topics of Practical Interest Related to Optical Fourier Transform

It is shown in Sect. 1.3.2 that an exact Fourier transform relationship exists between the field amplitude at the back focal plane of a thin lens and the object, which is placed in the front focal plane and illuminated by a plane wave. This Fourier transform result is further examined in this section in order to bring out some engineering design considerations of practical value, not only to optical Fourier spectrum analysis but also to coherent optical processing (Chap. 2). Specifically, we would like to investigate the effect of spherical (instead of plane) wave illumination on the Fourier transform, the accuracy required in determining the back focal plane position in order to observe the Fourier transform distribution, and desirable features associated with specially designed Fourier transform lenses.

1.7.1 Spherical Wave Illumination

If the input object is illuminated by a spherical wave, $(1/j\lambda\varrho)\exp[jk(\xi^2+\eta^2)/2\varrho]$, the input amplitude distribution will become

$$t_o^s(\xi,\eta)=(1/j\lambda\varrho)\exp[jk(\xi^2+\eta^2)/2\varrho]t_o(\xi,\eta), \tag{1.69}$$

where ϱ is the radius of curvature of the spherical wave [Ref. 1.16, pp. 109–110]. Furthermore, if the input object is at the front focal plane ($d_1=F$ in Fig. 1.6), the amplitude distribution at the back focal plane $u_v(x_v,y_v)$ will be

$$\begin{aligned}
u_v(x_v,y_v)&=C_3\hat{t}_o^s(v_x,v_y)\\
&=C_3\hat{t}_o(v_x,v_y)*\exp[-j\pi\lambda\varrho(v_x^2+v_y^2)]\\
&=C_3\hat{t}_o(x_v,y_v)*\exp[jk(x_v^2+y_v^2)/2(F^2/\varrho)]. \tag{1.70}
\end{aligned}$$

Since the factor $\exp[jk(x_v^2+y_v^2)/2(F^2/\varrho)]$ can be considered as the impulse response for diffraction effect over a distance (F^2/ϱ), the observed field in the back focal plane is an out-of-focused $\hat{t}_o(x_v,y_v)$; that is, the effect of spherical wave illumination is the same as smearing $t_o(x_{v_2}y_v)$ due to diffraction effects over a distance (F^2/ϱ). To observe an in-focused $\hat{t}_o(x_v,y_v)$ under spherical wave illumination, we should look at the plane at d_2 distance behind the lens:

$$d_2=F+F^2/\varrho. \tag{1.71}$$

1.7.2 Position Accuracy Requirement on Back Focal Plane

If the measured intensity distribution is to accurately represent the Fourier transform of the object, how small must the positioning error be for the back

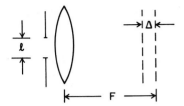

Fig. 1.25. Position accuracy requirement on back focal plane

focal plane? To determine this position accuracy requirement, let us consider a unit-amplitude, normal incident plane wave illuminating an object $u_1(x, y)$ of maximum linear dimension l, situated in front of a converging lens of focal length F (Fig. 1.25).

The field immediately behind the lens should be $u_1(x, y)$ $\cdot \exp[-jk(x^2+y^2)/2F]$. Applying the Fresnel diffraction formula of (1.4a) to this field over a distance $(F-\Delta)$, we obtain

$$u_2(x_v, y_v) = [1/j\lambda(F-\Delta)] \exp[jk(F-\Delta)] \iint u_1(x, y) \exp[-jk(x^2+y^2)/2F]$$
$$\cdot \exp\{jk[(x-x_v)^2 + (y-y_v)^2]/2(F-\Delta)\} dx dy$$
$$= [1/j\lambda(F-\Delta)] \exp[jk(F-\Delta)] \exp\{jk(x_v^2+y_v^2)/[2(F-\Delta)]\}$$
$$\cdot \iint u_1(x, y) \exp\left[-j\frac{k}{2}(x^2+y^2)\left(\frac{1}{F} - \frac{1}{(F-\Delta)}\right)\right]$$
$$\cdot \exp[jk(xx_v + yy_v)/(F-\Delta)] dx dy. \tag{1.72}$$

To accurately represent the Fourier transform pattern of the object, we need

$$\frac{k}{2}(x^2+y^2)_{max}\left[\frac{1}{F} - \frac{1}{(F-\Delta)}\right] \ll \pi$$

or

$$\Delta \ll 2\pi F^2/k(x^2+y^2)_{max}. \tag{1.73}$$

It is also noted in (1.72) that at $z=(F-\Delta)$, the scale of the Fourier transform pattern is slightly different from that at $z=F$ due to the factor $\exp[jk(xx_v + yy_v)/(F-\Delta)]$ inside the integral.

1.7.3 The Fourier Transform Lens

In order to satisfy the paraxial approximation for Fourier transform operation by a thin lens, the size of the lens aperture and the input object are frequently restricted. Obviously, it would be a desirable feature of a Fourier transform lens to be able to accept input objects of larger size, say sizes up to 75% of the lens.

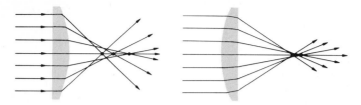

Fig. 1.26. SA for a planar-convex lens [Ref. 1.29, p. 177]

Since the size of a Fourier transform pattern is proportional to the focal length [$x_v = (\lambda v_x)F$], it would also be desirable for the Fourier transform lens to have a long effective focal length. However, the long effective focal length should be accompanied by a short physical distance between the exit pupil and lens vertex. Otherwise, it would not be convenient in working with the various planes of the optical transform system experimentally. Furthermore, it would be desirable for the Fourier transform plane to be flat.

To comprehend many of the considerations that go into designing a Fourier transform lens with the above-mentioned desirable features, we first need to discuss lens aberration briefly. Only those aberrations which affect Fourier transform lens design are touched upon here; they are spherical aberration, coma, astigmatism and field curvature.

a) Spherical Aberration

When the light rays passing near the edge of the lens (marginal rays) do not focus at the same plane as those passing near the center of the lens (central rays), spherical aberration is present (Fig. 1.26). The amount of spherical aberration (SA) when the aperture and focal length are fixed, varies with both the object distance and lens shape. A striking example is illustrated in Fig. 1.26 where simply turning the lens around markedly reduces the SA. When the object is at infinity, a simple concave or convex lens which has an almost, but not quite, flat rear side will suffer a minimum amount of spherical aberration. If the object and image distances are to be equal, the lens should be equiconvex to minimize SA. A combination of a converging and a diverging lens can also be utilized to dimish spherical aberration [Ref. 1.29, pp. 176–177].

b) Coma

Figure 1.27 illustrates the formation of the geometrical comatic image of an extra-axial point. Observe that each circular cone of rays whose endpoints (1–2–3–4–1–2–3–4) form a ring on the lens is imaged to a comatic circle on Σ_i. This case corresponds to a positive coma and so the larger the ring on the lens, the further its comatic circle will be from the axis. Like SA, coma is dependent on the shape of the lens. Thus, a strongly concave-meniscus lens with the object at infinity will have a large negative coma. Bending the lens so that it becomes

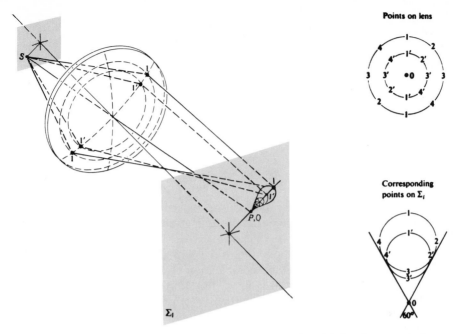

Points on lens

Corresponding points on Σ_i

Fig. 1.27. The geometrical coma image of a point. The central region of the lens forms a point image at the vertex of the cone [Ref. 1.29, p. 180]

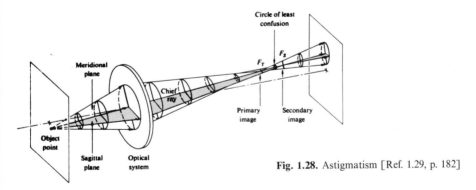

Fig. 1.28. Astigmatism [Ref. 1.29, p. 182]

planar-convex, then equi-convex, convex-planar and finally convex-meniscus will change the coma from negative, to zero, to positive. The fact that it can be made exactly zero for a single lens with a given object distance is quite significant. The particular shape is almost convex-planar and very nearly the configuration for minimum SA [Ref. 1.29, pp. 177–180].

c) Astigmatism

When an object point lies an appreciable distance from the optical axis, the incident cone of rays will strike the lens asymmetrically (Fig. 1.28). The light rays in a plane containing both the chief ray (i.e., the one passing through the center of the aperture) and the optical axis (the meridional plane) do not come to the same focus as the rays in the plane which contains the chief ray and is perpendicular to the meridional plane (the sagittal plane). It can be shown, using Fermat's principle, that the focal length difference depends effectively on the power of the lens (as opposed to the shape or index) and the angle at which the rays are inclined [Ref. 1.29, pp. 180–182]. This astigmatic difference increases rapidly as the rays become more oblique, i.e., as the object point moves further off the axis and is, of course, zero on axis.

d) Field Curvature

A planar object normal to the axis will be imaged approximately as a plane only in the paraxial region. For a finite lens aperture, the surface on which the image is formed is curved – the Petzval surface. The Petzval surface curves inward toward the object plane for a positive lens and it curves outward for a negative lens. Therefore, a suitable combination of positive and negative lenses will negate field curvature [Ref. 1.29, pp. 182–184].

Hence, to design a Fourier-transform lens for accepting large input objects we need to minimize spherical, coma and astigmatism aberrations. To provide a long effective focal length while maintaining controls over the optical system's physical length and aberrations, negative lens elements are frequently utilized. Moreover, to keep the Fourier plane flat we need to minimize the field curvature aberration [1.30].

At the moment, Fourier transform lenses are available commercially from Tropel. They comprise six lens elements each and, therefore, are quite expensive in comparison to the doublets offered by Space Optics Research Labs (model FX15/5) and Jaegers (6″ objectives).

References

1.1   L. Foucault: Ann. Obs. Imp. (Paris) **5**, 197 (1859)
1.2   E. Abbe: Arch. Mikros. Anat. **9**, 413 (1873)
1.3   A. B. Porter: Phil. Mag. **11**, 154 (1906)
1.4   F. Zernike: Z. Tech. Mag. **16**, 454 (1935)
1.5   P. M. Duffieux: Faculté des Sciences, Université de Besançon (1946)
1.6   P. Elias, D. S. Grey, D. Z. Robinson: J. Opt. Soc. Am. **42**, 127 (1952)
1.7   P. Elias: J. Opt. Soc. Am. **43**, 229 (1953)
1.8   E. L. O'Neill: IRE Trans. IT-**2**, 56 (1956)
1.9   A. Maréchal, P. Croce: Compt. Rend. **237**, 706 (1953)
1.10 L. J. Cutrona, E. N. Leith, L. J. Porcello, W. E. Vivian: Proc. IEEE **54**, 1026 (1966)
1.11 E. N. Leith: Proc. IEEE **59**, 1305 (1971)
1.12 A. B. Vander Lugt: IEEE Trans. IT-**10**, 130 (1964)
1.13 B. R. Brown, A. W. Lohmann: Appl. Opt. **5**, 967 (1966)
1.14 A. Sommerfeld: Optics, *Lectures on Theoretical Optics*, Vol. IV (Academic Press, New York 1954)

1.15 J.W.Goodman: *Introduction to Fourier Optics* (McGraw-Hill, New York 1968)

1.16 W.T.Cathey: *Optical Information Processing and Holography* (Wiley, New York 1974)

1.17 A.Papoulis: *The Fourier Integral and Its Applications* (McGraw-Hill, New York 1962)

1.18 R.N.Bracewell: *The Fourier Transform and Its Applications* (McGraw-Hill, New York 1965)

1.19 H.J.Caulfield: *Handbook of Optical Holography* (Academic Press, New York 1979)

1.20 J.B.DeVelis, G.O.Reynolds: *Theory and Applications of Holography* (Addison-Wesley, Reading, MA 1967)

1.21 M.J.Beran, G.B.Parrent, Jr.: *Theory of Partial Coherence* (Prentice-Hall, Englewood Cliffs, N.J. 1964)

1.22 G.B.Parrent, B.J.Thompson: *Physical Optics Notebook* (Society of Photo-optical Instrumentation Engineers, Bellingham, Washington 1969)

1.23 A.W.Lohmann: *Optical Information Processing*, Supplement to Vol. I (Course notes of University of California, San Diego 1974)

1.24 G.B.Brandt: Appl. Opt. **12**, 368 (1964)

1.25 W.Lukosz: J. Opt. Soc. Am. **56**, 1463 (1966)

1.26 A.W.Lohmann, D.P.Paris: Appl. Opt. **3**, 1037 (1964)

1.27 M.A.Grimm, A.W.Lohmann: J. Opt. Soc. Am. **56**, 1151 (1966)

1.28 J.L.Harris: J. Opt. Soc. Am. **54**, 941 (1964)

1.29 E.Hecht, A.Zajac: *Optics* (Addison-Wesley, Reading, MA 1973)

1.30 K.Von Bieren: Appl. Opt. **10**, 2739 (1971)

2. Coherent Optical Processing

S. H. Lee

With 23 Figures

The interesting and important fact that we can obtain the Fourier spectrum of a coherently illuminated object physically in the back focal plane of a lens has been discussed in Sect. 1.3. We will discuss in this chapter how to manipulate the Fourier spectrum directly, i.e., to perform spatial filtering [2.1–6] or coherent processing.

Figure 2.1 shows the configuration of a coherent optical processing system which is conceptually the simplest to understand. Light from the point source S is collimated by lens L_1. The input to be filtered is inserted as a space varying amplitude transmittance $f(\xi, \eta)$ in plane P_1. Lens L_2 Fourier transforms $f(\xi, \eta)$, producing an amplitude distribution $c_3 \hat{f}(v_x, v_y)$ across P_2, where c_3 is a complex constant. A filter is inserted in this plane to manipulate the amplitude and phase of the spectrum $\hat{f}$. Finally, the lens L_3 transforms the filtered spectrum to yield a filtered image acrosss P_3. Note that the coordinates (x', y') of P_3 are inverted with respect to (ξ, η) of P_1 because L_3 can only take forward Fourier transform and not inverse transform.

Using the coherent processing system of Fig. 2.1 we shall first discuss spatial filter synthesis, diffraction efficiencies and classifications of spatial filters in the following sections. Then, alternative spatial filtering systems and suppression of coherent noise in optical processing will be considered.

2.1 Spatial Filter Synthesis

2.1.1 Simple Filters

On the subject of spatial filter synthesis, let us first consider a few simple spatial filters which provide visually, quite dramatic filtered results, then proceed to more complicated filtering operations. Examples of simple filters which affect the amplitudes of object spectra are zero spatial frequency stop [2.7] and a horizontal slit [2.8] placed at the center of plane P_2 (Fig. 2.1). An example of a simple phase filter is one which retards the zero frequency component of the object spectrum by 90° [2.9].

When a filter of the zero spatial frequency stop is used, the average transmittance of an image will be eliminated yielding, for example, the high pass filtered result of Fig. 2.2b from the original image of Fig. 2.2a. When the filter of a horizontal slit is placed at the center of plane P_2 (Fig. 2.1), it will

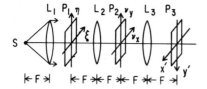

Fig. 2.1. Configuration of a coherent optical processing system

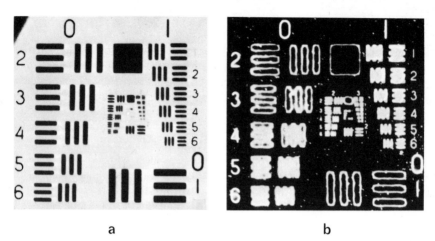

a   b

Fig. 2.2a, b. Original image of a resolution test target (**a**) and a high pass filtered image (**b**) [2.7]

transmit all the vertical lines of a mesh and eliminate all the horizontal lines (Fig. 2.3).

A 90° phase filter can help to visualize transparent objects. Let the amplitude transmittance of a transparent object be

$$f(\xi,\eta)=\exp[j\phi(\xi,\eta)] \tag{2.1a}$$

and assume, furthermore, that the phase $\phi(\xi,\eta)$ be less than 1 radian such that

$$f(\xi,\eta)\simeq 1+j\phi(\xi,\eta). \tag{2.1b}$$

When this spectrum of the transparent object is filtered by retarding the phase of its zero spatial frequency component by 90° relative to other frequency components, the amplitude and intensity of the filtered image will be, respectively,

$$g_o(x',y')=\exp[j(\pi/2)]+j\phi(x',y'), \tag{2.2a}$$

$$I(x',y')=|\exp[j(\pi/2)]+j\phi(x',y')|^2=|j(1+\phi)|^2$$
$$\cong 1+2\phi(x',y'). \tag{2.2b}$$

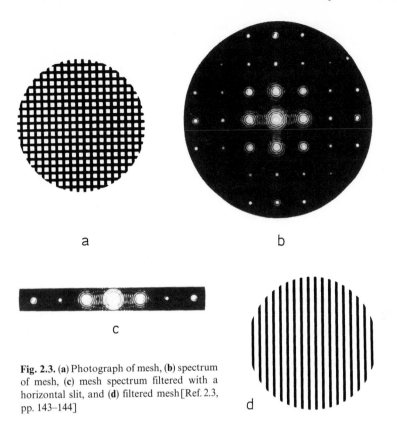

a   b

c

Fig. 2.3. (a) Photograph of mesh, **(b)** spectrum of mesh, **(c)** mesh spectrum filtered with a horizontal slit, and **(d)** filtered mesh [Ref. 2.3, pp. 143–144]

d

This amplitude or intensity variation is certainly visible.

There are many ways to produce the simple filters. The simplest way for producing the zero frequency stop and the slit filters is perhaps to expose a piece of photographic film with a small dot or a slit of light. The 90° phase filter can also be fabricated on film first by exposing a small dot, then bleaching it. Real-time interface devices which can be used for filter generation will be discussed in Chap. 4.

2.1.2 Grating Filters

Leading to the synthesis of more complicated spatial filtering operations, it will first be useful to understand the operation of simple grating filters because complicated filtering functions can generally be decomposed into the superposition of many sinusoidal gratings. With a sinusoidal grating filter it is a simple matter to show that an object placed on P_1 of Fig. 2.1 will be diffracted into three images on P_3 and that two spatially nonoverlapping objects are diffracted into six images, some of which may be overlapping. Let the transmission function of the grating be represented by

$$\hat{h}(v_x, v_y) = \tfrac{1}{2}\{1 + \exp[j(2\pi v_y b + \phi)] + \exp[-j(2\pi v_y b + \phi)]\}, \tag{2.3}$$

where ϕ is related to the grating position with respect to the optical axis. For the case of two nonoverlapping objects, $f_1(\xi, \eta - b')$ and $f_2(\xi, \eta + b')$ centered at $\pm b'$, the following expression will describe the light amplitude incident on the grating:

$$\hat{g}_i(v_x, v_y) = \hat{f}_1(v_x, v_y)\exp(-j2\pi v_y b') + \hat{f}_2(v_x, v_y)\exp(+j2\pi v_y b'). \tag{2.4a}$$

If the parameter b which characterizes the grating filter is equal to b' which is the off-axis distance of f_1 and f_2, the light amplitude behind the grating filter will be

$$\hat{g}_o(v_x, v_y) = \tfrac{1}{2}[\hat{f}_1(v_x, v_y)\exp(j\phi) + \hat{f}_2(v_x, v_y)\exp(-j\phi)]$$
$$+ 4 \text{ other terms}. \tag{2.4b}$$

After Fourier transforming $\hat{g}_o(v_x, v_y)$ by lens L_3, we obtain on P_3 for the output

$$g_o(x', y') = \tfrac{1}{2}[f_1(x', y') + f_2(x', y')\exp(-j2\phi)]\exp(j\phi)$$
$$+ 4 \text{ other terms off-axis}. \tag{2.4c}$$

When the grating has a maximum transmittance at the optical axis, $\phi = 0$ and optical addition results at the center portion of P_3. When the maximum transmittance of the grating is shifted by a quarter of a fringe spacing from the optical axis, $\phi = 90°$ and optical subtraction results. Experimental results of these addition and subtraction operations are shown in Fig. 2.4.

A different perspective on performing optical addition and subtraction by a sinusoidal grating can be obtained, if we recognize the result of an object spectrum $\hat{f}(v_x, v_y)$ filtered by $\hat{h}(v_x, v_y)$ as the convolution between $f(\xi, \eta)$ and $h(\xi, \eta)$. Figure 2.1 shows that the result of $\hat{f}(v_x, v_y)$ filtered by $\hat{h}(v_x, v_y)$ is

$$g_o(x', y') = \mathscr{F}\{\hat{f}(v_x, v_y)\hat{h}(v_x, v_y)\}, \tag{2.5a}$$

which is, according to the well-known convolution theorem,

$$g_o(x', y') = \int\int f(\xi, \eta)h(x' - \xi, y' - \eta)d\xi d\eta = f * h. \tag{2.5b}$$

When $\hat{h}(v_x, v_y)$ is a sinusoidal grating given by (2.3), the Fourier transform of the grating, or its impulse response, is the sum of three delta functions:

$$h(\xi, \eta) = [\delta(\xi, \eta) + \delta(\xi, \eta + b)e^{j\phi} + \delta(\xi, \eta - b)e^{-j\phi}]/2. \tag{2.6}$$

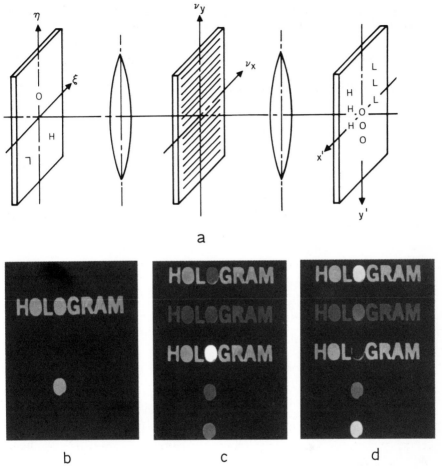

Fig. 2.4a–d. Complex amplitude addition and subtraction with gratings. (**a**) The optical system, (**b**) the two input pattern functions, (**c**) the pattern function "O" is added onto that of "HOLOGRAM" in the central region of the output plane, and (**d**) the pattern function "O" is subtracted from that of "HOLOGRAM" in the central region of the output plane [2.12]

The convolution of $[f_1(\xi, \eta - b) + f_2(\xi, \eta + b)]$ with $h(\xi, \eta)$ is

$$
\begin{aligned}
g_o(x', y') &= [f_1(\xi, \eta - b) + f_2(\xi, \eta + b)] \\
&\quad * \tfrac{1}{2}[\delta(\xi, \eta) + \delta(\xi, \eta + b)e^{j\phi} + \delta(\xi, \eta - b)^{-j\phi}] \\
&= \tfrac{1}{2}[f_1(\xi, \eta - b) * \delta(\xi, \eta + b)e^{j\phi} + f_2(\xi, \eta + b) * \delta(\xi, \eta - b)e^{-j\phi}] \\
&\quad + 4 \text{ other terms off-axis} \\
&= \tfrac{1}{2}[f_1(x', y') + f_2(x', y')e^{-j2\phi}]e^{j\phi} \\
&\quad + 4 \text{ other terms off-axis}.
\end{aligned}
\tag{2.7}
$$

Equation (2.7) is the same as (2.4c).

Composite gratings can be synthesized to perform other mathematical operations [2.10–12]. For example, a filtering function of two sinusoidal gratings which have slightly different frequencies and an impulse response with an off-axis component

$$h_1(\xi, \eta) = \delta[\xi + (b + \varepsilon), \eta] - \delta[\xi + b, \eta], \tag{2.8a}$$

can perform the differentiation operation $\partial f / \partial \xi$ because

$$\frac{\partial f}{\partial \xi} = \lim_{\varepsilon \to 0} \left(\frac{1}{\varepsilon}\right) \{f[\xi + (b + \varepsilon), \eta] - f[\xi + b, \eta]\}$$

$$= \lim_{\varepsilon \to 0} \left(\frac{1}{\varepsilon}\right) [f(\xi, \eta) * h_1(\xi, \eta)]. \tag{2.8b}$$

Similarly, the following differentiation operations can be performed by composite gratings whose off-axis components of their impulse responses are given, respectively, in the following equations:

$$\frac{\partial f}{\partial \eta} = \lim_{\varepsilon \to 0} \left(\frac{1}{\varepsilon}\right) [f(\xi, \eta) * h_2(\xi, \eta)]$$

$$h_2(\xi, \eta) = \delta[\xi, \eta + (b + \varepsilon)] - \delta(\xi, \eta + b), \tag{2.8c}$$

$$\frac{\partial f}{\delta \xi} + \frac{\partial f}{\partial \eta} = \lim_{\varepsilon \to 0} \left(\frac{1}{\varepsilon}\right) [f(\xi, \eta) * h_3(\xi, \eta)]$$

$$h_3(\xi, \eta) = \delta(\xi + b + \varepsilon, \eta) + \delta(\xi + b, \eta + \varepsilon) - 2\delta(\xi + b, \eta), \tag{2.8d}$$

$$\frac{\partial^2 f}{\partial \xi^2} + \frac{\partial^2 f}{\partial \eta^2} = \lim_{\varepsilon \to 0} \left(\frac{1}{\varepsilon^2}\right) [f(\xi, \eta) * h_4(\xi, \eta)]$$

$$h_4(\xi, \eta) = \delta(\xi + b + \varepsilon, \eta) + \delta(\xi + b - \varepsilon, \eta) + \delta(\xi + b, \eta + \varepsilon)$$
$$+ \delta(\xi + b, \eta - \varepsilon) - 4\delta(\xi + b, \eta), \tag{2.8e}$$

$$\frac{\partial^2 f}{\partial \xi \partial \eta} = \lim_{\varepsilon \to 0} \left(\frac{1}{\varepsilon^2}\right) [f(\xi, \eta) * h_5(\xi, \eta)]$$

$$h_5(\xi, \eta) = \delta(\xi + b + \varepsilon, \eta + \varepsilon) + \delta(\xi + b, \eta) - \delta(\xi + b, \eta + \varepsilon)$$
$$- \delta(\xi + b + \varepsilon, \eta). \tag{2.8f}$$

To produce the grating filters, the optical configuration of Fig. 2.5 can be used. Gratings are obtained by interfering the plane waves from the two focal points of L_2 and L_3. We can control the grating frequency by controlling the separation between L_2 and L_3. The relative phase differences between various grating components in a composite grating filter are recorded by changing the

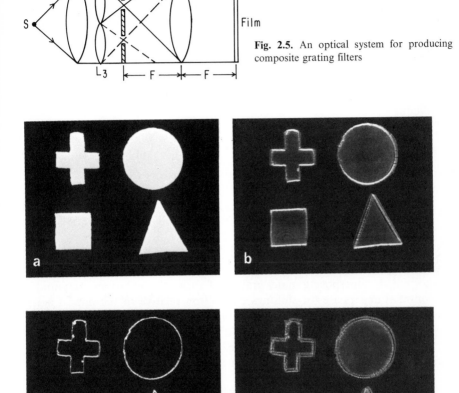

Fig. 2.5. An optical system for producing composite grating filters

Fig. 2.6a–d. Experimental results of optical differentiations with composite gratings. (a) The object pattern $f(\xi, \eta)$, (b) experimental result for $\partial f/\partial \eta$, (c) experimental result for $\partial f/\partial \xi + \partial f \partial \eta$, and (d) experimental result for $\partial^2 f/\partial \xi^2 + \partial^2 f/\partial \eta^2$

lateral position of the film before exposing each grating component. For example, to produce the composite grating filter (2.8e) for the Laplacian operation, five exposures are needed. The first four exposures are made by controlling the position of L_3 while the film is fixed. The fifth exposure is made by controlling both the positions of L_3 and the film. Samples of experimental results on using composite grating filters are illustrated in Fig. 2.6. These results appear off-axis in P_3 of Fig. 2.1 because the impulse responses described by (2.8a–f) are off-axis components.

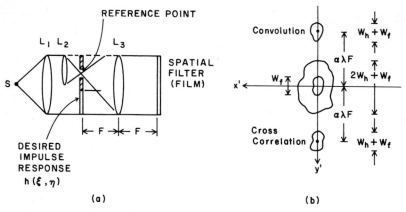

Fig. 2.7. (a) A modified Rayleigh interferometer system for producing complex spatial filters (the Vander Lugt filter) [2.13] and (b) location of the four terms of the processor output

2.1.3 Complex Spatial Filters

Complex spatial filters of filtering functions $\hat{h}(v_x, v_y)$ or $\hat{h}^*(v_x, v_y)$ can be synthesized by superposing many grating of various orientations, amplitudes (or contrast) and phase (or shiftings of position relative to the optical axis). When an input $f(\xi, \eta)$ is filtered by $\hat{h}(v_x, v_y)$, a convolution is obtained (2.5b), as explained in Sect. 2.1.2. When the input $f(\xi, \eta)$ is filtered by $\hat{h}^*(v_x, v_y)$, the following correlation function is obtained which has been proved to be very important for optical pattern recognition:

$$g(x', y') = \mathscr{F}\{\hat{f}(v_x, v_y)\hat{h}^*(v_x, v_y)\}$$
$$= \int\int f(\xi, \eta)h^*(\xi - x', \eta - y')d\xi d\eta = f \circledast h. \tag{2.9}$$

Figure 2.7a shows an interferometric system for producing complex spatial filters [2.13]. Each point (or pixel) of $h(\xi, \eta)$ gives rise to a tilted plane wave behind lens L_3 which will interfere with the tilted plane wave from the reference point to produce a sinusoidal grating. The amplitude and frequency of this grating depends on the pixel transmittance and the separation between the reference point and the pixel of $h(\xi, \eta)$, while the grating orientation will be perpendicular to the line joining the reference point and the pixel, as in the cases of producing composite gratings (Sect. 2.1.2). Superposing all sinusoidal gratings due to all pixels of $h(\xi, \eta)$, we obtain the spatial filter (transmittance) function of

$$t_a(v_x, v_y) \propto |Ae^{-j2\pi v_y b} + \hat{h}(v_x, v_y)|^2$$
$$= A^2 + |\hat{h}(v_x, v_y)|^2 + A\hat{h}^*(v_x, v_y)e^{-j2\pi v_y b}$$
$$+ A\hat{h}(v_x, v_y)e^{j2\pi v_y b}. \tag{2.10}$$

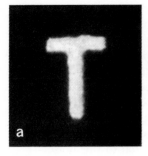
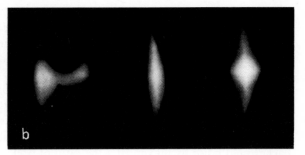

Fig. 2.8. (a) An important portion of the impulse response of a matched filter for the letter T and (b) correlation for the input letters CIT using a matched filter of the letter T

Note that the desired filtering function $\hat{h}^*(v_x, v_y)$ for correlation operation is contained in the third term of (2.10) and $\hat{h}(v_x, v_y)$ for convolution operation in the last term.

An input $f(x, y)$ filtered by the complex spatial transmittance of (2.10) yields an output described by

$$
\begin{aligned}
g(x', y') = \mathscr{F}\{f(v_x, v_y)t_a(v_x, v_y)\} \\
\propto A^2 f(x', y') + \mathscr{F}\{\hat{f}(v_x, v_y)\hat{h}(v_x, v_y)\hat{h}^*(v_x, v_y)\} \\
+ A\mathscr{F}\{\hat{f}(v_x, v_y)\hat{h}^*(v_x, v_y)e^{-j2\pi v_y b}\} \\
+ A\mathscr{F}\{\hat{f}(v_x, v_y)\hat{h}(v_x, v_y)e^{+j2\pi v_y b}\} \\
= A^2 f(x', y') + f(x', y') * h(x', y') * h^*(-x', -y') \\
+ A f(x', y') * h^*(-x', -y') * \delta(x', y' - b) \\
+ A f(x', y') * h(x', y') * \delta(x', y' + b).
\end{aligned}
\tag{2.11}
$$

The third term of (2.11) is the crosscorrelation of $f(x', y')$ and $h(x', y')$ centered at coordinates $(0, b)$ in the output plane P_3 of Fig. 2.1. The fourth term is the convolution of $f(x', y')$ and $h(x', y')$ centered at coordinates $(0, -b)$. The first two terms are of no particular use in the usual filtering operations and are centered at the origin.

If $f(x', y')$ and $h(x', y')$ have the spatial widths of W_f and W_h, respectively, it can be easily shown that the first two terms will occupy the spatial widths of W_f and $(W_f + 2W_h)$, respectively. However, both the third and fourth terms will occupy the same width of $(W_f + W_h)$ as shown in Fig. 2.7b.

As an example, Fig. 2.8a is a photograph of a portion of the impulse response from a complex filter, i.e., the fourth term of the Fourier transform of (2.10). Using this filter, one can observe on P_3 (Fig. 2.1) below the optical axis, the correlation output as shown in Fig. 2.8b when the input is CIT. Among the correlations of T and C, T and I, and T and T, the autocorrelation is identified as the bright spot on the right hand of Fig. 2.8b. The reason for obtaining the bright spot can be understood by observing the facts that the autocorrelation function is $\mathscr{F}\{\hat{f}_p\hat{h}_p^*\}=\mathscr{F}\{\hat{f}_p\hat{f}_p^*\}$, where $\hat{f}_p$ and $\hat{h}_p^*$ correspond to the Fourier transforms of the input function T and of the filter impulse response for T, respectively. Even if $\hat{f}_p$ is a complex function, $\hat{f}_p\hat{f}_p^*$ will still be a positive real-function with uniform phase. In taking the Fourier transform of $\hat{f}_p\hat{f}_p^*$, L_3 in Fig. 2.1 focuses this uniform phase wavefront into the bright spot.

A spatial filter, whose impulse response $h(\xi,\eta)$ contains $f^*(-\xi,-\eta)$, is frequently called a matched filter [2.14, 15]. From signal detection theory, a matched filter is useful for detecting known signals from additive random background noise by maximizing the ratio of peak signal to rms noise:

$$\hat{h}(v_x, v_y) = C\hat{f}^*(v_x, v_y)/N(v_x, v_y), \tag{2.12a}$$

where C is a constant and $N(v_x, v_y)$ the noise spectrum. If the stationary additive noise is white (i.e., the noise spectral density is uniform over the frequency range of interest), the optimum filter function of (2.12a) becomes

$$\hat{h}(v_x, v_y) = C'\hat{f}^*(v_x, v_y). \tag{2.12b}$$

Finally, the techniques used in synthesizing *complex* filters (discussed in this section) and those used in synthesizing *composite* grating filters (discussed in the last subsection) can be combined to synthesize the spatial filters for the combined operations of "subtraction-correlation", or "gradient correlation", etc. [2.16, 17].

2.1.4 Computer Generated Spatial Filters

In Sect. 2.1.3 we take note of the fact that a complex spatial filter can be synthesized from the superposition of many simple sinusoidal gratings. To simplify the tasks of producing complex spatial filters, the interferometric method of Fig. 2.7a can be used. However, even the interferometric method requires that the desire impulse response $h(\xi, \eta)$ of the complex filtering function be physically available. Since in many interesting and useful optical processing applications impulse responses are described in mathematical forms [2.18–20] but do not physically exist, the help of a digital electronic computer in producing the desired complex filter is welcomed.

The basic phenomena utilized in computer generated filters is optical diffraction by apertures (Fig. 2.9a, b). The important steps involved are (a) sampling the filtering function at periodic lattice points with a separation d, (b) determining the dimensions of the apertures so that the amount of light passing through them is proportional to the magnitudes of amplitude transmittance called for by the filtering functions at those sampled points, (c) shifting the position of each aperture with respect to its lattice point's position by an amount Δ proportional to the phase angle of the filtering function at that point,

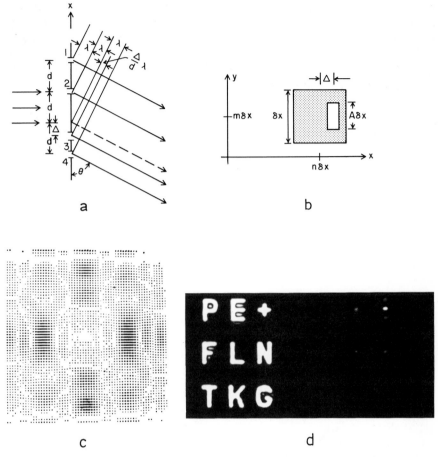

Fig. 2.9a–d. The "detour phase" principle of a computer generated spatial filter is illustrated in (**a**) and (**b**). An example of filter for gradient correlation is illustrated in (**c**). The result of gradient correlation of the letter *E* is shown in (**d**) [2.22]

and (d) photo-reducing the computer output [2.21, 22]. Figure 2.9c shows an example of such a filter for gradient correlation and Fig. 2.9d shows the result using this filter. More examples of computer generated filters are given in [2.21, 22]. Phase contrast filters, differential shearing interference filters and inverse filters are among the examples.

Several variations of this simple method of generating filters by computer have been suggested [2.23–25], each offering a different advantage. For example, some methods suggest the use of fixed position multi-apertures of variable size to synthesize a complex value instead of controlling the size and position of a single aperture to obtain the complex value. Some of these methods are more suited for implementation with a CRT, flatbed microdensito-

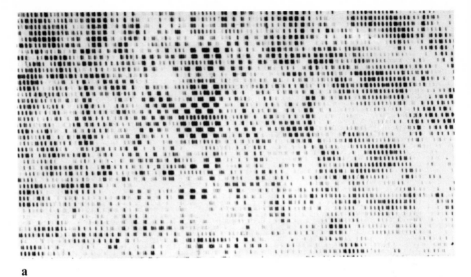

a

b   LASER SCANNING SYSTEM

Fig. 2.10. (a) Photomicrograph of the center of a computer generated filter. This filter consists of 786,432 rectangular apertures of varying transmittance. The size of the filter is 1.4 cm × 1.4 cm. Each aperture is approximately $10 \times 20\,\mu$ in size. (b) The laser beam scanning system is used to produce (a)

meter with writing capability or laser scanner instead of a CAL COMP plotter. Figure 2.10a shows an example of a computer generated filter based on three apertures of fixed positions but with variable transmittance to synthesize a complex value [2.18]. It was generated by the laser beam scanning system under computer control (Fig. 2.10b) and applied to performing Walsh-Hadamard transformations optically (Fig. 2.10c). The same laser scanner has recently also produced computer filters to perform Fukunaga Koontz transforms for classifying statistical patterns [2.19], while a flatbed microdensitometer with writing capability has been used under computer control to produce

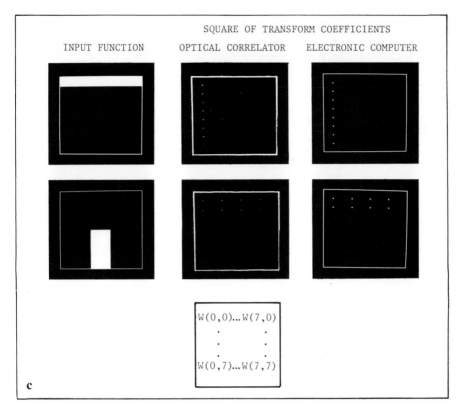

Fig. 2.10. (c) The output of the optical correlator (middle column) is compared to the solution obtained by an electronic computer (right column) for two input objects. The intensity of the spots is proportional to the square of the Walsh-Hadamard transform coefficients in both cases. The placement of coefficients is diagrammed in the bottom square [2.18]

filters for optical character recognition based on nonredundant correlation measurements [2.20]. The general consensus among researchers is that computer generated spatial filters possess the best potential for the future of optical processing.

2.1.5 Spatial Filtering Functions Based on Coherent Optical Feedback

Discussions on coherent processing up to this point have been concerned mainly with the system of Fig. 2.1, using a single spatial filter. The coherent optical feedback system of Fig. 2.11 would permit the use of two spatial filters and contribute a new class of filtering functions uniquely associated with feedback [2.26, 27].

The principles of operation of Fig. 2.11 are as follows. Lenses L_1 and L_2 image the input $f(\xi, \eta)$ telecentrically to the midplane P of the confocal Fabry-

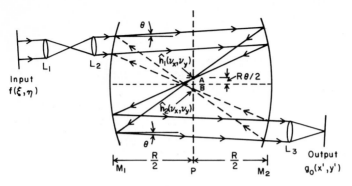

Fig. 2.11. A coherent optical feedback system using a confocal Fabry-Perot interferometer [2.29]

Perot system (CFP). The spherical mirrors M_1 and M_2 function both as reflecting and Fourier transforming elements. The mirror M_2 produces the Fourier transform $\hat{f}(v_x, v_y)$ of $f(\xi, \eta)$ centered at point A in the midplane. Because the zero spatial frequency component of $f(\xi, \eta)$ propagates at an angle θ to the optical axis, the point A is located a distance $R\theta/2$ above the optical axis, where R is the radius of curvature of the mirrors. $\hat{f}(v_x, v_y)$ is spatially filtered by $\hat{h}_1(v_x, v_y)$ at A and the result inverse transformed by mirror M_1. The part of this light (together with those parts from later reflections due to optical feedback) which is transmitted by M_2 constitutes the output $g_0(x', y')$ of the CFP. The output is imaged from the midplane to a vidicon by L_3. The rest of the light is reflected into the feedback path and Fourier transformed by M_2. This transform appears at point B and is spatially filtered by $\hat{h}_2(v_x, v_y)$. Finally, the feedback image is inverse transformed by M_1 and rejoins the original image with a phase shift β which depends in the mirror separation.

The coherent filtering function $\hat{h}_c(v_x, v_y)$ of the CFP is similar to an electronic feedback filtering function. By inspection of Fig. 2.11, an equation relating $\hat{g}_0(v_x, v_y)$ and $\hat{f}(v_x, v_y)$ is:

$$\hat{g}_0(v_x, v_y) = t^2 r^2 t_p^3 \hat{h}_1(v_x, v_y)\hat{f}(v_x, v_y)$$
$$+ r^4 t_p^4 \hat{h}_1(v_x, v_y)\hat{h}_2(v_x, v_y)e^{j\beta}\hat{g}_0(v_x, v_y), \tag{2.13a}$$

where t is the amplitude transmittance of the mirrors, r is the amplitude reflectance of the mirrors and t_p is the amplitude transmittance of the liquid gate holding the spatial filters in the midplane. Solving for the ratio $\hat{g}_0/\hat{f}$ gives

$$\hat{h}_c(v_x, v_y) \equiv \frac{\hat{g}_0}{\hat{f}} = \frac{t^2 r^2 t_p^3 \hat{h}_1}{1 - r^4 t_p^4 \hat{h}_1 \hat{h}_2 e^{j\beta}}$$

$$\propto \frac{\hat{h}_1}{1 - t_c \hat{h}_1 \hat{h}_2 e^{j\beta}}, \tag{2.13b}$$

where $t_c \equiv r^4 t_p^4$ is the amplitude transmittance of one round trip through the CFP. One advantage of the coherent optical feedback system over optical systems without feedback is that, even when $\hat{h}_1$ and $\hat{h}_2$ are real-valued functions, if $\beta = \pm \pi/2$, then $\hat{h}_c$ is complex-valued. Experimentally, this makes it possible to use amplitude transparencies fabricated on photographic film to implement a complex-valued filtering function.

A second advantage of the coherent optical feedback system is that the dynamic range d_r' of $\hat{h}_c$ is greater than the d_r of $\hat{h}$ achievable without feedback. To see this, notice that the spatial filters $\hat{h}_1$ and $\hat{h}_2$ can be considered to have a maximum amplitude transmittance of unity [since t_c in (2.13b) accounts for any residual loss in photographic film] and a minimum transmittance of $1/d_r$. Without feedback, the dynamic range of $\hat{h}$ implemented by using one of these filters is therefore d_r. With feedback, the maximum value of $\hat{h}_c$ can be found from (2.13b) by substituting $\hat{h}_1 = 1$, $\hat{h}_2 = 1$ and $\beta = 0$:

$$\hat{h}_{c\,max} = \frac{t^2 r^2 t_p^3}{1 - t_c}. \tag{2.14a}$$

The minimum value is found for $\hat{h}_1 = 1/d_r$, $\hat{h}_2 = -1$ and $\beta = 0$:

$$\hat{h}_{c\,min} = \frac{t^2 r^2 t_p^3/d_r}{1 + t_c/d_r}. \tag{2.14b}$$

The dynamic range d_r' of $\hat{h}_c$ is therefore:

$$d_r' = \frac{\hat{h}_{c\,max}}{\hat{h}_{c\,min}} = \frac{d_r(1 + t_c/d_r)}{1 - t_c}. \tag{2.14c}$$

Since $t_c < 1$ and $d_r \gg 1$, the use of feedback has increased the dynamic range by a factor of approximately $1/(1 - t_c)$. Experimentally, $d_r' = 8.3 d_r$ has been demonstrated. Furthermore, the feedback filtering of (2.13b) has been applied to image restoration, contrast control [2.28], analog solutions to partial differential equations and integral equations [2.28–31].

2.2 Diffraction Efficiencies and Classifications of Spatial Filters

In the discussions of spatial filter synthesis in Sects. 2.1.2–4, it has been noted that it is frequently the light diffracted into a certain off-axis direction by the spatial filter that contributes to the desired output. Diffraction efficiency is a quantitative measure of the percentage of light illuminating the spatial filter which is diffracted into the off-axis output [2.32]. Since all known detectors contribute noise to the output detected signals, it becomes important to consider the diffraction efficiencies of spatial filters to assure good signal-to-noise ratio in

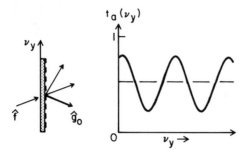

Fig. 2.12. Schematic diagram of a thin amplitude grating and the periodic variation of its transmittance $t_a(v_y)$

the output measurements. Specially important are diffraction efficiency considerations for multiplex filtering functions because more than one output signal would be diffracted into various regions in the output plane affecting the signal-to-noise ratio [2.18, 33].

2.2.1 Thin Amplitude Grating Filters

Spatial filtering functions can generally be decomposed into many grating components, as pointed out in Sects. 2.1.2, 3. A thin filter implies that the highest frequency grating component has a grating period larger than the thickness of the filter recording medium.

The amplitude transmittance of a thin grating filter can be expressed as

$$t_a(v_x, v_y) = \frac{1}{2}\left\{1 + \frac{m_1}{2}[\exp(j2\pi v_y b) + \exp(-j2\pi v_y b)]\right\}, \tag{2.15}$$

where m_1 is the modulation index of the grating. Illuminating the grating filter by a uniform plane wave of unit amplitude, the exponential terms in (2.15) will diffract light into ± 1 orders (Fig. 2.12). The amplitude of the diffracted light is $m_1/4$ and thus the diffraction efficiency η_e, which is defined as the intensity of diffracted light divided by the light intensity incident on the filter, is

$$\eta_e = (m_1/4)^2 = m_1^2/16. \tag{2.16}$$

The maximum diffraction efficiency occurs when m_1 is a maximum. Since both t_a and m_1 in (2.15) can have only the maximum value of one, $\eta_{e\,max} = 1/16$ (that is, a sinusoidal transmittance grating has an ultimate diffraction efficiency of 6.25 %).

Composite grating filters and complex filters which contain many grating components of various frequencies will have diffraction efficiencies less than 6.25 % because not all grating components can simultaneously have the same maximum modulation index of 1. On the other hand, computer generated filters especially, can take on binary forms – binary spatial filters. A special binary

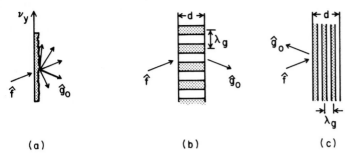

Fig. 2.13. (a) Thin phase grating, (b) thick transmission grating, and (c) thick reflection grating [2.32]

filter has the amplitude transmittance of a square-wave, e.g., a Ronchi grating. Since the first two terms of the Fourier series representation of the square wave are

$$t_a(v_x, v_y) = \frac{1}{2} + \frac{1}{\pi}[\exp(j2\pi v_x b) + \exp(-j2\pi v_x b)] + \ldots, \tag{2.17}$$

its diffraction efficiency is $(1/\pi)^2 = 10.1\%$.

2.2.2 Thin Phase Grating Filters

A thin phase filter has an amplitude transmittance of

$$t_a(v_x, v_y) = \exp[j\phi(v_x, v_y)]. \tag{2.18}$$

It absorbs no light, as indicated by the fact that $|t_a(v_x, v_y)|$ of (2.18) is unity and can be recorded on bleached photographic films [2.34, 35], dichromated gelatin [2.36, 37] or photo-plastic films [2.38–40].

The simplest thin phase filter provides a sinusoidal phase modulation:

$$t_a(v_x, v_y) = \exp[j\phi_1 \cos(2\pi v_y b)]$$
$$= \sum_{n=-\infty}^{\infty} j^n J_n(\phi_1) \exp(jn2\pi v_y b), \tag{2.19}$$

where J_n is the Bessel function of the first kind and nth order and ϕ_1 is a constant. When the sinusoidal phase grating is illuminated with an axial plane wave of unit amplitude, the diffracted amplitude in the -1 order is $J_1(\phi_1)$ (Fig. 2.13a). Since the maximum value of $J_1(\phi_1)$ is 0.582, the maximum diffraction efficiency is $(0.582)^2 = 33.9\%$.

Somewhat more light can be diffracted into the first order if the phase modulation varies as a square-wave function of v_y with $\phi = 0$ during half the square-wave period and $\phi = \pi$ during the remainder. The transmittance $t_a(v_x, v_y)$ in (2.18) is then $+1$ when $\phi = 0$ and -1 when $\phi = \pi$. This is analogous to the square-wave binary filter except that now the amplitude of the incident light diffracted into the first order is twice that for amplitude modulation. Hence, efficiency is four times as large or 40.4 %.

2.2.3 Thick Transmission Grating Filters

A thick grating filter implies that the grating period is small compared with the thickness of the recording medium. The dominant diffraction in thick grating filters occurs for light incident at or near the Bragg angle $[\theta_B = \sin^{-1}(\lambda/2n_o\lambda_g)$, where λ and λ_g are the wavelengths of light and grating, respectively, n_o is the index of refraction of the recording medium]. The fringes of a thick transmission grating are usually orthogonal to the surface of the filter (Fig. 2.13b). If the thick transmission grating filter is the phase type, no absorption will take place and the diffraction efficiency can reach 100 %. However, if it is the absorption type, the diffraction efficiency will be limited to 3.7 %. The analysis for the diffraction efficiencies of thick transmission filters involves the coupled wave theory (coupling between the incident and the diffracted waves) and can be found in [2.32]. Only recently do we find the interest in the thick transmission filters being applied to space variant processing [2.33].

2.2.4 Thick Reflection Grating Filters

Besides the transmission filters (thin or thick) discussed in Sects. 2.2.1–3, one can also produce filters of the thick reflection type. The fringes of a thick reflection grating are usually parallel to the surface of the filter (Fig. 2.13c). Otherwise, the characteristics of reflection gratings are very similar to those of the corresponding transmission gratings discussed in Sect. 2.2.3. The maximum diffraction efficiency of the phase and the absorption types of thick reflection gratings are 100 % and 7.2 %, respectively. Up to now, no work on applying thick reflection filters to coherent optical processing has been reported in the literature, perhaps because these filters are more difficult to produce. However, they may be able to offer the following advantages: (a) more compact processing systems and (b) improved signal-to-noise outputs (if there are any scattering centers in the recording medium contributing noise to the outputs, back scattering is usually not as strong as forward scattering). Table 2.1 summarizes the theoretical efficiencies for the various grating filters.

2.3 Alternative Designs of Optical Processors

Discussions up to here on coherent optical processing mainly concern the processing systems of Figs. 2.1 and 11. However, all optical imaging systems are

Table 2.1. Theoretical efficiencies for various grating filters

Thickness	Modulation	Mode of use	Max efficiency η_e
Thin	Absorption	Transmission	6.25%
Thin	Phase	Transmission	33.9%
Thick	Absorption	Transmission	3.7%
Thick	Phase	Transmission	100.0%
Thick	Absorption	Reflection	7.2%
Thick	Phase	Reflection	100.0%

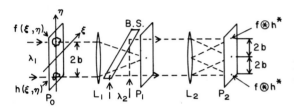

Fig. 2.14. A joint transform optical correlator configuration [2.41]

candidates for use as optical processors and there are alternatives to the coherent feedback system of Fig. 2.11. In this section some of these alternative processing systems are discussed.

2.3.1 Joint Transform Processor

The schematic diagram of a joint transform processor is shown in Fig. 2.14, in which both the input function $f(\xi, \eta)$ and the impulse response of the filter function $h(\xi, \eta)$ are placed side-by-side in the input plane [2.41]. The amplitude transmittance of plane P_0 is then

$$g_i(\xi, \eta) = f(\xi, \eta - b) + h(\xi, \eta + b). \tag{2.20}$$

where the center-to-center separation of the two nonoverlapping functions is $2b$. Lens L_1 forms the Fourier transform of $g_i(\xi, \eta)$ at plane P_1. The intensity of this transform $|\hat{g}_i(v_x, v_y)|^2$ is recorded. Assuming linear recording, the transmittance of P_1 will be proportional to $|\hat{g}_i|^2$:

$$\begin{aligned}
|\hat{g}_i(v_x, v_y)|^2 = &|\hat{f}(v_x, v_y)|^2 + |\hat{h}(v_x, v_y)|^2 \\
&+ \hat{f}(v_x, v_y)\hat{h}^*(v_x, v_y)\exp(-j4\pi v_y b) \\
&+ \hat{f}^*(v_x, v_y)\hat{h}(v_x, v_y)\exp(j4\pi v_y b).
\end{aligned} \tag{2.21}$$

Now, the writing beam is blocked and P_1 is illuminated by a normal plane wave (using the beam splitter BS). At plane P_2 we should then obtain the correlation

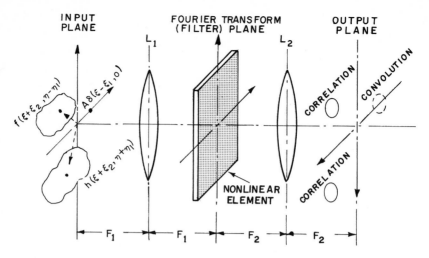

Fig. 2.15. An optical system that performs linear operations in real time with a nonlinear element [2.42]

of $f(x', y')$ and $h(x', y')$ at $y' = \pm 2b$:

$$
\begin{aligned}
g_o(x', y') &= \mathscr{F}[|\hat{g}_i|^2] \\
&= f(x', y') * f^*(x', y') + h(x', y') * h^*(x', y') \\
&\quad + f(x', y') * h^*(x', y') * \delta(x', y' - 2b) \\
&\quad + f^*(x', y') * h(x', y') * \delta(x', y' + 2b).
\end{aligned}
\tag{2.22}
$$

A couple of simple modifications to the joint transform processor as shown in Fig. 2.15 can help achieve real-time processing [2.42]. First, a nonlinear device with fast response time may be used in the filtering plane to provide the necessary mixing or multiplication of the Fourier spectra from the two inputs, $f(\xi, \eta)$ and $h(\xi, \eta)$. Nonlinear devices generally introduce some unwanted outputs also. To separate the convolution and correlation functions spatially from the unwanted outputs, a reference delta function is also incorporated into the input plane. Another useful role of the reference delta function is to provide a tilted plane wave, illuminating the nonlinear device and biasing it at the proper operating point.

2.3.2 Two-Lens Coherent Optical Processor

Figure 2.16 shows a two-lens coherent optical system design, which contains all the basic elements required for processing. A point source S is imaged to plane P_2 by lens L_1. A two-dimensional Fourier transform of the input at P_1 occurs at P_2. Since the input $f(\xi, \eta)$ is illuminated by a spherical wave, a quadratic phase

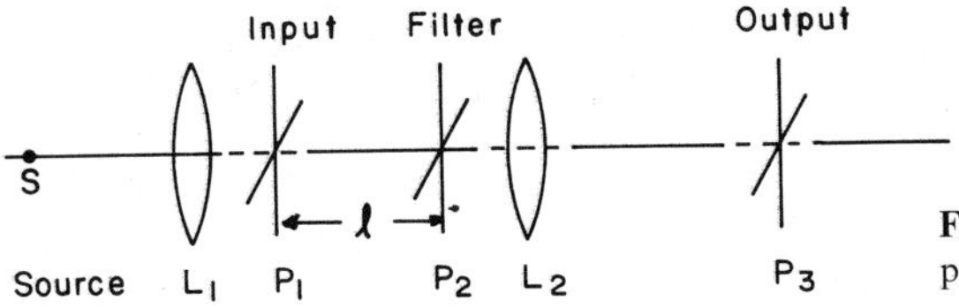

Fig. 2.16. A two-lens coherent optical processor [2.43]

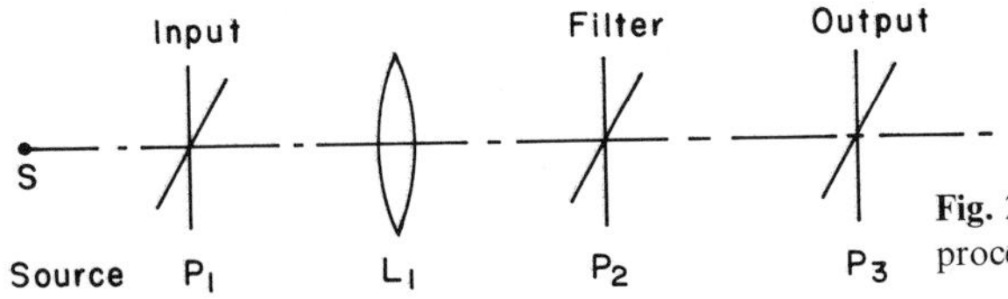

Fig. 2.17. A one-lens coherent optical processor [2.43]

factor will accompany $\hat{f}(v_x, v_y)$. A filter can be placed at P_2 for the optical process. Lens L_2 which satisfies the imaging condition from P_1 to P_3 produces the output image at P_3 [2.43].

This design provides certain advantages over the three lens design of Fig. 2.1. First, the distance l between the input at P_1 and the filter at P_2 can be adjusted so that a scaling of the input image to filter coordinates can be performed (Sect. 1.3.3). This is equivalent to tuning a filter to the frequency parameters of the input image. Second, the two-lens design also allows for adjusting the scale of the output. To obtain the desired output image scale, the P_1 to L_2 and L_2 to P_3 distances can be adjusted, as with any imaging system.

2.3.3 One-Lens Coherent Optical Processor

Figure 2.17 shows a one-lens coherent optical system design, which also satisfies all requirements for processing. The lens L, produces an image of the point source at P_2. The same lens produces an image of P_1 at P_3 [2.43]. The one-lens system offers similar flexibility of adjusting the scale between input and output and between the input and the filter.

Generally, a minimum of optical components should be used in a system on the basis of image quality and component costs. However, the complexity of filter design increases as the number of lenses in the processing system is reduced.

2.3.4 Alternative Coherent Feedback Processor

Figures 2.18–20 show three alternative coherent feedback processors using a various number of optical components. The system depicted in Fig. 2.18 and derived from [2.44] uses four lenses (L_2, L_3, L_5, and L_6), three mirrors (M_1, M_2, and M_3) and two beam splitters (B_1 and B_2). The input $f(\xi, \eta)$ is Fourier

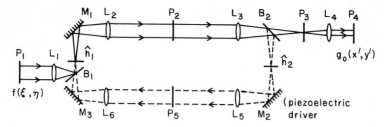

Fig. 2.18. A possible configuration for optical feedback [2.44]

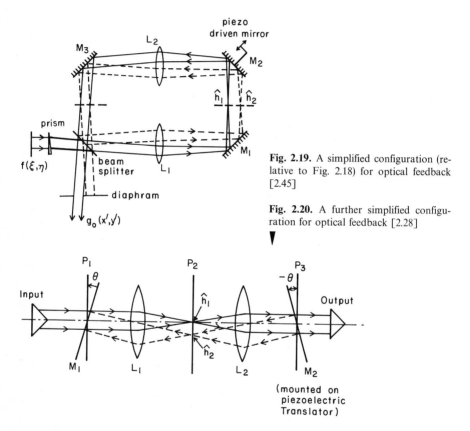

Fig. 2.19. A simplified configuration (relative to Fig. 2.18) for optical feedback [2.45]

Fig. 2.20. A further simplified configuration for optical feedback [2.28]

transformed by L_1 and fed into the feedback system by B_1. $\hat{f}$ is spatial filtered by $\hat{h}_1$ and the result Fourier transformed by L_2. The part of this light (together with those parts from later reflections due to optical feedback) which is telecentrically imaged by L_3 and L_4 from P_2 to P_4 and is transmitted by B_2, constitutes the output $g_o(x', y')$. The rest of the light is reflected into the feedback path by B_2 and Fourier transformed by L_3. This transform is filtered by $\hat{h}_2$, then telecentrically imaged by L_5 and L_6 to rejoin $\hat{f}$ with a phase shift β which depends on the path length of the entire feedback loop.

The system depicted in Fig. 2.19 and derived from [2.45] uses two lenses (L_1 and L_2), three mirrors (M_1, M_2, and M_3) and one beam splitter (B_1). The prism before the front focal plane of L_1 serves to split the beam inside the feedback loop into two separate components which add in amplitude at the output plane after *two* passes through the loop. One component corresponds to the forward (solid) path in Fig. 2.11, the other (dotted) the feedback path.

The system depicted in Fig. 2.20 uses two lenses (L_1 and L_2) and two mirrors (M_1 and M_2). By tilting the mirrors M_1 and M_2 at small angles $+\theta$ and $-\theta$, respectively, the Fourier transforms in the forward and feedback paths are spatially separated to facilitate filtering by $\hat{h}_1$ and $\hat{h}_2$, respectively.

The system depicted in Fig. 2.11 uses two spherical mirrors, each of which combine the functions of a lens and a plane mirror. Hence, this is the simplest system with a minimum number of optical components. Experimentally, it is found that the confocal Fabry-Perot system is easiest to align and most stable in operation.

Processing with feedback need not be restricted to purely optical systems. Currently there are in fact several advantages of performing feedback processing using a hybrid optical/electronic system. For example, amplification, thresholding and other nonlinear operations can readily be performed with electronics on the signal detected by a TV camera from the output of a coherent optical processor. After electronic processing of the optical output signal, the electronic output signal can be fed back to the input of a coherent processor using a high intensity CRT and liquid crystal light valve. To enhance the processing power, a microcomputer with a buffer memory in the electronics and a laser beam scanner under microcomputer control to generate spatial filters, can be included in the hybrid feedback system [2.46]. There is more discussion on real-time interface devices and hybrid systems in Chaps. 4 and 5, respectively.

2.4 Coherent Optical Noise Suppression

The dominant (spatial) noise in a coherent optical processor is scattering from film grains, when spatial filters are recorded on photographic films, or the diffraction rings generated by discrete bubbles and pits in the glass of lenses or by dust particles inside multielement lenses. To reduce scattering noise from film grain, extremely fine grain film should be used or the film should be replaced by some of the high quality real-time devices discussed in Chap. 4. To reduce the noise effects of diffraction rings caused by inhomogeneities in lens glass or by dust particles inside a multielement lens, the scheme depicted in Fig. 1.19 for reducing coherent imaging noise can also be applied here. This occurs because the input and output planes in a coherent processing system are basically imaging planes of each other.

Alternatively, the lens can be rotated around its optical axis to smear the noise effects of glass inhomogeneity and dust without affecting the Fourier

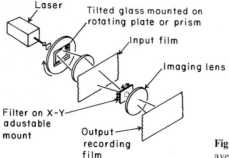

Fig. 2.21. Spatial filtering system with noise averaging device [2.48]

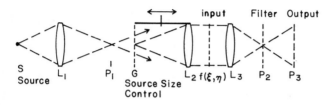

Fig. 2.22. An optical processor with a light source of controllable spatial coherence [2.49, 50]

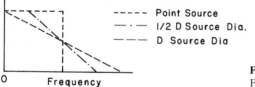

Fig. 2.23. MTF of the optical system in Fig. 2.22 [2.49, 50]

transform of the input ($\hat{f}$) or the processed image (g_o). If it is inconvenient to rotate the lens, a prism can be mounted on a rotating plate to deflect the direction of the light beam illuminating the input (Fig. 2.21). The rotating prism causes the smearing of coherent spatial noise; but the processed image in the output would not be smeared when the input object is kept stationary. Unfortunately, the rotating prism causes the origin of the Fourier transform of the input object to rotate also, thus complicating the filtering process [2.47, 48].

The spatial noise can normally be reduced to a tolerable level also by reducing the degree of spatial coherence of the light source. The laser illumination in Fig. 2.1 can be replaced by a mercury arc source with an interference filter which transmits the 5461 Å green line, for example. The arc is imaged onto a pinhole, and the degree of coherence is determined by the size of this pinhole. With a 25 µm pinhole, for example, the mercury arc system has approximately the same spatial coherence as the laser. As the pinhole is made larger, spatial coherence decreases and the spatial noise is reduced [2.48] (but, the transform plane resolution in the spatial filtering system is also lowered because the observed Fourier transform is the convolution of the object transform and the pinhole).

Figure 2.22 shows an interesting optical processing system with controllable spatial coherence. A small source S is imaged to a point at P_1 by L_1. A ground glass diffuser, G, is positioned on a slide together with L_2. When the diffuser is in plane P_1, the effective source is a small point. As the diffuser is moved from P_1, it intercepts a diverging field providing increasing source size with increasing distance. The input is transilluminated and the source is imaged into P_2, the back focal plane of L_3. The processed output is displayed on P_3 [2.49, 50]. Figure 2.23 illustrates the range of spatial frequency response for several source-to-lens diameter ratios. The dotted curve is obtained with a point source; the dashed curve is obtained with a source diameter equal to or greater than the lens diameter. The dot-dash curve is an intermediate response where the source diameter is one-half the lens diameter. Chapter 3 discusses further optical processing with (spatially) incoherent light.

Finally, the spatial noise in an optical processing system can be reduced by reducing the degree of temporal coherence of the light. White light processing is a subject of increasing importance, but it is also beyond the scope of this chapter. Interested readers are referred to [2.51–53].

References

2.1   D.K.Pollack, C.J.Koester, J.T.Tippett (eds.): *Optical Processing of Information* (Spartan Books, Baltimore, Md. 1965)
2.2   J.T.Tippett, D.A.Berkowitz, L.C.Clapp, C.J.Koester, A.Vanderburgh, Jr. (eds.): *Optical and Electro-optical Information Processing* (M.I.T. Press, Cambridge, Mass. 1965)
2.3   J.W.Goodman: *Introduction to Fourier Optics* (McGraw-Hill, New York 1968) Chap. 7
2.4   W.T.Cathey: *Optical Information Processing and Holography* (Wiley, New York 1974) Chap. 7
2.5   D.Casasent (ed.): *Optical Data Processing*, Topics in Applied Physics, Vol. 23 (Springer, Berlin, Heidelberg, New York 1978)
2.6   A.Vander Lugt: Proc. IEEE **62**, 1300 (1974)
2.7   G.Parrent, B.Thompson: J. SPIE **5** (1966)
2.8   A.B.Porter: Phil. Mag. (6)**11**, 154 (1906)
2.9   F.Zernike: Z. Tech. Phys. **16**, 454 (1935)
2.10 S.H.Lee: Pattern Recognition **5**, 21 (1973)
2.11 S.H.Lee: Opt. Eng. **13**, 196 (1974)
2.12 S.H.Lee: Appl. Phys. **10**, 203 (1976)
2.13 A.B.Vander Lugt: IEEE Trans. IT-**10**, 139 (1964)
2.14 A.Kozma, D.L.Kelly: Appl. Opt. **4**, 387 (1965)
2.15 G.L.Turin: IRE Trans. IT-**6**, 311 (1960)
2.16 K.J.Petrosky, S.H.Lee: Appl. Opt. **10**, 1968 (1971)
2.17 D.P.Jablonowski, S.H.Lee: Appl. Opt. **12**, 1713 (1973)
2.18 J.Leger, S.H.Lee: Opt. Eng. **18**, 518 (1979)
2.19 J.Leger, S.H.Lee: submitted to J. Opt. Soc. Am. (1981)
2.20 B.Braunecker, R.Hauck, A.W.Lohmann: Appl. Opt. **18**, 2746 (1979)
2.21 B.R.Brown, A.W.Lohmann: Appl. Opt. **5**, 967 (1966)

2.22 A.W.Lohmann, D.P.Paris: Appl. Opt. **7**, 651 (1968)
2.23 W.H.Lee: Appl. Opt. **9**, 639 (1970)
2.24 C.B.Burckhardt: Appl. Opt. **9**, 1949 (1970)
2.25 R.E.Haskell, B.C.Culver: Appl. Opt. **11**, 2712 (1972)
2.26 J.Cederquist, S.H.Lee: Appl. Phys. **18**, 311 (1979)
2.27 R.P.Akins, R.A.Athale, S.H.Lee: Opt. Eng. **19**, 347 (1980)
2.28 D.P.Jablonowski, S.H.Lee: Appl. Phys. **8**, 51 (1975)
2.29 J.Cederquist, S.H.Lee: J. Opt. Soc. Am. **70**, 944 (1980)
2.30 J.Cederquist, S.H.Lee: J. Opt. Soc. Am. **71** (June 1981)
2.31 J.Cederquist: J. Opt. Soc. Am. **71** (June 1981)
2.32 H.Kogelnik: Bell Syst. Tech. J. **48**, 2909 (1969)
2.33 J.P.Walkup: Opt. Eng. **19**, 339 (1980)
2.34 K.S.Pennington, J.S.Harper: Appl. Opt. **9**, 1643 (1970)
2.35 R.L.Lamberts, C.N.Kurtz: Appl. Opt. **10**, 1342 (1971)
2.36 L.H.Lin: Appl. Opt. **8**, 963 (1969)
2.37 T.A.Shankoff: Appl. Opt. **7**, 2101 (1968)
2.38 T.C.Lee, N.I.Marzwell, F.M.Schmit, O.N.Tufte: Appl. Opt. **17**, 2802 (1978)
2.39 W.S.Colburn, B.J.Chang: Opt. Eng. **17**, 334 (1978)
2.40 L.H.Lin, H.L.Beauchamp: Appl. Opt. **9**, 2088 (1970)
2.41 C.Weaver: Appl. Opt. **5**, 124 (1966)
2.42 S.H.Lee, K.T.Stalker: J. Opt. Soc. Am. **62**, 1366 (1972)
2.43 P.S.Considine, R.A.Gonsalves: Optical Image Enhancement and Image Restoration in *Optical Data Processing, Applications*, ed. by D. Casasent, Topics in Applied Physics, Vol. 23 (Springer, Berlin, Heidelberg, New York 1978) Chap. 3
2.44 E.S.Nezhvenko, B.I.Spektor: Autometriya **14** (1976)
2.45 E.Handler, U.Roder: Appl. Opt. **18**, 2787 (1979)
2.46 J.Leger, J.Cederquist, S.H.Lee: J. Opt. Soc. Am. **68**, 1414 (1978)
2.47 D.J.Cronin, A.E.Smith: Opt. Eng. **12**, 50 (1973)
2.48 C.E.Thomas: Appl. Opt. **7**, 517 (1968)
2.49 P.Considine: Opt. Eng. **12**, 36 (1973)
2.50 P.Jacquinot: "Apodization" in *Progress in Optics*, Vol. III, ed. by E. Wolf (North-Holland, Amsterdam 1964) Chap. 2
2.51 E.N.Leith, J.A.Roth: Appl. Opt. **18**, 2803 (1979)
2.52 E.N.Leith, J.A.Roth: Appl. Opt. **16**, 2565 (1977)
2.53 F.T.S.Yu, A.Tai: Appl. Opt. **18**, 2705 (1979)

3. Incoherent Optical Processing

W. T. Rhodes and A. A. Sawchuk

With 27 Figures

Most of the research and developmental effort in optical data processing in the 1960's and 1970's centered on coherent optical processing techniques, which depend on complex wave amplitudes to carry data. Work during this period was largely instigated by the development of the laser as a source of coherent wavefronts. However, optical information processing concepts predate the laser by many years. Much early processing was based on incoherent optical techniques, where information is carried by wave intensities. During the late 1970's, the pace of research on incoherent optical processing increased significantly, with a variety of new techniques being developed (for a review of early methods of incoherent optical processing, see [3.1]).

In this chapter we will analyze the basic characteristics of incoherent optical processing systems and look at several important classes. We begin with an overview and a brief comparison of coherent and incoherent methods. We then analyze the multichannel nature of incoherent optical processing systems and consider the consequences in terms of noise reduction and multichannel processing. In subsequent sections, we consider three basic approaches to incoherent optical data processing: incoherent spatial filtering, methods based on ray optics and plane-to-plane imaging, and, briefly, methods based on nonimaging ray optics or shadow casting.

3.1 Coherent vs Incoherent Processing

Coherent optical processing is conceptually simple and elegant. The Fourier transforming property of a spherical lens provides direct access to the spatial frequency spectrum of an input wavefield; linear, shift-invariant systems theory is directly applicable to a variety of information processing applications. However, most coherent optical processing schemes suffer from several distinct problems.

To begin with, coherent processing requires that the input distribution be present in the form of a complex wave amplitude distribution. This requirement precludes the use of CRTs or LED arrays as input devices. Inputs existing in photo transparency form can be used directly, but care must be taken to avoid unwanted phase modulation of the incident wave by emulsion and film base thickness variations. The use of a liquid gate is often necessary. If the information to be processed is not in suitable photo transparency form, then spatial light modulators (SLM) of high optical quality are required. Dynamic range requirements on the SLMs are often severe: a 100:1 range in input wave amplitude corresponds to a 10,000:1 range in input wave intensity, or a photographic density range of 4.0 for the SLM. Even with linearizing circuitry applied to the writing beam, such a dynamic range is often quite difficult or impossible to achieve.

A second problem with coherent optical processing is the coherent noise in the output distribution produced by dust, scratches, and other blemishes on optical components. A small speck of dust, if strongly illuminated, can produce scattered waves and a resulting interference pattern that largely obscures the desired output distribution. Lenses must be kept scrupulously clean; optical components that do not have high quality antireflection coatings may produce undesired interference patterns in the output through multiple reflections.

A third problem, somewhat more subtle than the preceeding two, is the nature of the output. In coherent optical processing, the information being processed is carried by complex wave amplitude distributions. This means not only that the input must be properly presented in wave amplitude form (sometimes necessitating nonlinear preprocessing of the input data), but also that the output distribution of direct concern is the output wave amplitude. However, what is actually measured in the output of the processor (except in rare circumstances where interferometric techniques are used) is the output wave intensity distribution. Phase information (or sign, if the distribution happens to be real valued) is lost. In some applications, matched filtering and image enhancement, for example, this loss of phase information is of little consequence. In other applications, however, for example, optical computation of spatial derivatives, loss of phase (or sign) information may seriously limit the applicability of the optical technique.

Incoherent optical processing does not suffer from these limitations. In incoherent processing, input and output information is conveyed not by complex wave amplitude distributions but by real-valued wave intensity distributions. The input can be a television picture, a 2-D LED array, or a self-luminous object. Photographic transparencies can be used without a liquid gate, since intensity transmittance of a photo transparency is unaffected by emulsion or film base thickness variations. More important perhaps, diffusely reflecting objects, such as printed pages, can be used as inputs to incoherent optical systems. No incoherent-to-coherent conversion via a SLM is necessary.

Incoherent optical processing systems exhibit a high tolerance for dust and blemishes on optical components. Figure 3.1 illustrates with the simple case of a two-lens imaging system. In Fig. 3.1a, the object transparency is illuminated coherently with light form a single source point. In Fig. 3.1b, object illumination is by an extended spatially incoherent source. Figure 3.1c shows the result of coherent imaging, Fig. 3.1d the result of incoherent imaging. In both cases, a microscope slide dusted with chalk was positioned near the pupil plane. The dust and blemishes that contribute distracting interference patterns and

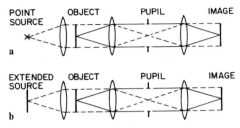

Fig. 3.1a–d. Coherent (a) and incoherent (b) imaging systems and respective images (c), (d). Microscope slide with chalk dust was placed near pupil plane to introduce noise

c

d

artifacts in the coherent imaging operation simply reduce the contrast of the incoherently produced image. As we shall show, this blemish resistance is due to the multichannel nature, or redundancy, of the incoherent operation.

Along with these advantages, there is one distinct disadvantage associated with incoherent processing. Since the carrier of information in the incoherent

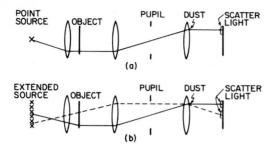

Fig. 3.2a, b. Effect of dust on image with coherent (a) and incoherent (b) illumination

system is light intensity, input and output distributions are restricted to be real and non-negative. This is in contrast to coherent optical processors, where the complex wave amplitudes convey both magnitude and phase information. If incoherent systems are to be used for bipolar or complex-valued processing operations, multiplexing schemes, usually involving some form of hybrid system, must be employed. Large bias distributions often accompany the desired information, and processor dynamic range may in some cases be severely limited. We consider these limitations later in the chapter when we consider three specific approaches to incoherent optical processing. First, however, we consider the important concept of redundancy and noise suppression in incoherent optical systems in greater detail.

3.2 Redundancy and Noise Suppression

As noted, one reason for turning to incoherent systems for optical processing is to achieve a reduction in noise and artifacts in the output distribution that result from dust and blemishes on optical components. This noise reduction results from the multiplicity of paths that light takes in traveling from input to output in an incoherent system. An instructive example, suggested by *Lohmann* [3.2], is shown in Fig. 3.2. This figure depicts an imaging system, (a) with coherent illumination and (b) with incoherent illumination. In the coherent case, light scattered by the dust speck is associated with a specific object point and, therefore, with a specific image point. In the incoherent case, light rays from different object points traverse the dust speck, producing scatter light in different regions of the image plane. Since these scatter distributions are ultimately associated with different mutually incoherent source points, they add incoherently producing a diffuse background or a reduction in image contrast. At the same time, of course, light from different source points illuminates the same object point, serving to reinforce its image.

If we ignore diffraction and go to the geometrical optics limit, light at a particular point anywhere in the system has, in the coherent case, a single direction of propagation. In the incoherent case, a given point in the system is

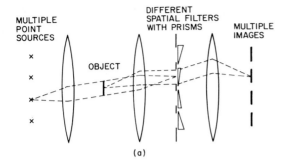

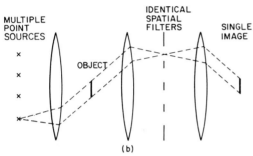

Fig. 3.3a, b. Multiple source imaging system configured (**a**) for parallel processing and (**b**) for redundancy

traversed by light rays traveling in a multiplicity of directions. As *Rogers* notes, the incoherent optical system has a multichannel nature that can be exploited to increase the number of processing operations that can be performed simultaneously (parallel processing) or to reduce the effects of system noise (redundancy) [Ref. 3.1, Sect. 3.3]. There is a continuous tradeoff between these two possibilities.

To illustrate this tradeoff, we consider the multichannel spatial filtering system shown in Fig. 3.3. By assumption, the four source points are mutually incoherent and quasimonochromatic. In Fig. 3.3a, we show the system configured for parallel processing. Each channel contains its own spatial filter; the output consists of four separate distributions, each corresponding to a different coherent spatial filtering operation. Figure 3.3b shows the system configured for maximum redundancy. In this case, the four spatial filters are identical, and all four outputs superpose. Since the sources are mutually incoherent, the four processed distributions add on an intensity basis. Should a lens blemish introduce noise in to one of the channels, the result is a degradation of the composite output image. The degradation is not nearly so severe, however, as it would be were there but a single noisy channel. Intermediate cases could be considered, for example, where two spatial filtering operations are performed in parallel, each with a two-fold redundancy (alternative methods for introducing redundancy to reduce noise affects have been considered; see, for example, [3.4]).

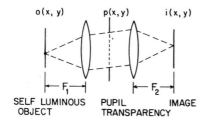

o(x, y)   p(x, y)   i(x, y)

SELF LUMINOUS   PUPIL   IMAGE
OBJECT   TRANSPARENCY

Fig. 3.4. Incoherent spatial filtering system with self-luminous object

3.3 Diffraction-Based Incoherent Spatial Filtering

3.3.1 Basic Theory

Of the three general methods of incoherent optical data processing discussed in this chapter, incoherent spatial filtering using the optical transfer function is the only method that depends on diffraction for its operation. Indeed, the other two methods operate best in the geometrical optics limit. An incoherent spatial filtering system is fundamentally an incoherent imaging system [Ref. 3.5, Chap. 6], but one for which the spatial impulse response (or, equivalently, the system transfer function) has been tailored for a specific data processing task. With reference to the system of Fig. 3.4, the image plane intensity distribution, $i(x, y)$, is given by a convolution integral,

$$i(x, y) = K \int\!\!\int_{-\infty}^{\infty} i_g(\xi, \eta) s(x - \xi, y - \eta) d\xi d\eta, \tag{3.1}$$

which we rewrite in shortened notation as

$$i(x, y) = K i_g(x, y) * s(x, y). \tag{3.2}$$

In this equation, $s(x, y)$ is the incoherent spatial impulse response or point-spread function (PSF) and $i_g(x, y)$ is the image distribution predicted by geometrical optics:

$$i_g(x, y) = (1/M)^2 o(-x/M, -y/M), \tag{3.3}$$

where $o(x, y)$ is the object plane intensity distribution, and where $M = F_2/F_1$ is the system magnification. The constant of proportionality K is determined in part by the precise definition of $s(x, y)$ used and in part by detailed characteristics of the object on a scale beyond the resolution of the imaging system. (An unknown constant of proportionality is consistent with our generally incomplete knowledge as to what fraction of the total light flux from the object propagates in the direction of the aperture.) It should be noted that, although Fig. 3.4 shows a self-luminous object, (3.1) also describes the input/output relationship for a transilluminated object, such as that shown in Fig. 3.1, so long as the spatial extent of the illumination source is sufficiently large [3.6].

In what follows, we assume unity magnification, i.e., $F_1 = F_2 = F$, and take the inversion of the image into account by reversing the direction of the $+x$ and $+y$ axis in the image plane. Equation (3.2) then assumes the convenient form

$$i(x, y) = K o(x, y) * s(x, y). \tag{3.4}$$

Taking the Fourier transform of both sides of this equation, we obtain the spatial frequency domain relationship

$$\hat{i}(v_x, v_y) = K \hat{o}(v_x, v_y) \hat{s}(v_x, v_y), \tag{3.5}$$

where $\hat{i}$, $\hat{o}$, and $\hat{s}$ are the 2-D Fourier transforms of i, o, and s. We refer to $\hat{s}(v_x, v_y)$ as the spatial frequency transfer function of the imaging system.

We gain physical insight into (3.4) if we consider the image distribution produced by a single point source object. Light from the source point, collimated by lens L_1, provides plane wave illumination of the pupil transparency. Lens L_2 acts on the resultant transmitted wave field and produces the Fraunhofer intensity pattern of the pupil function in its back focal plane. We can think of an extended spatially incoherent object as being made up of a collection of mutually incoherent point sources, each source producing a Fraunhofer pattern centered on its geometrical optics image point. Since the source points are mutually incoherent, the Fraunhofer pattern responses add on a wave intensity basis, and, so long as the Fraunhofer pattern does not change shape as a function of source point location (the shift-invariance condition), the superposition integral describing the input/output relationship assumes the convolution form of (3.4).

If we denote by $p(x, y)$ the pupil function of the system, i.e., the complex amplitude transmittance of the aperture, then the back focal plane Fraunhofer pattern that determines the PSF is given by [Ref. 3.5, Eq. (5.15)]

$$I_f(x, y) \propto s(x, y) \propto |\hat{p}(x/F, y/F)|^2, \tag{3.6}$$

where $\hat{p}(v_x, v_y)$ is the 2-D Fourier transform of $p(x, y)$:

$$\hat{p}(v_x, v_y) = \int\int_{-\infty}^{\infty} p(x, y) \exp[-j2\pi(v_x x + v_y y)] dx dy. \tag{3.7}$$

In order to simplify later analysis, we define a scaled pupil function, $s'(x, y)$, along with its 2-D Fourier transform, $\hat{p}'(v_x, v_y)$:

$$p'(x, y) = p(\lambda F x, \lambda F y), \tag{3.8}$$

$$\hat{p}'(v_x, v_y) = \int\int_{-\infty}^{\infty} p'(x, y) \exp[-j2\pi(v_x x + v_y y)] dx dy. \tag{3.9}$$

The function $p'(x, y)$, which we refer to as the *reduced coordinate pupil function*, is the actual pupil function scaled spatially by the numerical value of λF, where λ and F are measured in units of x and y. It is easily shown that $|\hat{p}'(x, y)|^2$ is proportional to $|\hat{p}(x/\lambda F, y/\lambda F)|^2$, i.e., to the point source response Fraunhofer pattern. Recalling that we have an arbitrary constant of proportionality K at our disposal, we *define* the PSF by

$$s(x, y) = |\hat{p}'(x, y)|^2, \tag{3.10}$$

i.e., the PSF is the squared modulus of the Fourier transform of the reduced coordinate pupil function. Taking the Fourier transform of (3.10) and invoking the autocorrelation theorem, we can write the transfer function $\hat{s}(v_x, v_y)$ in the form of an autocorrelation of the reduced coordinate pupil function:

$$\hat{s}(v_x, v_y) = p'(v_x, v_y) \circledast p'(v_x, v_y) = \int\limits_{-\infty}^{\infty} p'(\xi, \eta) p'^*(\xi - v_x, \eta - v_y) d\xi d\eta. \tag{3.11}$$

When normalized to unity at the origin, $\hat{s}(v_x, v_y)$ becomes the conventional optical transfer function, or (OTF), $\mathscr{H}(v_x, v_y)$, given by

$$\mathscr{H}(v_x, v_y) = \hat{s}(v_x, v_y)/\hat{s}(0, 0). \tag{3.12}$$

In the usual approach to describing incoherent imaging systems, $\hat{s}(v_x, v_y)$, $\hat{i}(v_x, v_y)$, and $\hat{o}(v_x, v_y)$ are all normalized by their zero frequency values. Later in this section, however, we describe hybrid spatial filtering systems where $\hat{s}(0, 0)$ and, therefore, $\hat{i}(0, 0)$ may be zero, and this normalization is thus impossible. Rather than adopting a convention we would later be forced to drop, we work consistently with unnormalized functions. To avoid a complication in terminology, we shall often refer to the unnormalized transfer function $\hat{s}(v_x, v_y)$ as the OTF. Strictly speaking, however, this term should be reserved for the normalized function.

As an example of transfer function calculation, assume a square pupil one cm on a side. If x and y are in millimeters, then the actual pupil function $p(x, y)$ is given by $p(x, y) = \text{rect}(x/10, y/10)$, where $\text{rect}(x, y)$ is the unit rectangle function [Ref. 3.7, Eq. (3.53)]. Assuming $F = 20$ cm and $\lambda = 500$ nm, the product λF (λ and F both in millimeters) has the numerical value $\lambda F = 10^{-1}$. Thus, the reduced coordinate pupil function is given by $p'(x, y) = \text{rect}(\lambda F x/10, \lambda F y/10) = \text{rect}(x/100, y/100)$. Calculating the Fourier transform of $p'(x, y)$ and substituting in (3.10), we obtain $s(x, y) = 10^4 \text{sinc}^2(100u, 100v)$, where $\text{sinc}(x, y) = (\sin \pi x/\pi x)(\sin \pi y/\pi y)$.

Although spectral bandwidth is usually of no concern in incoherent imaging, it is in spatial filtering operations. From (3.6) we see that the width of the PSF is proportional to the wavelength λ. If $s(x, y)$ is to remain the same for all wavelengths present in the object distribution, then a constraint must be

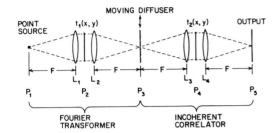

Fig. 3.5. Energy spectra correlator

imposed on the fractional spectral bandwidth $\Delta\lambda/\lambda$. The necessary constraint is easily derived if we note that the fractional change in the PSF width $\Delta W/W$ associated with a change in wavelength $\Delta\lambda$ satisfies the equation $\Delta W/W = \Delta\lambda/\lambda$. We therefore require that ΔW not exceed some reasonable fraction of a resolution cell of the PSF, or $\Delta W < \alpha W/N$, where N is the total number of resolution cells across the width W (the 1-D spacebandwidth product of the PSF), and where α is a constant in the range $0.1 \leq \alpha \leq 1.0$. Combining the two relationships, we obtain the basic condition

$$\Delta\lambda/\lambda < \alpha/N. \tag{3.13}$$

3.3.2 Energy Spectra Correlators

Incoherent spatial filtering was first discussed, in a paper by *Armitage* and *Lohmann*, as a means for correlating the energy spectra of alphanumeric characters in a character recognition system [3.8]. The basic system, which is a combination of coherent and incoherent optical subsystems, is illustrated in Fig. 3.5. The front half of the processor is a simple coherent optical Fourier transform system. Transparency $t_1(x, y)$ is coherently illuminated with quasi-monochromatic light of wavelength λ with a resultant light amplitude distribution in the back focal plane that varies as the spatial Fourier transform of t_1. The moving diffuser serves to destroy the coherence of the Fourier transform, producing an incoherent intensity distribution $I_f(x, y)$ proportional to the energy spectrum (Fraunhofer pattern) of t_1, i.e.,

$$I_f(x, y) \propto |\hat{t}_1(x/\lambda F, y/\lambda F)|^2, \tag{3.14}$$

where $\hat{t}_1(v_x, v_y)$ is the Fourier transform of $t_1(x, y)$. The remaining half of the system is an incoherent spatial filtering system. The output in plane P_5 is thus proportional to the convolution of the energy spectrum at the diffuser, $|\hat{t}_1|^2$, with the point spread function $s(x, y)$. Letting the (unscaled) pupil function be given by $t_2(x, y)$, we obtain from (3.6),

$$s(x, y) \propto |\hat{t}_2(x/\lambda F, y/\lambda F)|^2, \tag{3.15}$$

the energy spectrum of $t_2(x, y)$. Substituting in (3.4), we find the output of the system to be given by the convolution of the two energy spectra:

$$I_{out}(x, y) \propto \int\limits_{-\infty}^{\infty}\int |\hat{t}_1(\xi/\lambda F, \eta/\lambda F)|^2 |\hat{t}_2[(x-\xi)/\lambda F, (y-\eta)/\lambda F]|^2 d\xi d\eta. \qquad (3.16)$$

If either t_1 or t_2 is rotated through $180°$, the result is an integral of the form

$$I_{out}(x, y) \propto \int\limits_{-\infty}^{\infty}\int |\hat{t}(\xi/\lambda F, \eta/\lambda F)|^2 |\hat{t}_2[(\xi-x)/\lambda F, (\eta-y)/\lambda F]|^2 d\xi d\eta, \qquad (3.17)$$

which is the cross correlation of the two energy spectra. In most cases, t_1 and t_2 are real-valued. Under these circumstances, $|\hat{t}_1|^2$ and $|\hat{t}_2|^2$ are not only real but also symmetric, and the convolution of (3.16) and the correlation of (3.17) are equivalent under all circumstances.

An energy spectrum correlator of this type can be used for character recognition (although, as we shall discuss later, there are preferable methods). Its successful operation is based on the premise that the energy spectra associated with different characters differ from one another. This is, for the most part, true, although the differences between some spectra may be subtle, as between the spectra for the letters o and Q, for example. (In fact, characters that are identical to within a $180°$ rotation, e.g., the numerals 6 and 9, have identical energy spectra.) In a character recognition operation, the energy spectrum of the unknown character $t_u(x, y)$ is correlated with the different energy spectra of a standard set of characters $\{t_i(x, y)\}$. For each correlation, the light intensity on the optical axis in the output plane is measured. The measured values are the cross-correlation coefficients, defined by

$$c_{u,j} = \int\limits_{-\infty}^{\infty}\int |\hat{t}_u(\xi/\lambda F, \eta/\lambda F)|^2 |\hat{t}_i(\xi/\lambda F, \eta/\lambda F)|^2 d\xi d\eta, \qquad (3.18)$$

associated with the different correlations. It can be shown using the Schwartz inequality [3.9], that the correlation coefficient is maximum when the input character and the pupil plane character are the same. For each unknown input character, the entire set of possible characters is placed in the pupil plane of the incoherent spatial filtering system and the resultant correlation coefficient measurements compared. Alternatively, a number of spatial filtering systems, each with its own output plane detector, can operate side by side with the same input energy spectrum.

The energy spectrum correlator has several attractive features, among them the invariance to the position of the unknown character in the input plane. A translation of the input character introduces a linear phase factor in its amplitude spectrum, but this has no effect on the energy spectrum. An obvious advantage is the simplicity of the pupil plane transparency: it is simply the

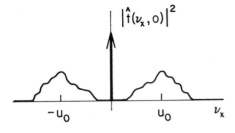

Fig. 3.6. Cross section of energy spectrum of a bandpass signal

character (or other distribution) to be recognized itself. A major disadvantage is that there are great similarities between the energy spectra of alphanumeric characters. What is needed is a system that correlates on specific distinguishing features of the input character energy spectrum or, better yet, on features of the character itself. We shall return to this point shortly.

A natural extension of the basic Armitage-Lohmann scheme is to be found in the correlation of the energy spectrum of an unknown analog signal waveform with a number of energy spectra of "standard" (preclassified) signals [3.10]. Assume the unknown signal is recorded on a SLM in a one-dimensional format. We represent the transmittance function of the SLM by

$$t_u(x, y) = [t_b + Kf(x)],\tag{3.19}$$

where $f(t)$ is the temporal signal input to the recorder and t_b and K are constants.

The recording is placed in the input plane, P_2, of the processor. If we temporarily assume that the processor aperture is sufficiently large, the associated energy spectrum can be adequately represented by

$$|\hat{t}_u(v_x, v_y)|^2 = a\delta(v_x, v_y) + E(v_x)\delta(v_y),\tag{3.20}$$

where $E(v_x)$ denotes the energy spectrum of $f(t)$:

$$E(v_x) = \left| \int_{-\infty}^{\infty} f(t)\exp(-j2\pi v_x t)dt \right|^2.\tag{3.21}$$

For reasons that will become clear, it is generally necessary that $f(t)$ be a narrowband signal. This is not really a handicap, for any lowpass signal can be converted to a bandpass signal without changing the underlying spectral characteristics by multiplying the lowpass signal by $\cos \omega t$, where ω is chosen to be greater than the cutoff frequency of the lowpass signal. Under such circumstances, $|\hat{t}_u(v_x, v_y)|^2$ assumes the appearance suggested in Fig. 3.6. The symmetry inherent in the spectrum of a real bandpass signal can be indicated explicitly by rewriting (3.20) in the form

$$|\hat{t}_u(v_x, v_y)|^2 = a\delta(v_x, v_y) + [\tilde{E}(-v_x - u_0) + \tilde{E}(v_x - u_0)]\delta(v_y),\tag{3.22}$$

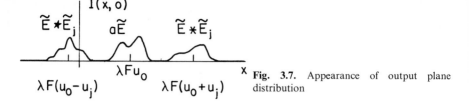

Fig. 3.7. Appearance of output plane distribution

where $\tilde{E}(v_x)$ is the one-sided, or positive frequency spectral distribution measured about its center frequency u_0.

In plane P_3 of the processor we place a similar recording of a "standard" signal, with a resultant pointspread function governed by the energy spectrum

$$|\hat{t}_j(v_x, v_y)|^2 = a\delta(v_x, v_y)$$
$$+ [\tilde{E}_j(-v_x - u_j) + \tilde{E}_j(v_x - u_j)]\delta(v_y). \qquad (3.23)$$

Here the subscript j designates the jth "pattern" energy spectrum, chosen so as to be characteristic of a particular signal class.

The output plane correlation pattern given by $|\hat{t}_u|^2 \circledast |\hat{t}_j|^2$, contains a total of nine terms, some of which, in general, overlap. Of interest to us for signal classification purposes is the signal term $\tilde{E}(v_x - u_0) \circledast \tilde{E}_j(v_x - u_j)$, the correlation of the one-sided power spectrum of the unknown signal with the one-sided spectrum of the standard signal. Because of the symmetry of the energy spectra, the output distribution can be greatly simplified while still retaining this term by placing a stop in the diffuser plane that blocks the zero and negative spatial frequency portion of the spectrum. The result is an intensity distribution in the output plane of the form

$$I(x, 0) = a\tilde{E}(v_x - u_0)$$
$$+ \tilde{E}(v_x - u_0) * \tilde{E}_j(-v_x - u_j)$$
$$+ \tilde{E}(v_x - u_0) \circledast \tilde{E}_j(-v_\delta - u_j), \qquad (3.24)$$

where $v_x = x/\lambda F$. Such a distribution is sketched in Fig. 3.7.

Because of the bandpass nature of the signals, it is possible to guarantee a physical separation in the output plane of the desired cross correlation term from the other two terms by maintaining u_0 and u_j sufficiently large relative to the respective signal bandwidths. This done, unambiguous measurements can be made on a number of such cross correlations, the measured values serving as input to appropriate pattern recognition algorithms.

3.3.3 PSF Synthesis with Holographic Pupil Transparencies

In 1968, *Lohmann* and, independently, *Lowenthal* and *Werts*, greatly extended the range of capabilities of incoherent spatial filtering systems by recognizing

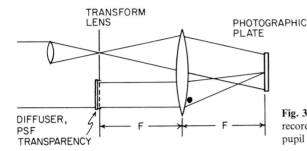

Fig. **3.8.** Fourier transform hologram recording geometry for holographic pupil transparency

that holographic pupil transparencies could be used to realize arbitrary non-negative real PSFs [3.11, 12].

To see how this might be done, assume that the desired PSF, designated by $s_d(x, y)$, exists as the intensity transmittance of a photographic transparency, possibly computer generated. This transparency, in contact with a diffuser, is placed in the input plane of a Fourier transform hologram recording setup, as shown in Fig. 3.8. The complex wave amplitude in the input plane is represented by

$$u_1(x, y) = \sqrt{s_d(x, y)}\, \exp[j\phi(x, y)] + \delta(x - x_0, y),\tag{3.25}$$

where $\phi(x, y)$ is a random phase factor associated with the diffuser. The photographic plate in the back focal plane of the transform lens records the energy spectrum of $u_1(x, y)$, such that the resultant pupil transparency, expressed in reduced coordinates, is given by the complex amplitude transmittance

$$p'(x, y) = |\hat{u}_1(x, y)|^2 w'(x, y),\tag{3.26}$$

where $\hat{u}_1$ denotes the Fourier transform of u_1, and where $w'(x, y)$ is the aperture function, also in reduced coordinates. Evaluating the Fourier transform of $p'(x, y)$ and substituting in (3.10), we obtain the following expression for $s(x, y)$:

$$s(x, y) = |[u_1(x, y) \circledast u_1(x, y)] * \hat{w}'(x, y)|^2,\tag{3.27}$$

where $\hat{w}'(v_x, v_y)$ is the Fourier transform of $w'(x, y)$. *Substituting for* $u_1(x, y)$, we obtain

$$
\begin{aligned}
s(x, y) = |\hat{w}'(x, y) \\
+ [(\sqrt{s_d(x, y)}\, \exp[j\phi(x, y)]) \circledast (\sqrt{s_d(x, y)}\, \exp[j\phi(x, y)])] * \hat{w}'(x, y) \\
+ (\sqrt{s_d(x, y)}\, \exp[j\phi(x, y)]) * \delta(x + x_0, y) * \hat{w}'(x, y) \\
+ (\sqrt{s_d(-x, -y)}\, \exp[-j\phi(-x, -y)]) * \delta(x - x_0, y) * \hat{w}'(x, y)|^2.
\end{aligned}\tag{3.28}
$$

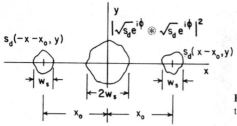

Fig. 3.9. Widths and relative locations of terms in (3.29)

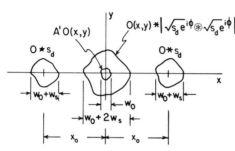

Fig. 3.10. Widths and relative locations of output distribution terms

If x_0 is sufficiently large, the third and fourth terms are spatially isolated from the other two terms. Further, if the aperture is sufficiently large, both $\hat{w}'(x, y)$ and its squared modulus can be treated as delta functions. Under these circumstances, to within a proportionality constant determined by the aperture area, (3.28) assumes the form

$$s(x, y) = a\delta(x, y)$$
$$+ |(\sqrt{s_d(x, y)}\, \exp[j\phi(x, y)]) \circledast (\sqrt{s_d(x, y)}\, \exp[j\phi(x, y)])|^2$$
$$+ s_d(x + x_0, y) + s_d(-x + x_0, -y). \tag{3.29}$$

The regions occupied by these different terms are shown in Fig. 3.9, where w_s designates the width in the x direction of $s_d(x, y)$. When object $o(x, y)$ is placed in the input plane of the spatial filtering system, the desired convolution term $s_d(x, y) * o(x, y)$ appears off axis. If this term is to be physically separated from the other distributions, its displacement from the origin x_0 must be sufficiently large, as suggested in Fig. 3.10. In this figure, w_o designates the width in the x direction of $o(x, y)$. The widths of the four contributions to Fig. 3.10 have been obtained by noting that the widths w_s and w_o add under convolution. The desired distribution $o(x, y) * s_d(x, y)$ is separated from the on-axis distributions if x_0 satisfies the condition

$$x_0 > w_o + (3/2)w_s. \tag{3.30}$$

With holographically recorded pupil functions it is possible to synthesize more or less arbitrary non-negative pointspread distributions. *Lohmann* and *Werlich* made holograms of alphanumeric characters and were thus able to correlate the characters themselves, rather than their energy spectra [3.13]. *Maloney* took oscilloscope traces as his input object for real time signal analysis [3.14], and showed that the incoherent spatial filtering could be performed without lenses if the lens elements were effectively included in the holographic recording itself [3.15]. Other variations on the basic technique are discussed in [3.16–18].

3.3.4 Bipolar Spatial Filtering with Two-Pupil OTF Synthesis

In the form discussed above, incoherent spatial filtering systems are marked by a serious limitation. Whereas coherent processors operate with complex valued impulse responses and input/output distributions, the incoherent optical system must operate with real, non-negative quantities. The restriction to non-negative real inputs is seldom a handicap; much information suitable for optical processing is of this form anyway. The restriction to a non-negative real impulse response is, however, quite serious, since many linear shift-invariant 2-D processing operations of great practical importance, e.g., deblurring, code translation, Wiener filtering, and differentiation (including gradients and other higher order differentiation operations), require impulse responses that are bipolar. A key to the practical success of incoherent spatial filtering lies in the implementation of hybrid systems that allow a bipolar processing capability. In this section we describe a broad class of such systems, referred to generically as two-pupil spatial filtering systems, that are again based on OTF synthesis concepts. We illustrate the idea with a simple example: bandpass spatial filtering.

True bandpass spatial filtering, where low spatial frequency content is totally removed, cannot be performed directly with an incoherent spatial filtering system. The reason can be explained in terms of the autocorrelation function nature of the OTF (this autocorrelation function nature is, of course, consistent with the non-negative real nature of the PSF). As discussed by *Lukosz*, the OTF is fundamentally a lowpass structure [3.19]. Thus, if a two-opening pupil function of the form shown in Fig. 3.11a is used, the resultant OTF, Fig. 3.11b, although containing distinct passbands, is also characterized by large values at low spatial frequencies. Consider, however, the operation suggested in Fig. 3.12. We first perform a spatial filtering operation with the pupil and transfer function of Figs. 3.12a and b. The image obtained with this system is recorded. The pupil function is then modified by a 180° phase shift across one opening (using, for example, a half wave plate) and a second image recorded. For this operation, the pupil and transfer functions have the form shown in Figs. 3.12c and d. The final step is the subtraction of the two images. Denoting the two images by $i_1(x, y) = K o(x, y) * s_1(x, y)$ and

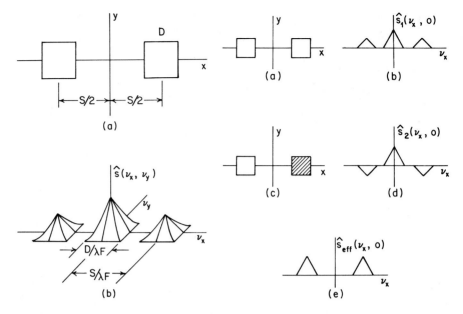

Fig. 3.11a, b. Two-opening pupil functions (a) and associated OTF (b)

Fig. 3.12a–e. Pupil functions (a), (c) and corresponding OTFs (b), (d) used to synthesize OTF shown in (e). Crosshatching in (c) denotes 180° phase shift

$i_2(x, y) = K o(x, y) * s_2(x, y)$, where $s_1(x, y)$ and $s_2(x, y)$ are the two PSFs, the difference image $i(x, y)$ is given by

$$i(x, y) = K o(x, y) * s_1(x, y) - K o(x, y) * s_2(x, y)$$
$$= K o(x, y) * s_{eff}(x, y), \tag{3.31}$$

where

$$s_{eff}(x, y) = s_1(x, y) - s_2(x, y) \tag{3.32}$$

is the synthesized PSF. Taking Fourier transforms, we obtain the frequency domain relationship for the synthesized OTF:

$$\hat{s}_{eff}(v_x, v_y) = \hat{s}_1(v_x, v_y) - \hat{s}_2(v_x, v_y). \tag{3.33}$$

The effective transfer function for the bandpass filtering example is shown in Fig. 3.12e. Various stages of a bandpass spatial filtering operation are shown in Figs. 3.13a–d.

It is clear from the preceeding discussion that bipolar spatial filtering, if it is to be performed incoherently, requires some method for representing negative

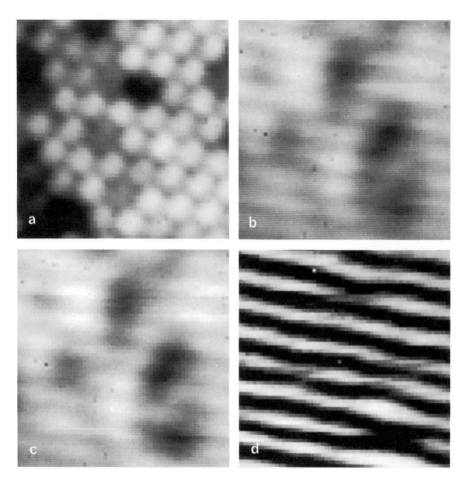

Fig. 3.13a–d. Bandpass spatial filtering operation: (**a**) conventional image of test object, an optical fiber bundle; (**b**) $\theta = 0°$ image; (**c**) $\theta = 180°$ image; (**d**) bandpass image obtained by subtracting 180° image from 0° image (bias added for display)

values with non-negative light intensity distributions. One approach is to use wavelength or polarization to encode bipolar information. Alternatively, bipolar intensity distributions can be represented numerically or electronically in a hybrid system.

The bandpass filtering example is representative of a class of hybrid spatial filtering techniques investigated by *Rhodes* et al. [3.20–22]. The method employed, referred to as phase-switching synthesis of OTFs, has its basis in the phase switching methods commonly employed in processing signals from radio telescope interferometers [3.23]. Consider the imaging system illustrated in Fig. 3.14. This system, characterized by an extended pupil region, has two pupil

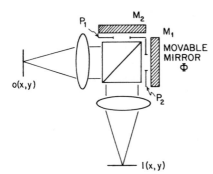

Fig. 3.14. Two-pupil (interferometric) incoherent spatial filtering system

functions, p_1 and p_2, which are effectively superposed by the beamsplitter. With the optical path lengths of the two arms of the system equal, the overall system pupil function is given by the sum of p_1 and p_2. (In this particular system, light traverses each pupil transparency twice. Thus p_1 and p_2 are in fact the squares of the actual complex transmittance functions of the pupil transparencies.) If the pathlength in one arm is changed slightly, however, for example, by moving mirror M_2 a small distance, a phase factor is introduced in one of the component pupil functions, with the result (using reduced coordinate pupil functions)

$$p'(x, y) = p'_1(x, y) + p'_2(x, y) \exp(j\phi). \tag{3.34}$$

By changing the value of ϕ, it is possible to change $p'(x, y)$ and, thereby, the PSF.

In a typical PSF synthesis operation, ϕ is switched from $\phi = 0°$ to $\phi = 180°$. The corresponding PSF component states are given by, using (3.34) and (3.10),

$$s_1(x, y) = |\hat{p}'_1(x, y) + \hat{p}'_2(x, y)|^2, \tag{3.35}$$

$$s_2(x, y) = |\hat{p}'_1(x, y) - \hat{p}'_2(x, y)|^2. \tag{3.36}$$

Evaluating these expressions and substituting in (3.32), we obtain

$$s_{\text{eff}}(x, y) = 2[\hat{p}'_1(x, y)\hat{p}'^*_2(x, y) + \hat{p}'^*_1(x, y)\hat{p}'_2(x, y)] \tag{3.37a}$$

$$= 4\text{Re}\{\hat{p}'_1(x, y)\hat{p}'^*_2(x, y)\} \tag{3.37b}$$

$$= 2|\hat{p}'_1(x, y)\hat{p}'_2(x, y)|\cos[\theta_1(x, y) - \theta_2(x, y)], \tag{3.37c}$$

where Re denotes the real part and where $\theta_i(x, y) = \arg[\hat{p}'_i(x, y)]$. The corresponding transfer function is given by

$$\hat{s}_{\text{eff}}(v_x, v_y) = 2[p'_1(v_x, v_y) \circledast p'_2(v_x, v_y) + p'_2(v_x, v_y) \circledast p'_1(v_x, v_y)]. \tag{3.38}$$

Since $s_{eff}(x, y)$ is bipolar, it is possible that the effective output image is itself negative for certain values of x and y. If the difference image is retained in numerical form, negative values are perfectly acceptable. If, on the other hand, the difference image is to be displayed, something must be done to insure only positive display values. A bias must be added, the signal must be rectified (full or half wave), or some other such operation must be performed.

If pupil transparencies $p_1(x, y)$ and $p_2(x, y)$ are recorded holographically (natural or computer generated), arbitrary PSFs can be synthesized using the 0–180° phase switching operation. In some situations, however, the pupil transparencies may be constrained in form, for example, p_1 and p_2 may correspond to pure phase transparencies. In this case, greater flexibility can be gained by switching ϕ between 90° and 270° states as well. By taking a linear combination of the image distribution obtained with $\phi = 0°$, 90°, 180°, and 270° (actually, three scans taken with phase differences of 120° suffice), a final output distribution can be obtained which corresponds to an effective PSF given by

$$s_{eff}(x, y) = \alpha[|\hat{p}_1'(x, y)|^2 + |\hat{p}_2'(x, y)|^2]$$
$$+ \beta\,\mathrm{Re}\{\hat{p}_1'(x, y)\hat{p}_2'^*(x, y)\}$$
$$+ \gamma\,\mathrm{Im}\{\hat{p}_1'(x, y)\hat{p}_2'^*(x, y)\}, \qquad (3.39)$$

where Im denotes the imaginary part. The phase ϕ can be monitored indirectly by measuring light intensity in the pupil plane: maximum pupil plane intensity corresponds to $\phi = 0°$, minimum to $\phi = 180°$. Mirror positions for $\phi = 90°$ and $\phi = 270°$ are easily extrapolated from the 0° and 180° positions. Two independent PSFs can be synthesized if $p_1(x, y)$ and $p_2(x, y)$ are sufficiently general (e.g., holographically recorded). Thus, two separate spatial filtering operations can be performed at the same time. Alternatively, the two independent PSFs can be identified with the real and imaginary parts of a complex-valued PSF for complex processing operations. See [3.24] for additional discussions along these lines.

The problem of pupil function specification still remains. *Stoner*, in connection with a closely related bipolar spatial filtering method discussed below, has suggested a number of clever solutions for special optical processing operations [3.25]. Computer generated pupil transparencies determined by iterative algorithms can serve in more general applications such as matched filtering. A detailed analysis of pupil function specification is presented in [3.26]. An important consideration noted in that reference is the contrast of the processed component images. With reference to the PSF synthesis of (3.32), for example, it can be shown that the choice

$$s_1(x, y) = \begin{cases} s_{eff}(x, y), & s_{eff}(x, y) \geq 0 \\ 0, & \text{else} \end{cases} \qquad (3.40)$$

$$s_2(x, y) = \begin{cases} -s_{eff}(x, y), & s_{eff}(x, y) < 0 \\ 0, & \text{else} \end{cases} \qquad (3.41)$$

minimizes the bias in the component output distributions and thus allows optimum utilization of the overall system dynamic range.

As an alternative to discrete phase switching, the bipolar PSF and image information can be placed on a temporal frequency carrier. In this method, the phase ϕ is made to vary as $\phi = \phi(t) = \omega t$. Methods for varying the phase in this manner are discussed in [3.21]. The effective reduced coordinate pupil function is then

$$p'(x, y) = p'_1(x, y) + p'_2(x, y) \exp(j\omega t). \tag{3.42}$$

The corresponding time varying PSF can be written in the form

$$s(x, y, t) = |\hat{p}'_1(x, y)|^2 + |\hat{p}'_2(x, y)|^2 + s_\phi(x, y, t), \tag{3.43}$$

where, letting $\Delta\theta = \theta_1 - \theta_2$,

$$s_\phi(x, y, t) = 2|\hat{p}'_1(x, y)\hat{p}'_2(x, y)| \cos[\omega t - \Delta\theta(x, y)]. \tag{3.44}$$

For object distribution $o(x, y)$, the resultant image distribution $i(x, y, t)$ is given by

$$i(x, y, t) = K o(x, y) * s_1(x, y) + K o(x, y) * s_2(x, y)$$
$$+ K o(x, y) * s_\phi(x, y, t). \tag{3.45}$$

If the image $i(x, y, t)$ is scanned with a nonintegrating detector, e.g., an image dissector, the first two terms of (3.45) produce lowpass signals at the detector output. The third term is bandpass in nature and, assuming ω is sufficiently high, can be isolated by a bandpass filter and demodulated to yield a bipolar scan signal as output. For convenience, we write $s_\phi(x, y, t)$ of (3.44) in the form

$$s_\phi(x, y, t) = 2s_R(x, y) \cos\omega t + 2s_I(x, y) \sin\omega t, \tag{3.46}$$

where

$$s_R(x, y) = |\hat{p}'_1(x, y)\hat{p}'_2(x, y)| \cos\Delta\theta(x, y)$$
$$= \mathrm{Re}\{\hat{p}'_1(x, y)\hat{p}'^*_2(x, y)\}, \tag{3.47}$$

$$s_I(x, y) = |\hat{p}'_1(x, y)\hat{p}'_2(x, y)| \sin\Delta\theta(x, y)$$
$$= \mathrm{Im}\{\hat{p}'_1(x, y)\hat{p}'^*_2(x, y)\}. \tag{3.48}$$

A quadrature demodulation system, like the one shown in Fig. 3.15, can be used to separate two components from the input scan signal, one proportional to $o(x, y) * s_R(x, y)$, the other to $o(x, y) * s_I(x, y)$. A weighted sum of these can be

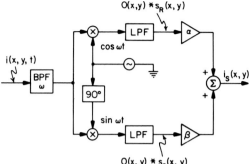

Fig. 3.15. Schematic diagram of a quadrature demodulation system that yields desired image distribution from scan signal

taken for the final output image, given by

$$i(x, y) = K o(x, y) * [\alpha s_R(x, y) + \beta s_I(x, y)]. \tag{3.49}$$

Since both $s_R(x, y)$ and $s_I(x, y)$ are, in general, bipolar and for unconstrained pupil functions independent of one another, any real, bounded, bandlimited, bipolar impulse response can be synthesized in this way.

In a particularly convenient scheme ϕ is varied sinusoidally in time, for example, by moving the mirror of Fig. 3.14 sinusoidally. Setting $\phi = A \sin \omega t$ in (3.34) and calculating the time-varying PSF, we obtain an expression like (3.44), but with $s_\phi(x, y, t)$ now given by

$$s_\phi(x, y, t) = 2|\hat{p}'_1(x, y) \hat{p}'_2(x, y)| \cos[A \sin \omega t - \Delta\theta(x, y)]. \tag{3.50}$$

Using a Bessel series expansion for the time-varying term, we can write $s_\phi(x, y, t)$ in the form

$$\begin{aligned}
s_\phi(x, y, t) = &\, 2s_R(x, y) J_0(A) \\
&+ 4s_I(x, y)[J_1(A) \sin \omega t + J_3(A) \sin 3\omega t + \dots] \\
&+ 4s_R(x, y)[J_2(A) \cos 2\omega t + J_4(A) \cos 4\omega t + \dots],
\end{aligned} \tag{3.51}$$

where $s_R(x, y)$ and $s_I(x, y)$ are as defined before. At the output of a nonintegrating image scanner, terms varying at odd harmonics of ω are governed by $s_I(x, y)$; terms varying at even harmonics of ω are governed by $s_R(x, y)$. A general processing system using both $s_R(x, y)$ and $s_I(x, y)$, analogous to the system of Fig. 3.15, is shown in Fig. 3.16. The outputs of the two demodulator sections are $o(x, y) * 2J_2(A)s_R(x, y)$ and $o(x, y) * 2J_1(A)s_I(x, y)$, as indicated. A reasonable choice for A is $A = 2.65$, for at that value, $J_1(A) = J_2(A)$. The corresponding peak-to-peak excursion of the mirror is approximately 4/10 of a wave.

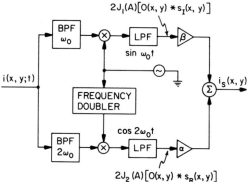

Fig. 3.16. Demodulation system for use with sinusoidal phase modulation

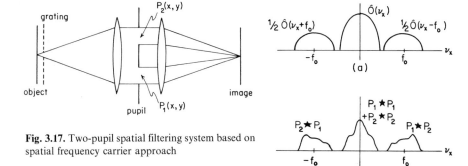

Fig. 3.17. Two-pupil spatial filtering system based on spatial frequency carrier approach

Fig. 3.18a, b. Spatial frequency spectrum of grating-modulated object (a) and transfer function of two-pupil system (b)

Closely related to the temporal carrier method is a spatial carrier method for incoherent processing, proposed independently by *Lohmann* [3.24] and *Stoner* [3.27], and investigated extensively by the latter [3.25, 28]. In this method, two pupil transparencies are again used, but this time they are placed side by side in the pupil plane of an imaging system as in Fig. 3.17. Writing the total pupil function in the form

$$p'(x, y) = p'_1(x, y) * \delta(x - f_0, y) + p'_2(x, y) * \delta(x + f_0, y), \tag{3.52}$$

we obtain, on evaluating the Fourier transform $\hat{p}'(v_x, v_y)$ and substituting in (3.10),

$$s(x, y) = |\hat{p}'_1(x, y)|^2 + |\hat{p}'_2(x, y)|^2$$
$$+ 2|\hat{p}'_2(x, y)\hat{p}'_2(x, y)| \cos[\omega x - \Delta\theta(x, y)], \tag{3.53}$$

where $\omega = 2\pi f_0$. This is virtually the same expression we had for the temporal carrier case, (3.43), but the temporal carrier has been replaced by a spatial carrier. In order to use this kind of processing system, it is necessary that the object distribution be on a spatial carrier also. This is accomplished by placing the object in contact with a sinusoidal grating with intensity transmittance

$$t_g(x, y) = (1/2)(1 + \cos 2\pi f_0 x). \tag{3.54}$$

The effective input object is then $o(x, y)t_g(x, y)$, with a corresponding spatial frequency domain representation

$$\hat{o}(v_x, v_y)\hat{t}_g(v_x, v_y) = (1/2)\hat{o}(v_x, v_y)$$
$$+ (1/4)\hat{o}(v_x - f_0, v_y) + (1/4)\hat{o}(v_x + f_0, v_y). \tag{3.55}$$

An exemplary distribution is shown in Fig. 3.18, along with the transfer function of the system, given by

$$\hat{s}(v_x, v_y) = p_1'(v_x, v_y) \circledast p_1'(v_x, v_y)$$
$$+ p_2'(v_x, v_y) \circledast p_2'(v_x, v_y)$$
$$+ [p_1'(v_x, v_y) \circledast p_2'(v_x, v_y)] * \delta(v_x - f_0, v_y)$$
$$+ [p_2'(v_x, v_y) \circledast p_1'(v_x, v_y)] * \delta(v_x + f_0, v_y). \tag{3.56}$$

The important point is that the grating modulation shifts the object spectrum up to the center frequency of the bandpass portion of the transfer function. The corresponding output image distribution thus has the form

$$i(x, y) = K o(x, y) * |\hat{p}_1'(x, y)|^2 + K o(x, y) * |\hat{p}_2'(x, y)|^2 + K(1/2)o(x, y) * s_{\omega x}(x, y), \tag{3.57}$$

where

$$s_{\omega x}(x, y) = 2|\hat{p}_1'(x, y)\hat{p}_2'(x, y)| \cos[\omega x - \Delta\theta(x, y)]$$
$$= 2s_R(x, y) \cos \omega x + 2s_I(x, y) \sin \omega x, \tag{3.58}$$

where again $s_R(x, y)$ and $s_I(x, y)$ are as defined before. If the image is scanned line by line by a detector that moves with constant velocity in the x direction, the spatial carrier is converted into a temporal carrier; the output scan signal can be demodulated and otherwise processed as in the temporal carrier case. Note that with the spatial carrier, an integrating detector such as a vidicon can be used with an attendant gain in performance at low light levels. The necessary local oscillator signal for demodulation can be obtained if a double frequency sinusoidal "pilot" distribution, $\cos 2x$, is added to the object [3.28]. The corresponding double frequency temporal sinusoid at the output of the scanning detector is halved in frequency to yield the desired local oscillator waveform. An alternative approach, satisfactory if preservation of the sign of the bipolar output distribution is not essential, is to use an envelope detector on the bandpass portion of the output scan signal waveform.

As noted by *Lohmann*, the spatial carrier method can be used to process bipolar (or even complex-valued) input distributions incoherently [3.24], the basic idea being to encode both the magnitude and sign (phase) of the bipolar input on a spatial carrier. The object distribution then assumes the form

$$o(x, y) = b(x, y) + |f(x, y)| \cos[2\pi f_0 x + \pi \operatorname{sgn}[f(x, y)]], \qquad (3.59)$$

where $f(x, y)$ is the spatial distribution to be processed and $b(x, y)$ is a lowpass function chosen to guarantee that $o(x, y) \geqq 0$, as required of an intensity distribution. *Furman* and *Casasent*, using hardclipped computer generated transparencies, have investigated this method in connection with optical pattern recognition applications [3.29].

The two-pupil OTF synthesis methods discussed in this subsection have been described in [3.26] as pupil interaction syntheses because of their dependence on interference of light from the two pupils p_1 and p_2. These methods are relatively complicated analytically and somewhat difficult to implement because of the interferometric nature of their operation. They are nevertheless attractive because of the great flexibility they allow in the synthesis of relatively general PSFs with pupil transparencies that operate in the zero order (i.e., nonholographic.) A complementary noninteraction regime of OTF synthesis schemes, investigated initially by *Chavel* and *Lowenthal*, is conceptually and analytically more straightforward and more easily implemented. In this regime, the effective PSF is given by the weighted difference of two conventional PSFs:

$$s_{\text{eff}}(x, y) = |\hat{p}'_1(x, y)|^2 - |\hat{p}'_2(x, y)|^2 . \qquad (3.60)$$

Several methods for implementing this kind of synthesis have been considered [3.30–32]. The most straightforward is a direct subtraction of two images of $o(x, y)$, one obtained with PSF $|\hat{p}'_1(x, y)|^2$, the other with PSF $|\hat{p}'_2(x, y)|^2$. The two images can be obtained sequentially with intermediate storage, or simultaneously using parallel optical systems. If general PSFs are to be synthesized, pupil transparencies operating in the zero order are inadequate; holographic pupil functions must be used, with an attendant loss in the usable system spacebandwidth product. Further discussions of incoherent spatial filtering are to be found in [3.33], and some early examples of incoherent spatial filtering in [3.34, 35].

3.4 Geometrical Optics-Based Processing: Imaging

In this section we present several different methods for incoherent optical processing that rely on geometrical optics-based imaging between various planes of the system. There is no direct spatial filtering of wavefronts or OTF modification as in the systems of Sect. 3.3. The systems described in Sect. 3.5 also have no spatial filtering or OTF modification, but they do not form images in the conventional sense. Instead they rely on principles generally known as "shadow-casting", and can be completely analyzed in terms of geometrical optics. Although the effects of diffraction are neglected in the description of the systems in Sects. 3.4, 5, in practice, diffraction effects limit the space-bandwidth product and processing capabilities of these systems. The influence of diffraction is generally more severe in the shadow-casting systems of Sect. 3.5 than in the imaging systems of Sect. 3.4. A brief discussion of the problem is given in Sect. 3.5.

In a review paper, *Monahan* et al. [3.36] have discussed many types of geometrical optics-based incoherent imaging and shadow-casting systems. In this section, we distinguish between spatial scanning and temporal scanning systems, generally following their classification. We make no attempt at an exhaustive list of all published techniques, but instead we concentrate on those having the greatest potential and the most interesting applications. For additional references the reader is referred to [3.36].

The operations performed by the systems in Sects. 3.4 and 5 are generally linear transformations of the one-dimensional form

$$g(u) = \int f(x)s(u;x)dx ,\tag{3.61}$$

or the two-dimensional form

$$g(u, v) = \iint f(x, y)s(u, v; x, y)dxdy ,\tag{3.62}$$

or discrete versions in which the integrals are replaced by summations. Most of the limits on integrations and summations will be omitted for clarity in this discussion. The general operations in (3.61, 62) include a large variety of useful linear processing functions, including convolution (linear shift-invariant superposition), correlation, shift-variant filtering, matrix-vector multiplication, and transforms such as Fourier or Walsh-Hadamard. In Sects. 3.4, 5 we change our notation, denoting inputs by $f(x)$ or $f(x, y)$ and outputs by $g(u)$ or $g(u, v)$, because many of these systems perform processing of optical or electronic signals that are not necessarily objects or images. We retain the notation $s(u;x)$ and $s(u, v; x, y)$ for general shift-variant impulse responses or point-spread functions.

It is important to note once again that the information-carrying physical quantity in all these systems is the irradiance and not a complex field amplitude. The irradiance is a non-negative real quantity and special processing techniques are needed to handle bipolar or complex functions. One way of achieving bipolar operation is to add a positive bias term sufficient to make all negative signals non-negative. This method has the disadvantages of reducing the usable dynamic range and introducing offset terms which must be measured

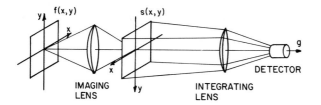

Fig. 3.19. System for obtaining the integral of the product of two functions

and cancelled in the final output. Another approach is to separately perform calculations on positive and negative parts of the signals with a multiplexing system, then combining them later with analog or digital subtraction. This approach can be extended to calculations involving real complex numbers. In fact these approaches are special cases of a general approach discussed by *Lohmann* and *Rhodes* [3.26]. A specific type of temporal scanning/spatial integrating system discussed later in Sect. 3.4.4 uses several types of coding to perform complex calculations [3.37–39]. Other techniques of circumventing the problem of non-negative signals have been described by *Piety* [3.40] and *Rogers* [3.1, 41]. Since these difficulties and the methods of solution are common to all the systems here, we assume that all functions in the sections to follow are non-negative real, with the understanding that extension to bipolar real or complex is possible.

Because the geometrical optics imaging systems in Sect. 3.4 can be made achromatic and the fact that they do not rely on diffraction for the modification of information, these systems operate under a much wider range of source spectral bandwidths than the diffraction based systems described previously.

3.4.1 Basic Processing Variables

The simplest kind of geometrical optics-based imaging system for incoherent processing is shown in Fig. 3.19. The system consists of an imaging lens which transfers the input function $f(x, y)$ to the plane of a transmitting filter function $s(x, y)$. The combination of an integrating lens and detector produces an output of the form

$$g = g(0, 0) = \iint f(x, y)s(x, y)dxdy. \tag{3.63}$$

The input $f(x, y)$ is an incoherent two-dimensional irradiance function which could be a photographic transparency, a CRT or other incoherent display, or simply a real-world scene viewed by the optics. The transmittance $s(x, y)$ could also be a photographic transparency or mask, or a real-time erasable electro-optical device which is optically or electronically controlled. Such systems have been used for character recognition by *Goldberg* [3.42] and for two-dimensional area measurement by *Tea* [3.43].

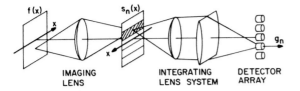

IMAGING LENS   INTEGRATING LENS SYSTEM   DETECTOR ARRAY

Fig. 3.20. Multichannel system for one-dimensional processing

The system shown in Fig. 3.19 is fundamental to most of the techniques described in the following sections. By spatial scanning of $f(x, y)$, $s(x, y)$, the imaging lens or some combination of the above, many two-dimensional and one-dimensional linear processing operations can be achieved. Parallel arrays which replicate the system of Fig. 3.19 are also described. In spatial scanning systems the integration takes place over spatial variables. Another major type of linear processing system combines the basic system of Fig. 3.19 with some type of temporal modulation of the light source or the input $f(x, y)$. These systems generally require mechanical or electronic scanning to collect the output so that the integration takes place over time. It is important to note that all these systems assume perfect imaging in the system of Fig. 3.19. Thus, any diffraction, aberrations, scattering due to film-grain noise, stray light, etc., will cause errors in the processing.

3.4.2 Spatial Scanning Processors

The simplest type of spatial scanning linear processor can be used to compute one-dimensional convolution or correlation integrals. The system consists of the image casting setup shown in Fig. 3.19 along with a mechanism for mechanical translation of $f(x, y)$, $s(x, y)$ or the imaging lens in the x dimension alone. If the input is shifted by an amount x_0, then we obtain a one-dimensional cross correlation

$$g(x_0) = \int f(x - x_0) s(x) dx, \tag{3.64}$$

with the inverted coordinate system shown in Fig. 3.19. If the mask function $s(x, y)$ is inverted and shifted by x_0, then

$$g(x_0) = \int f(x) s(x_0 - x) dx \tag{3.65}$$

which is a one-dimensional convolution. *Field* [3.44] achieved such a system by putting the functions on moving film strips and *Ferre* [3.45] used a scanning mirror between f and s for relative displacement. Many similar systems using acousto-optic modulators for real-time modulation and translation of f and/or s have been described by [3.46–52].

Because of the two dimensions available, the spatial convolver/correlator can be multiplexed to provide simultaneous processing of a one-dimensional input $f(x)$ with N independent channels. The system is shown schematically in Fig. 3.20. The imaging lens transfers $f(x)$ to an array of filter functions $s_n(x)$, $n = 1, 2, ..., N$. The second imaging system is an astigmatic spherical/cylindrical combination which images each channel in the y dimension and integrates in the x dimension onto N separate detectors. With no scanning, we have an N channel multiplexed line imaging system which produces

$$g_n = \int f(x)s_n(x)dx \qquad n = 1, 2, ..., N.\tag{3.66}$$

Dillard [3.53] has proposed the use of such a system for real-time pattern recognition of the spectral characteristics of speech. The input $f(x)$ is the short-time power spectrum of speech supplied to the system using a CRT. The detector outputs g_n are compared to find the best match to a stored reference $s_n(x)$.

By translating $f(x)$ in Fig. 3.20, the $s_n(x)$ array or the imaging lens in x, an n channel cross correlator which performs

$$g_n(x_0) = \int f(x - x_0)s_n(x)dx \qquad n = 1, 2, ..., N.\tag{3.67}$$

is achieved. The use of such systems in seismic signal analysis has been described by *Piety* [3.40]. The most general use for the system in Fig. 3.20 is to cross correlate or convolve N different inputs $f_n(x), n = 1, 2, ..., N$, with N different filter functions $s_n(x), n = 1, 2, ..., N$. By translating the $f_n(x)$ array along the x axis, the system performs the operation

$$g_n(x_0) = \int f_n(x - x_0)s_n(x)dx \qquad n = 1, 2, ..., N.\tag{3.68}$$

The system of Fig. 3.19 can perform a one-dimensional cross correlation on a two-dimensional signal with integration in the y dimension of the form

$$g(x_0) = \int\int f(x - x_0, y)s(x, y)dxdy\tag{3.69}$$

if f, s or the lens is translated in the x dimension over distance x_0. By spatially reversing f or s, a similar convolution is obtained. *Ator* [3.54] used such a system as an aircraft velocity sensor. In this system, the translating input was the irradiance of the ground as viewed from moving aircraft. This system produces an effective time-varying output because of the velocity conversion from position to time. However, the fundamental type of integration is spatial.

By translating $f(x, y)$ in Fig. 3.19 in two-dimensions, a cross correlation/convolution of the form

$$g(x_0, y_0) = \int\int f(x - x_0, y - y_0)s(x, y)dxdy\tag{3.70}$$

is achieved. *Shack* [3.55] and *Swindell* [3.56] have assembled systems operating on this principle for image processing. Image restoration to remove the effects

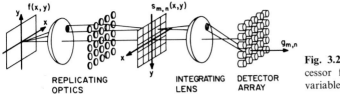

Fig. 3.21. Multichannel processor for functions of two variables

of motion blur and lens defocus has been achieved by convolving the degraded input $f(x, y)$ with an appropriate deconvolution operator $s(x, y)$. Their system achieves effective two-dimensional translation by scanning the input on a rotating drum and displaying the corrected output on a synchronously moving drum. Unfortunately, most deconvolution filters have bipolar impulse responses [3.57, 58] so special coding techniques must be used. *Shack* and *Swindell* used two orthogonally polarized illumination systems to independently convolve $f(x, y)$ with the positive and negative portions of the impulse response $s(x, y)$. These separate scanned outputs were electronically subtracted before being transferred to the output drum. In a similar system *Barrett* and *Swindell* [3.59] used this type of processing to decode information recorded in a transaxial tomogram. Other applications are in infrared target detection systems which use scanning or rotating masks [3.60–63].

The final extensions of spatial scanning use "fly's-eye" replicator arrays to provide many multiplexed imaging systems operating on the same input. A schematic diagram of this arrangement is shown in Fig. 3.21. The replicating optics relays a copy of $f(x, y)$ to individual masks $s_{m,n}(x, y)$, and an integrating lens collects the individual integrated products onto an array of detectors. There are MN total detectors and the output of one of them is

$$g_{m,n} = \iint f(x, y) s_{m,n}(x, y) dx dy, \quad \begin{matrix} m = 1, 2, ..., M \\ n = 1, 2, ..., N. \end{matrix} \tag{3.71}$$

Jackson [3.64] used an LED array as the input $f(x, y)$ and constructed an associative memory for binary computer words using such a principle, and *Babcock* et al. [3.65] have used such a system for pattern recognition. The most ambitious extension of the "fly's-eye" system is to provide one or two-dimensional shifting of $f(x, y)$ or the $s_{m,n}(x, y)$ array to perform many channel parallel convolution/correlations with the stored references. The relatively poor quality and uniformity of available fly's-eye lens systems may limit the practical application and development of this scheme.

3.4.3 Temporal Scanning Processors

In all the systems described previously, the input information is introduced by fixed or mechanically moving spatial functions so that the effective integration

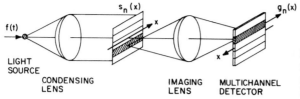

Fig. 3.22. Multichannel system for one-dimensional processing using a modulated light source

LIGHT SOURCE

CONDENSING LENS

IMAGING LENS

MULTICHANNEL DETECTOR

is over spatial variables. In this section we describe several systems which perform linear convolution and correlations of the form in (3.65) except that the input, output and integration variables are time functions. In the next section we describe a multichannel temporal-spatial processor which performs general linear vector-matrix spatial superpositions over time varying functions at a very high rate.

Most temporal scanning processors combine several parallel channels which operate on a single input time function $f(t)$. A general schematic diagram of such a system is shown in Fig. 3.22. The modulated light source $f(t)$ could be an LED, an incandescent source, or a source/modulator combination. The integration operation is performed in parallel over N channels by effectively scanning the filter functions $s_n(x), n = 1, 2, ..., N$, by physically moving the imaging lens to translate the images of $s_n(x)$, or by moving or electronically scanning the detector array. The condensing lens collects light from the source and uniformly illuminates all the masks $s_n(x)$. The imaging lens transfers the product of $f(t)$ and $s_n(x)$ to the detector array, which time integrates over horizontal strips stacked in the y dimension. All N strips have a parallel output $g_n(x)$ and are selected to have an integration time equal to the desired processing time interval over $f(t)$. The translation of $s_n(x)$, the imaging lens or the scanning of the detector takes place during this integration time so that the output of the nth channel of the detector can be written as

$$g_n(x) = \int f(t) s_n(x - vt) dt \qquad n = 1, 2, ..., N \tag{3.72}$$

where v is a velocity constant. A further subdivision of temporal scanning into scanning-mask and scanning-detector systems can be made.

In a temporal scanning-mask implementation, the $s_n(x)$ function in Fig. 3.22 is a physically moving optical mask which moves in the x direction [3.66, 67]. *Talamini* and *Farnett* [3.68] and others [3.66, 67, 69] have described systems in which the mask is a rotating disk. The *Talamini* and *Farnett* system used 241 channels mounted on a 40 cm disk spinning at 7000 rpm for radar matched filtering operations. *Parks* [3.70, 71] and *Skenderoff* et al. [3.66, 67] described systems using a rotating cylindrical drum. *Bromley* [3.72] used a mechanically scanning mirror to move the product of f and s over a detector array. *Strand* and *Persons* [3.73] developed a more general version of the system in Fig. 3.22 which had several input channels. Their system used a parallel array of sources, $f_n(t), n = 1, 2, ..., N$, arranged so that each illuminated only the corresponding

$s_n(x)$. The detector array output is then

$$g_n(x) = \int f_n(t) s_n(x - vt) dt \qquad n = 1, 2, ..., N .$$ (3.73)

Recently, *Sprague* and *Koliopoulos* [3.74] and *Kellman* [3.75, 76] have described a single-input version of the system in which the moving mask is replaced by an acousto-optic delay line driven by a time signal $s(t)$. The system performs the operation of (3.72) without any moving parts. By introducing complex modulation on the time signals, large time-bandwidth spectral analyses and other types of signal processing are possible. One reason for these improved capabilities is that the integration interval is set by the detector array and not by spatial limitations of the input devices as in spatial integration systems. *Turpin* [3.77] and *Kellman* [3.75, 76] have also designed a two-dimensional extension of the acousto-optic delay line system which uses two one-dimensional delay lines arranged orthogonally. The readout is performed by a two-dimensional matrix of time integrating detectors. Such a system performs operations of the form

$$g(x, y) = \int f(t) a(t) s_1(x - vt) s_2(y - vt) b(x, y) dt ,$$ (3.74)

where $g(x, y)$ is the two-dimensional output, s_1 and s_2 are general filter functions, a is a multiplicative time function and b is a multiplicative spatial function. By appropriate choice of these functions, versions of the system that compute ambiguity functions for radar signal processing and perform two-dimensional spectral analysis have been designed.

As an alternative to the scanning mask technique, the multichannel detector in Fig. 3.22 may be temporally scanned to produce $g_n(x)$. A system described by *Faiss* [3.78] used moving photographic film as the integrating detector. With recent advances in microelectronics, detector arrays that accomplish the time integration on scanned discrete elements are available. These devices enable detection to take place with no physical movement of the array. *Talamini* and *Farnett* [3.68] envisioned such a system for radar signal processing, and *Monahan* et al. [3.79] implemented the idea using scanning charge-coupled device (CCD) array.

3.4.4 Temporal/Spatial Processors

Recently a new temporal/spatial approach to incoherent diffraction-limited optical processing has been explored by *Goodman* et al. [3.37–39]. Their system accepts N parallel temporal input channels f_n, $n = 1, 2, ..., N$, and performs spatial integration to produce M parallel multichannel temporal output channels. The system performs discrete linear matrix-vector operations of the form

$$g_m = \sum_{n=1}^{N} s_{mn} f_n \qquad m = 1, 2, ..., M$$ (3.75)

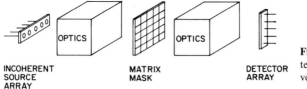

INCOHERENT
SOURCE
ARRAY

MATRIX
MASK

DETECTOR
ARRAY

Fig. 3.23. Basic system for temporal/spatial matrix-vector multiplication

or in matrix notation

$$g = Sf, \tag{3.76}$$

where s_{mn} are arbitrary entries in the matrix S, f is an N element vector and g is an M element vector. In its most general form, (3.75) is a discrete version of the space-variant integral in (3.61).

The *Goodman* system is shown schematically in Fig. 3.23. The input data is entered in parallel on a modulated array of incoherent sources (LED's, for example). The first set of astigmatic optics images the sources in the horizontal direction onto the matrix mask, but acts as a condenser in the vertical direction to spread the light into a uniformly illuminated line. The matrix entries in the mask attenuate the light and the second optical system images the outputs in the vertical dimension onto the detector array while integrating the outputs across the horizontal dimension. Collecting the light in the horizontal dimension performs the summation in (3.75). An advantage of this system is that the data is entered and extracted as parallel time signals; thus the potential throughput rates are much higher than many of the previous systems.

One application of the system in Fig. 3.23 is in the computation of the one-dimensional discrete Fourier transform (DFT). In this case, the inputs and outputs are generally complex bipolar signals which must be encoded into non-negative quantities for incoherent processing [3.37]. One method for doing this is to represent the complex numbers in terms of four non-negative real numbers, specifically the positive and negative components of the real and imaginary parts. Another more efficient representation in terms of three non-negative unit complex vectors has been described by *Burckhardt* [3.80] and employed in computer-generated holograms. The *Burckhardt* approach decomposes the complex vector f as

$$f = f^{(0)} + f^{(1)} \exp(j2\pi/3) + f^{(2)} \exp(j4\pi/3), \tag{3.77}$$

where the f superscripts are real and non-negative. A final approach uses biased real and imaginary parts; the relative merits of each have been compared and analyzed for the temporal/spatial processor [3.39].

Figures 3.24 and 25, taken from *Goodman* et al. [3.38], illustrate an experimental implementation of the temporal/spatial processor of Fig. 3.23 for performing DFT's. The input f, matrix S and output g are all decomposed using the three unit vector *Burckhardt* method of (3.77). To perform a 10 point

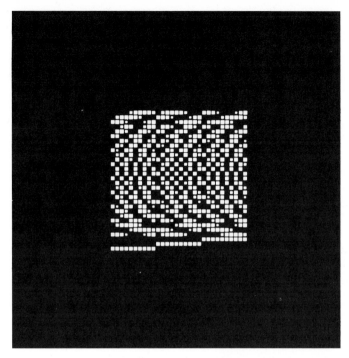

Fig. 3.24. Matrix mask for 10 point DFT with *Burckhardt* decomposition

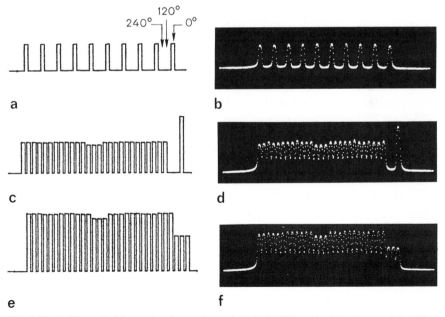

Fig. 3.25a–f. Theoretical (**a, c, e**) and experimental (**b, d, f**) DFT results (*Goodman et al.* [3.38]).

DFT, a 30×30 matrix mask is needed as shown in Fig. 3.24. The various weights are encoded as area modulation on the mask, and the three output components $g_k^{(0)}$, $g_k^{(1)}$ and $g_k^{(2)}$ for the kth Fourier coefficient are designed to appear side by side. Figure 3.25 shows theoretical and experimental outputs for several different input sequences. Although the output functions were obtained using a mask-modulated incandescent source in this experiment, the basic principle is the same as if a time modulated source array as in Fig. 3.23 had been used. In Figs. 3.25a and b, the input function is the sequence (1, 0, 0, 0, 0, 0, 0, 0, 0, 0). The DFT of this sequence should be completely real and of constant magnitude. As shown in Fig. 3.25b, the nonzero components appear only at $0°$ and have equal magnitude. Other components at $120°$ and $240°$ are zero. In Figs. 3.25c and d, the input sequence is real and constant. The DFT shows a large real zero-frequency component (at the right) with all other triplets at equal magnitude. Any set of triplets of equal magnitude corresponds to a zero, thus the only nonzero component is at zero frequency. Figures 3.25e and f show the results with uniform input illumination (all zero input sequence). The output has triplets of equal magnitude, showing that the DFT of a zero input sequence is a zero output sequence.

Psaltis et al. [3.81] have demonstrated an extension of temporal/spatial processing using three LED sources of different colors (nonoverlapping spectral outputs). Complex time signal inputs are decomposed into three components by the *Burckhardt* method and are applied to the system of Fig. 3.23 in parallel. A colored mask in the system separately modulates these components, and a grating system between the matrix mask and detectors spatially separates the spectral components so that they are detected by parallel strips of detectors stacked horizontally. *Psaltis* et al. have also described an optical iteration scheme for solving sets of simultaneous linear equations. An electronic accumulator at the output of the temporal/spatial processor stores updated outputs and provides feedback to the input for recursion.

Several extensions of the basic one-dimensional temporal/spatial processor to two dimensions are possible [3.39]. Two-dimensional inputs to be linearly processed can be lexicographically ordered [Ref. 3.58, Chap. 5] by stacking rows or columns of input arrays into long vectors for input to the system. Another extension of the spatial-temporal processor uses electrical analog devices such as CCD's or surface acoustic wave (SAW) devices to do time signal filtering in parallel on the input and/or output lines shown in Fig. 3.23. Capabilities of the system using currently available LED's and detectors for computing DFT's are that it has a processing rate of 50 complex data vectors in 10 ns and a throughput rate of $5 \cdot 10^9$ samples per second.

3.5 Geometrical Optics-Based Processing: Shadow Casting

The geometrical optics-based incoherent processing systems described in this section are often called "shadow-casting" systems and generally do not form

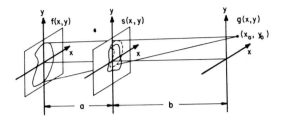

Fig. 3.26. Basic system for "shadow-casting" geometrical optics incoherent processing

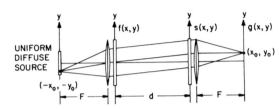

Fig. 3.27. System with lenses for "shadow-casting" geometrical optics incoherent processing

images. Historically, such systems were among the earliest for performing incoherent processing, and they have the advantage of requiring no mechanical or electronic scanning to perform their operations. The inherent redundancy of incoherent systems provides many parallel channels of processing on the input information. This information is collected and integrated by the system detector to provide the output [3.1, 41]. As noted earlier, a limitation of geometrical optics-based shadow casting systems is that inputs with large space-bandwidth product introduce diffraction effects, thus violating the assumptions on which the system is based [Ref. 3.5, Chap. 7] and [3.82]. A discussion of this problem is given at the end of this section.

There are two basic configurations for geometrical optics systems which perform signal processing without image formation. These are shown in Figs. 3.26 and 27. In both systems the function $f(x, y)$ can be an incoherent self-luminous input (a diffuse object, a CRT or other electronic display) or a diffusely illuminated transparency. The important point is the diffuse nature of $f(x, y)$, which must spread the light from all source points uniformly in the direction of the readout plane where the output $g(x, y)$ is measured. Assuming the propagation laws of geometrical optics, all rays from every source point travel in straight lines to the readout plane and are simply attenuated by intermediate transparencies. The effects of scattering, diffraction and oblique distortion due to the viewing angle are neglected. Many other configurations for shadow casting are possible. *Knopp* and *Becker* [3.83] have presented a general model that includes the two major configurations described here as special cases.

3.5.1 Lensless Shadow Casting Systems

Referring to Fig. 3.26, the rays converging to (x_0, y_0) in the readout plane can be extended back to $f(x, y)$ through the filter function $s(x, y)$ which is a transparency placed in an intermediate plane. By simple geometrical analysis, the shadow of $f(x, y)$ can be shown to be displaced at the filter plane $s(x, y)$ by $[ax_0/(a+b)]$, $[ay_0/(a+b)]$ where a and b are shown in Fig. 3.26. Also, the shadow of $f(x, y)$ is minified by a factor $b/(a+b)$. The product of the $f(x, y)$ shadow and $s(x, y)$ is integrated at (x_0, y_0) and the intensity in the readout plane becomes

$$g(x_0, y_0) = \int\int f\left[\frac{x - ax_0/(a+b)}{b/(a+b)}, \frac{y - ay_0/(a+b)}{b/(a+b)}\right] s(x, y) dx dy \qquad (3.78)$$

which is a cross-correlation operation. By inverting either f or s, a convolution is obtained. One important feature of this system is that the integral is evaluated in parallel at all output points simultaneously without the need for spatial or temporal scanning.

The system shown in Fig. 3.26 has had many practical applications. *Meyer-Eppler* [3.84] described the use of the system for general linear signal processing, and *Robertson* [3.85] and *Bragg* [3.86] used a similar configuration to study the lattice structure of crystals. The system has been used for image processing and pattern recognition by *McLachlan* [3.87], *Wilde* [3.88], and *Hawkins* and *Munsey* [3.89]. In some of these systems, the preprocessing of image data was done by several parallel optical processors, followed by electronic thresholding for decision making. *Maure* [3.90–92] has described a version of the system in Fig. 3.26 which uses discrete modulated LED arrays for $f(x, y)$ and a discrete detector array to measure $g(x, y)$. He proposed that the system be used for logic function applications, including read-only memories. *Fomenko* [3.93] has described how one-dimensional vectors of long length (N^2) could be rearranged into a 2-D array of size $N \times N$ by inverse vector stacking for one-dimensional signal processing.

The shadow-casting system of Fig. 3.26 has been combined with spatial scanning as described in Sect. 3.4.2 to perform convolution or correlation in the form of (3.64) [3.94, 95]. In these setups, lenses are not needed or are not used to form images as in the Fig. 3.20 system. Instead, relative motion between the f and s functions of Fig. 3.26 causes the output convolution to be measured on a single detector or small number of detectors in the output plane.

Other discrete generalizations of the Fig. 3.26 system have been used for vector-matrix operations of the form of (3.75). As described in Sect. 3.4, these operations are not necessarily restricted to convolution and correlation. *Mengert* and *Tanimoto* [3.96] developed a shadow casting setup like that of Fig. 3.26 to do matrix-vector multiplication with a discrete array of detectors and sources. The system was used for solving linear systems of equations and feedback was employed to generalize the system. *Ullmann* [3.97] has presented many pattern recognition applications using matrix-vector (generally space-variant) operations performed on systems similar to that in Fig. 3.26.

As pointed out by *Monahan* et al. [3.36], the space-variant Fourier transform linear operation

$$g(v_x, v_y) = \iint f(x, y) e^{-j2\pi(xv_x + yv_y)} dx dy \tag{3.79}$$

is extremely useful in two-dimensional signal processing. Special attention to this particular operation has been given by a number of groups. *Leifer* et al. [3.98] used a shadow casting system to do Fourier analysis of inputs to a character recognition system. The input is cross correlated with binary square-wave masks of different spatial frequency and complex Fourier components are extracted from the contrast and position of the output. Eq. (3.79) can be rearranged in the form

$$g(v_x, v_y) = e^{-j\pi(v_x^2 + v_y^2)} \iint f(x, y) e^{-j\pi(x^2 + y^2)} \cdot e^{j\pi[(x - v_x)^2 + (y - v_y)^2]} dx dy, \tag{3.80}$$

in which the input $f(x, y)$ is first multiplied by $\exp[-j\pi(x^2 + y^2)]$, convolved with a quadratic phase "chirp" function, and finally multiplied by another quadratic phase $\exp[-j\pi(v_x^2 + v_y^2)]$. *Mertz* [3.99] has described a shadow casting scheme based on the Fig. 3.26 system using Fresnel zone plates (FZP) to encode the quadratic phase multiplicative and convolutional functions. The complete system is called a "Fresnel sandwich". The major complexity is associated with the non-negative nature of the incoherent processor and the need to process complex data in (3.80). *Stephens* and *Rogers* [3.100] and *Richardson* [3.101] described an alternate way of evaluating (3.79) by performing two sequential cross correlations of $f(x, y)$ with chirp functions whose spatial frequency increases in opposite directions. As before, FZP's can be used to provide the complex masks needed.

3.5.2 Shadow Casting Systems with Lenses

Figure 3.27 shows an alternate way of realizing a shadow casting processor [Ref. 3.5, Chap. 7] and [3.102, 103]. Considering a point $(-x_0, -y_0)$ on the diffuse source, all the divergent rays are collimated by the first lens, located one focal distance away. The collimated beam is modulated by the input $f(x, y)$, which projects to the filter function $s(x, y)$ and is collected by the second lens to be detected at the output point (x_0, y_0). The lenses eliminate the minification present in the Fig. 3.26 system, so that the overall operation can be expressed as

$$g(x_0, y_0) = \iint f\left(x - \frac{dx_0}{F}, y - \frac{dy_0}{F}\right) s(x, y) dx dy. \tag{3.81}$$

A variation of this setup was used by *Trabka* and *Roetling* [3.104] with a real-time input device to do incoherent signal processing for pattern recognition.

Jackson [3.105] also used the system for image spatial filtering and enhancement. *Kovasznay* and *Arman* [3.103] and *Horwitz* and *Shelton* [3.106] used a modified version of the Fig. 3.27 system for autocorrelation. By spatially folding the input $f(x, y)$ with mirrors, the need for two separate transparencies of $f(x, y)$ is eliminated and the autocorrelation is produced.

A discrete modification of the system in Fig. 3.27 has been presented by *Schneider* and *Fink* [3.107]. Their system performed multiplications of the form

$$g_{mp} = \sum_{n=1}^{N} s_{mn} f_{np}, \quad \begin{array}{l} m = 1, 2, ..., M \\ p = 1, 2, ..., P. \end{array} \tag{3.82}$$

where g, s and f are general matrix entries. Their system was a generalization of that in Fig. 3.27 using a displaced, segmented spherical lens system. They experimentally demonstrated 5×5 matrix multiplications.

Krivenkov et al. [3.108, 109] have developed a hybrid incoherent imaging/shadow-casting system for performing several matrix multiplications of the form of (3.82). Their system has several cascaded stages which perform imaging in one dimension and shadow casting in the orthogonal dimension using an astigmatic lens system. The input is a two-dimensional mask function and the output is detected by a two-dimensional array. They have performed 8×8 matrix multiplications with the system and have used it to perform Walsh-Hadamard transforms. This transform has positive and negative entries in the matrices, thus parallel encoding of the four-quadrant output of multiplied bipolar signals is used.

As a final note, some comments on the limitations of these geometrical optics processors is needed. An important limitation of these systems is that diffraction effects may be significant under certain conditions when functions having a large space-bandwidth product are to be processed [Ref. 3.5, Chap. 7] and [3.82]. As discussed earlier, the space-bandwidth product is a measure of the complexity or degrees-of-freedom of an input function. The goal of all the optical processing systems described in this chapter is to maximize the effective space-bandwidth product.

Unfortunately, the systems in Sect. 3.5 suffer most from diffraction effects that arise when the input functions have a high space-bandwidth product. As the spatial detail on the input increases, diffraction effects become more significant and the laws of geometrical optics are increasingly inaccurate. Thus, the principles on which these systems are based are no longer valid. In particular, diffraction occurs in the system of Fig. 3.26 over the distance a, so that a filtered version of $f(x, y)$ will be present at the $s(x, y)$ plane. Similar diffraction effects occur over the distance b in Fig. 3.26 and over the distance d in Fig. 3.27. These diffraction effects limit the space-bandwidth product over which these systems can operate accurately. The systems described in Sect. 3.4 are also affected by diffraction, but to a much lesser extent because they are imaging systems which are in focus according to the principles of geometrical optics. In spite of these limitations, such systems have found many practical uses. *Green* [3.82] has analyzed several types of incoherent geometrical optics systems and has shown that the maximum number of resolvable points in the system of Fig. 3.27 varies directly as the square of the width of the input function and inversely as the separation d. Using Green's theory, *Monahan* et al. [3.36] have quoted the following example: for input functions of width 4 cm and separation 16 cm, the system space-bandwidth product is at least 10^4.

3.6 Summary and Conclusions

In this chapter we have analyzed the basic characteristics of incoherent optical processing, considering three major classes of systems: (1) systems that rely on diffraction; (2) systems that rely on plane-to-plane imaging in the geometrical optics sense; and (3) systems that rely on diffractionless geometrical optics "shadow casting" for their operation. Incoherent systems are often characterized by a redundancy and immunity to noise not associated with coherent optical systems. However, the non-negative real nature of the information-bearing irradiance distributions precludes direct implementation of incoherent systems in many signal processing applications, and various tricks must be employed. Dynamic range limitations with incoherent systems are an area of active study, and the relative advantages of incoherent systems over coherent systems are not known conclusively.

Acknowledgements. W. T. Rhodes gratefully acknowledges the support of the U.S. Army Research Office for research in incoherent optical processing. *A. A. Sawchuk* gratefully acknowledges the support of the Air Force Office of Scientific Research, the National Science Foundation and the Joint Services Electronics Program/AFSC at USC for research in optical information processing. Both authors are grateful to *P. Chavel* and *T. C. Strand* for their careful review of the manuscript.

References

3.1   G.L.Rogers: *Noncoherent Optical Processing* (Wiley, New York 1977)
3.2   A.W.Lohmann: Private communication
3.3   G.L.Rogers: "The Importance of Redundancy in Optical Processing", in *Proceedings of ICO-11 Conference, Madrid, Spain,* 1978 (Optica Hoy *y* Mañana), ed. by J.Bescos, A.Hidalgo, L.Plaza, J.Santamaria (Sociedad Española de Optica, Madrid 1978) pp. 307–310
3.4   G.Brandt: Appl. Opt. **12**, 368 (1973)
3.5   J.W.Goodman: *Introduction to Fourier Optics* (McGraw-Hill, New York 1968)
3.6   M.Brousseau, H.H.Arsenault: Opt. Commun. **15**, 389 (1975)
3.7   J.D.Gaskill: *Linear Systems, Fourier Transforms, and Optics* (Wiley, New York 1978)
3.8   J.D.Armitage, A.W.Lohmann: Appl. Opt. **4**, 461 (1965)
3.9   R.Bracewell: *The Fourier Transform and Its Applications* (McGraw-Hill, New York 1965)
3.10   W.T.Rhodes, W.R.Limburg: "Coherent and Noncoherent Optical Processing of Analog Signals", in *Proceedings of the 1972 Electro-Optical Systems Design Conference* (Industrial and Scientific Conference Management, Chicago 1972) pp. 314–320
3.11   A.W.Lohmann: Appl. Opt. **7**, 561 (1968)
3.12   S.Lowenthal, A.Werts: C. R. Acad. Sci. Ser. B **266**, 542 (1968)

3.13   A.W.Lohmann, H.W.Werlich: Appl. Opt. **10**, 670 (1971)
3.14   W.T.Maloney: Appl. Opt. **10**, 2554 (1971)
3.15   W.T.Maloney: Appl. Opt. **10**, 2127 (1971)
3.16   R.A.Gonsalves, P.S.Considine: Opt. Eng. **15**, 64 (1976)
3.17   J.Fleuret, H.Maitre, E.Cheval: Opt. Commun. **21**, 361 (1977)
3.18   H.H.Barrett, M.Y.Chiu, S.K.Gordon, R.E.Parks, W.Swindell: Appl. Opt. **18**, 2760 (1979)
3.19   W.Lukosz: J. Opt. Soc. Am. **52**, 827 (1962)
3.20   W.T.Rhodes: Appl. Opt. **16**, 265 (1977)
3.21   W.T.Rhodes: "Temporal Frequency Carriers in Noncoherent Optical Processing", in *Proceedings of the 1978 International Optical Computing Conference (Digest of Papers)* (IEEE No. 78CH1305-2C, 1978) pp. 163–168
3.22   J.M.Florence, H.E.Pettit, W.T.Rhodes: "Feature Enhancement Using Noncoherent Optical Processing", in *Optical Pattern Recognition*, SPIE Proc., Vol. 201, ed. by D.Casasent (1979) pp. 95–99
3.23   M.Ryle: Proc. R. Soc. London Ser. A **211**, 351 (1952)
3.24   A.W.Lohmann: Appl. Opt. **16**, 261 (1977)
3.25   W.Stoner: Appl. Opt. **17**, 2454 (1978)
3.26   A.W.Lohmann, W.T.Rhodes: Appl. Opt. **17**, 1141 (1978)
3.27   W.Stoner: Appl. Opt. **16**, 1451 (1977)
3.28   W.Stoner: "Optical Data Processing with the OTF", in *Proceedings of the 1978 International Optical Computing Conference (Digest of Papers)* (IEEE No. 78CH1305-2C, 1978) pp. 172–176
3.29   A.Furman, D.Casasent: Appl. Opt. **18**, 660 (1979)
3.30   P.Chavel, S.Lowenthal: J. Opt. Soc. Am. **66**, 14 (1976)
3.31   D.Görlitz, F.Lanzl: Opt. Commun. **20**, 234 (1977)
3.32   B.Braunecker, R.Hauck: Opt. Commun. **20**, 234 (1977)
3.33   W.T.Rhodes: Opt. Eng. **19**, 323 (1980)
3.34   K.Sayanagi: Ind. Photogr. **30**, 172 (1959)
3.35   M.Lasserre, R.W.Smith: Opt. Commun. **12**, 260 (1974)
3.36   M.A.Monahan, K.Bromley, R.P.Bocker: Proc. IEEE **65**, 121 (1977)
3.37   J.W.Goodman, L.M.Woody: Appl. Opt. **16**, 2611 (1977)
3.38   J.W.Goodman, A.R.Dias, L.M.Woody: Opt. Lett. **2**, 1 (1978)
3.39   J.W.Goodman, A.R.Dias, L.M.Woody, J.Erickson: "Some New Methods for Processing Electronic Image Data Using Incoherent Light", in *Proceedings of ICO-11 Conference, Madrid, Spain, 1978* (Optica Hoy y Mañana), ed. by J.Bescos, A.Hidalgo, L.Plaza, J.Santamaria (Sociedad Española de Optica, Madrid, 1978) pp. 139–145
3.40   R.G.Piety: "Optical Computer", US Patent 2,712,415 (1955)
3.41   G.L.Rogers: Opt. Laser Tech. **7**, 153 (1975)
3.42   E.Goldberg: "Statistical Machine", US Patent 1,838,389 (1931)
3.43   P.L.Tea: "Area Measuring Device", US Patent 2,179,000 (1939)
3.44   H.S.Field: "Method and Approaches for the Optical Cross-Correlation of Two Functions", US Patent 3,283,133 (1966)
3.45   M.C.Ferre: "Computing Apparatus", US Patent 3,030,021 (1962)
3.46   H.L.Barney: "Transversal Filter", US Patent 2,451,465 (1948)
3.47   L.Slobodin: Proc. IEEE **51**, 1782 (1963)
3.48   M.Arm, L.Lambert, I.Weissman: Proc. IEEE **52**, 842 (1964)
3.49   E.B.Felstead: IEEE Trans. AES-**3**, 907 (1967)
3.50   C.Atzeni, L.Pantani: Proc. IEEE **57**, 344 (1969)
3.51   R.M.Wilmotte: "Method and Apparatus for Optically Processing Information", US Patent 3,111,666 (1963)
3.52   R.Sprague: Opt. Eng. **16**, 467 (1977)
3.53   H.E.Dillard: Private communication with M.A.Monahan, K.Bromley, R.P.Bocker, in Ref. 3.36
3.54   J.T.Ator: Appl. Opt. **5**, 1325 (1966)

3.55 R.V.Shack: Pattern Recognition **2**, 123 (1970)

3.56 W.Swindell: Appl. Opt. **9**, 2459 (1970)

3.57 H.C.Andrews, B.R.Hunt: *Digital Image Restoration* (Prentice-Hall, Englewood Cliffs, NJ 1977)

3.58 W.K.Pratt: *Digital Image Processing* (Wiley-Interscience, New York 1978)

3.59 H.H.Barrett, W.Swindell: Proc. IEEE **65**, 89 (1977)

3.60 G.F.Aroyan: Proc. IRE **47**, 1561 (1959)

3.61 J.A.Jamieson, R.H.McFee, G.N.Plass, R.H.Grube, R.G.Richards: *Infrared Physics and Engineering* (McGraw-Hill, New York 1963) Chap. 12

3.62 J.Alward: "Spatial Frequency Filtering", in *Handbook of Military Infrared Technology* (Office of Naval Research, Washington, D.C. 1965) Chap. 16

3.63 R.D.Hudson, Jr.: *Infrared Systems Engineering* (Wiley, New York 1969) Chap. 6

3.64 A.S.Jackson: "A New Approach to Utilization of Optoelectronic Technology", in *Proceedings of the 1974 IEEE Computer Conference (Digest of Papers)* (IEEE Computer Society 1974) pp. 251–254

3.65 T.R.Babcock, R.C.Friend, P.Heggs: "Linear Discrimination Optical-Electronic Implementation Techniques", in *Optical Processing of Information*, ed. by D.K.Pollock, C.J.Koester, J.T.Tippett (Spartan Books, Baltimore, MD 1963) Chap. 12

3.66 C.Skenderoff et al.: "Radar Receiver Having Improved Optical Correlation Means", US Patent 3,483,557 (1969)

3.67 C.Skenderoff et al.: "Optical Correlation Systems for Received Radar Signals in Pseudo-Randomly Coded Radar Systems", US Patent 3,526,893 (1970)

3.68 A.J.Talamini, Jr., E.C.Farnett: Electronics **38**, 58 (1965)

3.69 D.E.Jackson: Sperry Eng. Rev. **19**, 15 (1966)

3.70 J.K.Parks: J. Acoust. Soc. Am. **37**, 268 (1965)

3.71 J.K.Parks: J. Aircraft **3**, 278 (1966)

3.72 K.Bromley: Opt. Acta **21**, 35 (1974)

3.73 T.C.Strand, C.E.Persons: "Incoherent Optical Correlation for Active Sonar": Tech. Rpt. TR 1887, Naval Electronics Lab. Center, San Diego, CA (1973)

3.74 R.Sprague, C.Koliopoulos: Appl. Opt. **15**, 89 (1976)

3.75 P.Kellman: "Detector Integration Acousto-Optic Signal Processing", in *Proceedings of the 1978 International Optical Computing Conference (Digest of Papers)* (IEEE No. 78CH1305-2C, 1978) pp. 91–95

3.76 P.Kellman: "Time Integrating Optical Signal Processing", Ph. D. Thesis, Dep. Elec. Eng., Stanford University (1979)

3.77 T.Turpin: "Time-Integrating Optical Processors", in *Real-Time Signal Processing*, SPIE Proc., Vol. 154, ed. by T.F.Tao (1978) pp. 196–203

3.78 R.D.Faiss: "Coherent Simultaneous Cross-Correlating Signal Separator", US Patent 3,486,016 (1969)

3.79 M.A.Monahan, R.P.Bocker, K.Bromley, A.Louie: "Incoherent Electrooptical Processing with CCD's", in *Proceedings of the 1975 International Optical Computing Conference (Digest of Papers)* (IEEE No. 75CH0941-5C, 1975) pp. 25–34

3.80 C.B.Burckhardt: Appl. Opt. **9**, 1949 (1970)

3.81 D.Psaltis, D.Casasent, M.Carlotto: Opt. Lett. **4**, 348 (1979)

3.82 E.L.Green: Appl. Opt. **7**, 1237 (1968)

3.83 J.Knopp, M.F.Becker: Appl. Opt. **17**, 984 (1978)

3.84 W.Meyer-Eppler: Optik **1**, 465 (1946)

3.85 J.M.Robertson: Nature London **152**, 511 (1943)

3.86 L.Bragg: Nature London **154**, 69 (1944)

3.87 D.McLachlan, Jr.: J. Opt. Soc. Am. **52**, 454 (1962)

3.88 H.J.Wilde: "Generation of Two-Dimensional Optical Spatial Auto- and Cross-Correlation Functions"; Rept. 744, US Navy Underwater Sound Laboratory, New London, CN (1966)

3.89 J.K.Hawkins, C.J.Munsey: "A Natural Image Computer", in *Optical Processing of Information*, ed. by D.K.Pollock, C.J.Koester, J.T.Tippett (Spartan Books, Baltimore, MD 1963) Chap. 17

3.90   D.R.Maure: "Read-Only Memory", US Patent 3,656,120 (1972)
3.91   D.R.Maure: "Optical Memory Apparatus", US Patent 3,676,864 (1972)
3.91   D.R.Maure: "Optical Logic Function Generator", US Patent 3,680,080 (1972)
3.93   S.M.Fomenko: "Signal Processing System", US Patent 3,211,898 (1965)
3.94   T.S.Gray: J. Franklin Inst. **212**, 77 (1931)
3.95   F.K.Preikschat: "Optical Correlator With Endless Grease Belt Recorder", US Patent 3,358,149 (1967)
3.96   P.Mengert, T.T.Tanimoto: "Control Apparatus", US Patent 3,525,856 (1970)
3.97   J.R.Ullmann: Opto-Electronics **6**, 319 (1974)
3.98   I.Leifer, G.L.Rogers, N.W.F.Stephens: Opt. Acta **16**, 535 (1969)
3.99   L.Mertz: *Transformations in Optics* (Wiley, New York 1965) pp. 94–95
3.100   N.W.F.Stephens, G.L.Rogers: Phys. Educ. **9**, 331 (1974)
3.101   J.M.Richardson: "Device for Producing Identifiable Sine and Cosine (Fourier) Transforms of Input Signals by Means of Noncoherent Optics", US Patent 3,669,528 (1972)
3.102   F.B.Berger: "Optical Cross-Correlator", US Patent 2,787,188 (1957)
3.103   L.S.G.Kovasznay, A.Arman: Rev. Sci. Instrum. **28**, 793 (1957)
3.104   E.A.Trabka, P.G.Roetling: J. Opt. Soc. Am. **54**, 1242 (1964)
3.105   P.L.Jackson: Appl. Opt. **6**, 1272 (1967)
3.106   L.P.Horwitz, G.L.Shelton, Jr.: Proc. IRE **49**, 175 (1961)
3.107   W.Schneider, W.Fink: Opt. Acta **22**, 879 (1975)
3.108   B.Krivenkov, P.E.Tverdokhleb, Yu.V.Chugui: Appl. Opt. **14**, 1829 (1975)
3.109   B.Krivenkov, S.V.Mikhlyaev, P.E.Tverdokhleb, Yu.V.Chugui: "Noncoherent Optical System for Processing of Images and Signals", in *Optical Information Processing*, ed. by Yu.E.Nesterikhin, G.W.Stroke, W.E.Kock (Plenum, New York 1976) pp. 203–217

4. Interface Devices and Memory Materials

G. R. Knight

With 32 Figures

The requirements for fast, efficient interface devices operating in coherent optical computers are quite different from the requirements on optical storage materials. The storage materials can have applications to devices and systems completely removed from optical processing and can be erasable or archival. The interface device must act as a real-time transparency and be useable over a large number of input output cycles to be useful in an optical processor. This chapter is, therefore, divided into two separate sections. Section 4.1 covers interface devices and compares the usefulness of recent implementations. Section 4.2 covers optical storage materials for a variety of applications.

4.1 Interface Devices

If optical computers are to achieve their potential for high-speed parallel data throughput, then real-time, transparency-type devices must become available. A real-time transparency device can, in fact, be a transmissive or reflective spatial light modulator which can be electrically or optically accessed and modulated. A spatial light modulator is required in the data input plane of the coherent optical computer to allow an imagewise phase or amplitude modulation of a coherent light wavefront. In most optical computing applications, it is also required that a spatial light modulator act as a two-dimensional optical filter in the Fourier transform plane of the optical system. The requirements on the data input plane and spatial filter plane interface devices can be quite different. For example, the spatial filter light modulator must, in general, be a very high resolution device since holographic type filters are usually required in a coherent optical computer.

Table 4.1 lists the spatial light modulators that will be discussed in this section along with their operating parameters. The actual figures listed are those available at the time this chapter was written; where active development is still occurring, the parameters should continually improve. The parameters in Table 4.1 allow comparison of the various interface devices. The relative importance of each parameter is discussed below; the physical mechanisms and operating characteristics of the individual devices are also described in detail.

An interface device may be electrically or optically accessible, and recent survey articles [4.1–3] have separated the available devices into these two classifications. With good, high-resolution TV vidicons and monitors available

Table 4.1. Spatial light modulators

	Optical write energy	Electrical write/ maintain energy	Contrast ratio	Resolution	Activation time
1. Liquid crystal light valve	24 ergs/cm² (0.3 ergs/cm² threshold)	Several mW at 10 kHz	>100:1	>100 l/mm	10 ms
2. e-Beam KDP (Titus tube)	—	100 µA beam current	60:1	40 l/mm	1/30 s full frame
3. PROM	50 ergs/cm²	4×10^{-4} J/frame	10,000:1	500 l/mm (limiting)	10 ns
4. Photo-Titus (KDP):	100 ergs/cm²	40 W (incl. cooling)	70:1	40 l/mm	10 µs
$(Bi_4Ti_3O_{12})$	2.5×10^{-3} J/cm²		>100:1	90 l/mm	50 ms
5. Ceramic ferro-electrics (strain biased)	—	50–300 V	<55:1	50 l/mm	10–50 µs switching
(scattering)	—	50–300 V	1,000:1	40 l/mm	10–50 µs switching
6. Deformable surface tubes					
a) GE	—	1250 W			1 ms
b) IBM (deformogra-phic tube)	—	6×10^{-8} coul./cm²	>100:1	15 l/mm	20 ms
c) CBS (Lumatron)	—	7.5 kV		71 l/mm	Few ms
d) Eidophor		20 µA beam current	<400:1	32 l/mm	<13 ms
7. Ruticon	10–50 ergs/cm²	200 V bias		>100 l/mm	<17 ms
8. Folded acousto-optic	—	Few W		0.5 l/mm	350 µs
9. Membrane light modulator	—	80 V		20 l/mm	0.1–1 µs

today, it is not terribly important whether an interface device is optically or electrically accessible. Good optical processing systems can be configured around either or both types of devices.

However, there are several parameters in Table 4.1 which are critical for the realization of efficient, high speed, high quality operation in an optical processor. Cycle speed, which involves both activation and decay or erase times, must be compatible with available input signal sources. This generally implies operating at TV rates, i.e., one frame every 33 milliseconds (ms). If the

Table 4.1 (continued)

	Decay time	Uniformity	Life	Coherent image	Comp. erase	Size	Oper. temp.
1. Liquid crystal light valve	15 ms	$<2\lambda$		Yes	Yes	2.5×2.5 cm	Room
2. e-Beam KDP (Titus tube)	Semi-permanent erase <1 ms with flood e-beam	Needs improvement		Yes	Yes (at $-52\,°C$)	5.1×5.1 cm	$-52\,°C$
3. PROM	Several hours; erase $<6\,\mu s$ optical flood	$<\lambda/6$	Long	Yes	Yes	4 cm diameter	Room
4. Photo-Titus (KDP): ($Bi_4Ti_3O_{12}$)	>1 h erase $<10\,\mu s$ flood light	$<\lambda/2$		Yes / Yes	Yes / Yes	4 cm diameter / 1 cm $\times$ 1 cm	$-50\,°C$
5. Ceramic ferro-electrics (strain biased)	$10–50\,\mu s$ switching	(Poor) local strain variations	Long ($>10^8$ cycles)	Yes	Yes	>3.25 cm diameter	Room
(scattering)	$10–50\,\mu s$ switching		Long	No	Yes	>3.25 cm diameter	Room
6. Deformable surface tubes							
a) GE	5–300 ms	6λ		Yes	Yes		Room
b) IBM (deformo-graphic tube)	20 ms to min		$>20 \times 10^6$ cycles	Yes	Yes and no	12.5 cm diameter	Room
c) CBS (Lumatron)	20 months; <1 s erase		$>3 \times 10^4$ cycles	Yes	Yes	3.8×3.8 cm	Room
d) Eidophor	adjustable			Yes	Yes		Room
7. Ruticon	>5 min; 10 ms erasure	$\lambda/4$	$>10^6$ cycles	Yes	Yes		Room
8. Folded acousto-optic	$350\,\mu s$	Excellent	∞	Yes (2-D) no (3-D)	Yes	15 cm diameter	Room
9. Membrane light modulator	$0.1–1\,\mu s$	$\lambda/10$	$\sim \infty$ in vacuum	Yes	Yes	5 mm $\times$ 5 mm	Room

interface device is optically accessed with a focused image (from a TV monitor), then the activation time plus erase time can be as long as 33 ms. If the device is addressed by a scanned electron beam, scanned laser beam, or an electrical matrix, then the device must respond to signals in the 3 to 30 megahertz (MHz) bandwidth region, depending upon device resolution. This implies a single element activation time in the 30 to 300 nanosecond (ns) range, with the additional requirement for a memory capability of 1/30 s frame storage while the total image is written into the device. Image decay time must be long enough to allow the total image to be formed and optically processed (i.e., 1/30s for TV rates); however, the image decay can be much longer and not cause any problems as long as an erase capability exists in the device.

4.1.1 Liquid Crystal Light Valve

Many types of liquid crystals in several device configurations have been employed as spatial light valves [4.3]. This section will not attempt to review all of these devices but will, instead, describe two of the more recent and successful light valve implementations using liquid crystals.

The major problem which faced most liquid crystal light valve devices was that of a very poor lifetime due to *dc* electrochemical degradation of the liquid crystal [4.4]. The device described below is a special adaptation of an *ac* photoactivated liquid crystal light valve [4.5–7]. The configuration of the device is shown in Fig. 4.1. The light valve consists of a number of continuous thin film layers sandwiched between two optical glass substrates. The device, as shown in Fig. 4.1, operates in reflection. A low voltage (5 to $10\,V_{rms}$) audio frequency power supply is connected to the two outer, thin film indium-tin-oxide transparent electrodes and thus across the entire thin film sandwich. The device is activated optically by a writing light beam onto a 15 micron (μm) thick cadmium sulfide photoconductor. The input writing light can be coherent or incoherent. A dielectric mirror and a cadmium telluride light-blocking layer optically separate the photoconductor from the readout light beam. The light-blocking layer and dielectric mirror are fundamental to the operation of the device. They enable simultaneous writing and reading of the device without regard to the spectral composition of the two light beams (with the photoconductor chosen to respond to the write wavelength). Because of the insulating nature of the dielectric mirror, no *dc* current can flow, and *ac* operation is necessary to establish a charge pattern across the liquid crystal. The dielectric mirror consists of alternating $\lambda/4$ films of high and low index materials and can be designed to provide greater than 90 % reflectivity across the visible spectrum or can be tuned to reflect a specific spectral region. The CdTe layer provides greater than 4 ND isolation between the photoconductor and the readout light. The dielectric mirror and the counterelectrode are overcoated with sputter-deposited films of SiO_2. These films provide surfaces in contact with the liquid crystal that are both chemically and electrochemically inert. This configuration, along with the *ac* mode of operation, provides for long device lifetime. The SiO_2 films also provide a suitable surface for aligning the liquid crystal. The SiO_2 surface is ion-beam etched at an angle of less than 20° to the surface to provide highly uniform homogeneous (parallel) or twisted homogeneous (twisted nematic) alignment. With the additional use of chemical bonding of aliphatic compounds to the ion-beam etched surface, a unidirectional, tilted homeotropic (perpendicular) alignment is possible. Thus, a complete range of align-

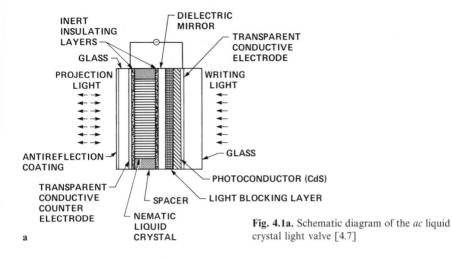

Fig. 4.1a. Schematic diagram of the *ac* liquid crystal light valve [4.7]

Fig. 4.1b. Photograph of the liquid crystal light valve

ment modes is possible allowing the electro-optic performance of the device to be optimized.

The liquid crystal layer, the counterelectrode, and the liquid crystal spacer complete the light valve structure. The spacer is fabricated from sputtered SiO_2 or thermally deposited Al_2O_3 in the form of a border around the device on the counterelectrode. The spacer is used to accurately define the liquid crystal layer thickness. The counterelectrode glass is 1/2 inch thick to minimize distortion when the light valve is mounted in the holder. The counterelectrode glass and substrate film glass are all optically polished to $\lambda/10$ flatness for good low optical noise performance. An antireflection coating is deposited on the outer

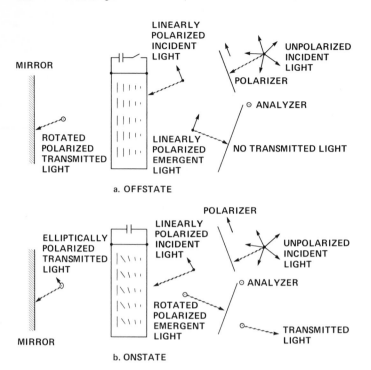

Fig. 4.2a, b. Operation of the hybrid field effect device: **(a)** the off-state; **(b)** the on-state [4.6]

surface of the counterelectrode to reduce the contrast of interference fringes caused by multiple reflections between the dielectric mirror and counterelectrode surfaces. The indium-tin-oxide film on the inner surface of the counterelectrode is deposited to an optical half-wave thickness to provide optical matching between the liquid crystal and the glass.

Liquid crystals exhibit several different electro-optic effects: dynamic scattering, and two electric field effects; optical birefringence and the twisted nematic effect. The light valve shown in Fig. 4.1b uses a hybrid field effect mode; one that uses the conventional twisted nematic effect in the off-state (no voltage on the liquid crystal) and the pure optical birefringence effect of the liquid crystal in the on-state (voltage on the liquid crystal). The liquid crystal molecules at the electrodes are aligned with their long axes parallel to the electrode surfaces. In addition, they are aligned to lie parallel with each other along a preferred direction that is fabricated into the surfaces. The twisted alignment configuration is obtained by orienting the two electrodes so that the directions of liquid crystal alignment of the two electrode surfaces make an angle of 45° with respect to each other. In conventional twisted nematic devices, the twist angle is 90°. The device described here cannot use the 90° twist configuration because it operates in the reflection mode. To obtain maximum modulation of the light beam in reflection, the liquid crystal molecules are twisted through 45°. This twisted alignment configuration, combined with the intrinsic optical birefringence of the liquid crystal, causes each pass of the linearly polarized light beam to rotate exactly through the twist angle.

The operation of the hybrid field effect light valve is shown schematically in Fig. 4.2. The off-state of the light valve is shown in Fig. 4.2a. A polarizer is placed in the incident beam and an analyzer is placed in the reflected beam. This provides a dark off-state, because after its first pass through the liquid crystal layer the direction of the linearly polarized incident light is rotated through 45°. However, upon reflection from the dielectric mirror, the light passes through the liquid crystal a second time and its polarization is rotated back to the direction of the incident light where it is blocked by the crossed analyzer. Thus, the off-state of the device is determined entirely by the twisted nematic effect.

The on-state of the device is shown schematically in Fig. 4.2b. If a voltage is applied across the liquid crystal, the molecules rotate to the homeotropic alignment in which the long axis of the molecules is oriented perpendicular to the electrode surfaces. For complete rotation, the polarization of the light would be unaffected by the liquid crystal and a dark on-state would result. Between full rotation and the off-state, however, a voltage regime exists where the device will transmit light. The molecules begin to tilt toward the homeotropic alignment (Fig. 4.2b) and the optical birefringence of the molecules affects the polarization of the light. The light that emerges from the device after reflection from the mirror is no longer linearly polarized so that some transmission occurs. The effect of the voltage is to destroy the twist spiral of the molecules. With the voltage on, half of the molecules in the layer tend to adopt the preferred alignment direction associated with one electrode; the other half adopt the alignment direction associated with the other electrode. With the preferred directions of the two electrodes at an angle of 45° with respect to each other, the polarization of the light will make an angle of 45° with respect to the extraordinary axis of the layer. This optimizes the transmission of the device. In Fig. 4.3a, the transmission of the device versus voltage is plotted for the hybrid field effect. The two curves correspond to twist angles of 90° and 45°, respectively, between the preferred directions on the electrodes. The data of Fig. 4.3a were taken with a $2 \mu m$ thick layer of an ester nematic liquid crystal.

A sensitometry curve of a typical light valve designed for application in optical data processing is shown in Fig. 4.3b. The threshold sensitivity occurs at about $3.3 \mu W/cm^2$. At about $100 \mu W/cm^2$, full contrast ($> 100:1$) is obtained.

A typical modulation transfer function (MTF) curve for the light valve is shown in Fig. 4.4. The 50% modulation point occurs at a resolution of 60 lines/mm. Devices 2.5 cm square have been built corresponding to 1500 resolvable lines across the aperture. The light-valve limiting resolution is in excess of 100 lines/mm.

Light valve time response measurements show a rise time of about 15 ms. These times are very fast for a liquid crystal and are attributable to the thinness ($\sim 2 \mu m$) of the crystal layer used in the device. The response times are

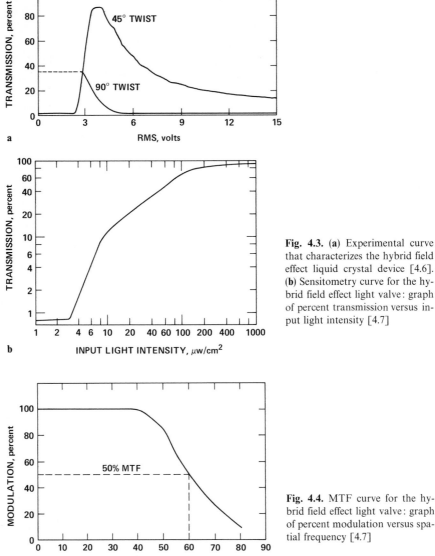

Fig. 4.3. (a) Experimental curve that characterizes the hybrid field effect liquid crystal device [4.6]. (b) Sensitometry curve for the hybrid field effect light valve: graph of percent transmission versus input light intensity [4.7]

Fig. 4.4. MTF curve for the hybrid field effect light valve: graph of percent modulation versus spatial frequency [4.7]

essentially independent of the duration of the input light pulse. The storage time is typically 10 to 20 ms in this device and erasure occurs by decay.

This liquid crystal device is still undergoing development. One area of development is improvement in cosmetic quality and optical flatness. The cosmetic quality suffers principally from point defects caused by inclusions in the thin films which underlay the mirror. The optical flatness of the device is less than optimum because internal stresses in the thin films bow the substrate glass. Fabrication techniques and photosensor responses are also being improved so that a tuned liquid crystal thickness of 1.37 µm can be utilized for optimum 632.8 nm operation. The thinner liquid crystal will also allow for less than 10 ms response time.

Another liquid crystal device of importance is an electroded liquid crystal structure [4.8] used as a computer-addressed adaptive Fourier plane filter. The device's format consists of a wedge filter with forty 9° segments and a ring filter with 20 concentric rings in a 2.5 cm diameter area. The optical processing operations with a wedge/ring detector are discussed in Sect. 5.4.2. The wedge and ring filters are made separately and then oriented and cemented before mounting. The dynamic scattering effect is used, but by controlling the polarization within specific scattering cones, high (10,000:1) contrast ratios and a high (50%) on-state transmission are obtained. Any of the filter segments can be separately activated. The wedge pattern permits filtering by pattern orientation while the ring pattern enables conventional filtering based on spatial frequencies present in the image to be achieved.

The transmittance of any of the 40 possible filter segments (oppositely oriented wedges are externally connected) can be controlled by a 7-bit computer word. 30 V signals are required for operation and once a location is addressed by an ON command it remains latched until it receives an OFF signal. The low resolution of this binary filter is a limitation but the structure has been used to remove the zero-order term in the Fourier transform (by activating the first ring of the ring filter) and one arm of a triangle (by activating the appropriate pair of wedge sections).

4.1.2 Electron-Beam Addressed DKDP

The interface device described in this section is an electron-beam addressed electro-optic crystal variously described in the literature as an Ardenne tube [4.9, 10], a Titus tube [4.11], or a Pockels effect tube [4.12, 13]. There is extensive literature describing the applications of this device to optical data processing as well as the most recent performance parameters [4.14–20].

A diagram of one version of this interface device is shown in Fig. 4.5. The device is a light modulator operated in the transmission mode utilizing the Pockels effect. The input information in electrical form as a video signal modulates the grid of a special purpose, off-axis writing electron gun, thus varying its beam current. The resulting spatial charge pattern developed across the target results in an electric field distribution across the target surface that is a controllable point-by-point function of the input information. By the Pockels effect, this electric field distribution causes a collimated input laser beam to be phase and polarization modulated point-by-point. The phase modulation can be converted to amplitude modulation after propagation through a crossed polarizer/analyzer.

The input electrical signal is conventionally scanned onto the target crystal during the frame time of standard 525 line television and the information is

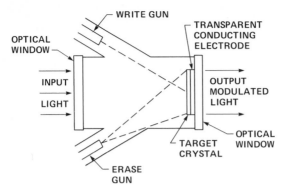

Fig. 4.5. Titus tube light modulator block diagram [4.14]

optically read during the first portion of the vertical retrace interval. A second off-axis electron gun with a reduced accelerating potential emits a defocused flood of low-energy electrons which results in erasure because of a secondary emission ratio greater than one.

The target crystal used is DKDP (Potassium Dideuterium Phosphate), one of the isomorphs of KDP. It belongs to the $\overline{4}2$ m point group and exhibits the linear longitudinal, electro-optic or Pockels, effect. The crystal is typically $5.1 \times 5.1 \times 0.025$ cm thick and is a $0°$ Z-cut basal plane section. Polarized light incident on the crystal travels along two orthogonal axes; these two components are the ordinary and extraordinary rays. The crystal's indices of refraction along these induced axes are proportional to the electric field developed across the crystal's thickness at each point. The electrically induced index of refraction change results in a phase retardation between the component optical waves. At the crystal's half-wave voltage, destructive interference between the component waves creates a full $\pi/2$ radians of phase modulation. If the polarization of the incident light is oriented at $45°$ to the induced crystal axes, the phase shift causes the emerging light to be elliptically polarized. With a crossed analyzer following the crystal, the transmitted light intensity at each point varies with the imagewise voltage distribution, and intensity modulation results.

A transparent conducting layer of cadmium oxide is vacuum deposited on the rear face of the crystal and serves as the conducting ground reference anode for the depressed cathode mode electron gun. The assembly is mounted on a calcium fluoride substrate for support. Each element of the crystal can be viewed as a parallel-plate capacitor of diameter, d, (the size of the uniform spot of charge deposited by the electron gun) at a potential V, with respect to the grounded transparent electrode. Because of field fringing, the diameter of the equipotential $V/2$ surfaces is $d' = d + 0.44b$ where d is the electron-beam spot size and b is the crystal thickness. Thus, thin crystals and small electron-beam spot sizes are required for high-resolution imaging. With a $25 \, \mu m$ spot size, resolution is limited to only 7.4 lines/mm which is inadequate for most applications.

Improved performance results from reducing the crystal operating temperature. An effective crystal thickness b' has been described [4.13] as $b/\sqrt{\varepsilon_3\varepsilon_1}$, where ε_1 and ε_3 are components of the diagonalized dielectric constant tensor. At room temperature $\sqrt{\varepsilon_3/\varepsilon_1} \simeq 1$, while near the Curie point T_c of DKDP ($-52\,^\circ$C), it can exceed 30. Then b' is negligible compared to d, and the crystal resolution then quals the electron beam resolution (40 lines/mm).

Operation of the crystal near its Curie temperature also improves image uniformity. The time constant of charge decay varies from 30 ms at room temperature to over 10^{10} s at T_c. Thus a charge storage mode results at the Curie temperature. It has been experimentally determined that various sections of the stored image decay before others due to point-by-point variations in the crystal RC time constant. At room temperature, this quickly results in nonuniform image decay. In the charge-storage mode near T_c, there is effectively no image decay, hence, excellent image uniformity.

Another advantage of Curie point operation is the availability of complete erasure with the flood electron gun. At room temperature, the Pockels tube suffers from incomplete erasure because the target crystal surface potentials involved are comparable to the crystal's second crossover point at which the erase gun cathode is biased. Local erasure of low-potential regions and a charge redistribution, rather than a charge removal, then occurs. At the Curie temperature, the half-wave potential of the crystal is reduced to 200 V and the erase gun then has an adequate accelerating potential for complete target erasure.

The three modes of operation, write, read, and erase, are synchronized to one television frame time. The leading edge of the vertical synchronization pulse signals the end of the write portion of a cycle. This pulse activates a modulator which pulses the readout laser during which time all optical processing operations are performed. The trailing edge of this pulse then activates a 1 ms pulse from the erase gun.

The electron guns and the target crystal are housed in a vacuum chamber. A two-stage Peltier cooling system is employed for keeping the crystal near its Curie temperature. Using 5.1×5.1 cm^2 crystals, optical processing operations with 1000×1000, well-resolved spots are possible. Typical contrast ratios of 60:1 are obtainable, but are a function of the degree of collimation of the laser beam. The major remaining problem area is obtaining good optical quality of the total interface device.

4.1.3 Pockels Readout Optical Modulator (PROM)

The Pockel's Readout Optical Modulator (PROM) is an electro-optical interface device developed at the Itek Corporation [4.21–29]. The PROM is a single-active component device constructed from a thin slice of bismuth silicon oxide ($Bi_{12}SiO_{20}$). $Bi_{12}SiO_{20}$ is a cubic crystal exhibiting the Pockels effect, photoconductivity when illuminated with blue light, and resistivity sufficient to

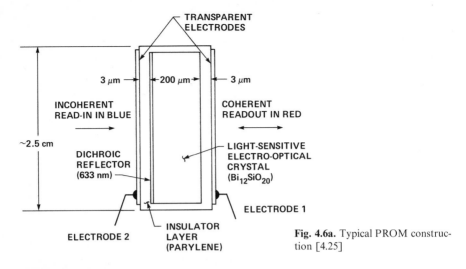

Fig. 4.6a. Typical PROM construction [4.25]

Fig. 4.6b. Photograph of the PROM

allow up to two hours of charge storage in the dark. A typical PROM construction is shown in Fig. 4.6. The crystal is cut and the faces polished flat with an orientation normal to the (100) crystalline plane. The crystal thickness is normally between 0.2 mm and 1.0 mm and a 3 μm to a 10 μm insulating layer of parylene is coated onto the crystal. Typical outside dimensions of the crystal are 25 to 30 mm square. Transparent conducting electrodes of indium oxide (InO_2) are evaporated onto the outside surfaces of the parylene. When the device is to be operated in a reflection mode, a dichroic layer reflecting red light but transmitting blue, is deposited on one side of the crystal before it is coated

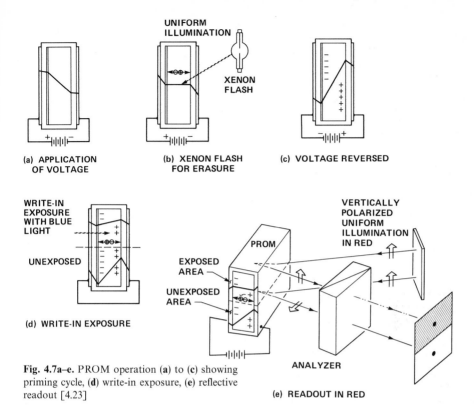

Fig. 4.7a–e. PROM operation (a) to (c) showing priming cycle, (d) write-in exposure, (e) reflective readout [4.23]

with parylene. A slight angle is usually introduced between the crystal surfaces to allow multiple reflections inside the device to be separated.

A typical operation cycle of the device is given in Fig. 4.7. A voltage V_0 (about 1200 V), is applied between the electrodes at a time when the crystal is fully insulating and the voltage is divided between the various layers (Fig. 4.7a). The device is now flooded with light from a short duration Xenon flash which creates mobile electrons. The mobile carriers drift to the crystal-parylene interfaces until the electric field in the crystal has been cancelled and potential differences of $V_0/2$ are established across each parylene layer. Any previous image in the device has now been erased (Fig. 4.7b). The erasing light is turned off and the voltage V_0 is reversed (Fig. 4.7c). A voltage of the order of $2V_0$ now appears across the crystal; its exact value depends on the voltage division between the $Bi_{12}SiO_{20}$ crystal and the parylene. The device is now primed for the image exposure step in blue light (Fig. 4.7d). The blue light reduces the stored field in the illuminated areas of the crystal. But, since electrons are the mobile carriers and blue light is largely absorbed by the crystal, then greater sensitivity is obtained (~ 50 ergs/cm^2) when the image is written on the negative side of the device. Finally, the image is read out in red light (Fig. 4.7e). Readout can be by transmission or, with the addition of the dichroic layer mentioned above, by reflection. The readout light is polarized parallel to the (100) crystal plane and is analyzed by a crossed polarizer which also serves to reject light reflected from the front surface.

Readout efficiency is a function of the voltage ($\sim 2V_0$) established across the crystal and the half-wave voltage of the crystal. In transmission, the readout light passes once through the crystal so that $2V_0$ must approach the full half-wave voltage of 3900 V at 633 nm for full modulation. In reflection, only 1950 V is required for full modulation so that the reflection mode is preferred. In addition, optical activity, which is wavelength dependent, is cancelled in this mode.

The contrast of the image can be arbitrarily modified by reapplying a voltage to the PROM while reading out a stored image. The intensity at a given image point is determined by the magnitude of the electric field at that point. The electric field, however, has two possible signs which correspond to positive and negative phase differences between the light propagation modes (ordinary and extraordinary rays) in the crystal. Thus, the application of a uniform electric field to the crystal can either add or subtract a constant amplitude from the entire image. If the modulation of the image is small, then the contrast can be enhanced by subtraction of a constant voltage. If the substracted voltage is equal to, or greater than the maximum voltage element in the image, the contrast of the image is reversed. Finally, if the subtracted voltage has some intermediate value, all image points having that voltage are dark, and a constant intensity contour appears in the image. The ability to modify the image contrast has been termed baseline subtraction [4.27] and can be quite useful in coherent optical processing operations.

Performance parameters of a typical PROM were measured [4.27] and are given in Figs. 4.8, 9. The results were obtained with a device having a 900 μm thick $Bi_{12}SiO_{20}$ crystal, 5 μm thick parylene layers, 100 nm InO_2 electrodes, and an applied voltage of 2000 V. Figure 4.8 shows sensitivity measurements taken at various recording light wavelengths. It can be seen that maximum sensitivity is reached at approximately 440 nm over much of the exposure range except for very heavy exposures where the shorter wavelengths are more effective. Figure 4.8 also demonstrates that a 10^4:1 contrast ratio is obtainable from the device in parallel readout illumination. Figure 4.9 shows the energy required to reduce the output light intensity to $1/e$ and the gamma of each sensitivity curve, both of which as a function of recording wavelength. The sensitivity at the usual readout wavelength (633 nm) is down by a factor of 400 compared to the peak sensitivity at 440 nm allowing readout with almost no erasure at low readout levels. The exposure gamma (the straight-line portion of each exposure curve) varies from 1.5 at 589 nm to 4.8 at 365 nm.

The modulation transfer function for the PROM is shown in Fig. 4.10. The measurements were taken in an interferometer with collimated light beams. In a thick recording crystal such as that used in the PROM, the MTF will be reduced considerably from the values in Fig. 4.10 if high numerical aperture

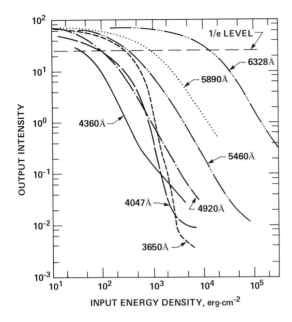

Fig. 4.8. Sensitivity as a function of recording wavelength [4.27]

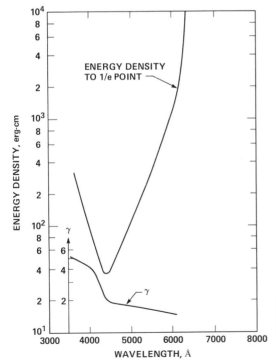

Fig. 4.9. Exposure energy and γ as a function of recording wavelength [4.27]

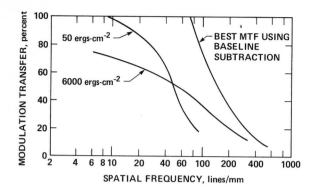

Fig. 4.10. PROM MTF for two exposure energies [4.27]

imaging cones are used, since the linear medium response causes recording throughout the light cone in the medium. The results shown in Fig. 4.10 are for two exposure levels; 50 ergs/cm² average exposure which yielded a highly linear recording as well as the best MTF out to around 50 cycles/mm and 6000 ergs/cm² which gave a higher MTF above 50 cycles/mm. The figure also shows the best MTF curve using baseline subtraction. Baseline subtraction allows the contrast to be electrically modified. This adjustment of contrast, however, can only be determined to an exact value at one spatial frequency. The natural MTF of the device then determines the effect at other spatial frequencies. Thus, only one frequency can be optimized in an image having a continuous frequency range if the image intensity is to remain undisturbed. The erasure-exposure readout cycle can be repeated indefinitely, with typical operation at TV frame rates.

The quality of surface polishing and the growing of successively more strain-free crystals has resulted in one-tenth wave performance over a 25 mm diameter crystal (such devices are offered for purchase from Itek). A remaining problem is piezoelectric bending of the crystal. Since the $Bi_{12}SiO_{20}$ crystal is slightly piezoelectric, it has a tendency to bend by an amount depending upon the nonuniformity of the stored electric field. This can cause several waves of astigmatism in a coherent reflected readout beam. Bending of the transmission PROM does not introduce aberrations since refracting effects at the two surfaces cancel in a thin plate. A system that works for reflective readout places a transmission PROM mounted with viscous oil onto a dichroic-coated glass flat [4.27].

4.1.4 Single Crystal Ferroelectrics (Photo-Titus)

This section describes single-crystal, ferroelectric-photoconductor, optically addressed light valves (Photo-Titus). A typical construction of this type of light valve and its use as a projection display device are shown in Fig. 4.11. Two types of ferroelectric materials can be used in a Photo-Titus light valve: materials operated above their Curie temperature which are electrically

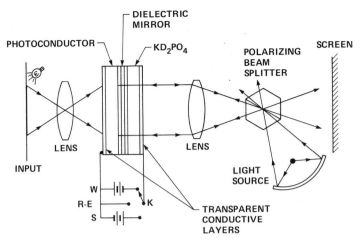

Fig. 4.11. Schematic structure and projection system of Photo-Titus. The different positions of the switch K correspond to the writing (W), reading or erasing (R–E), and subtracting (S) modes. The readout is effected in the reflection mode, the calcite polarizing beam splitter acting as two crossed polarizers [4.32]

monostable (such as KD_2PO_4), and materials operated below their Curie temperature which are electrically bistable (such as $Bi_4Ti_3O_{12}$). The electrically monostable light valves are well adapted to analog processing functions while the electrically bistable devices are better adapted to digital and storage functions [4.30]. The properties of both types of Photo-Titus light valves are discussed below.

First, let us consider a ferroelectric-photoconductor light valve employing KD_2PO_4 (DKDP), an electrically monostable crystal best operated just above its Curie temperature of $-52\,°C$. The operation of DKDP in an optical access mode is similar to the electron-beam accessed DKDP (Titus tube) described in Sect. 4.1.2. The Photo-Titus assembly (Fig. 4.11) uses a DKDP crystal operated near its Curie temperature and is light addressed onto the photoconductor [4.30–33]. The physical target is similar to that employed in the Titus tube with the addition, over the dielectric mirror, of an amorphous selenium photoconductive layer and a second transparent conductive layer of platinum. The structure is cooled to near $-52\,°C$ by a two-stage Peltier cell. The operation of the cell in a vacuum has the advantage of protecting the selenium layer and the DKDP crystal against moisture from the air.

The principal of operation of Photo-Titus is shown in Fig. 4.11. In the write-in step, an image is projected onto the photoconductor while a *dc* voltage is applied between the two transparent electrodes. This causes electric charges to move toward the dielectric mirror, the amount of charge being proportional to the exposure. During the readout cycle, a polarized projection light source (coherent or noncoherent) illuminates the DKDP crystal, reflects off the dielectric mirror, passes again through the crystal and is then passed through an analyzer to create an intensity image of the image input. During the readout cycle, the electrodes are short-circuited to prevent any electric field from appearing by capacitive coupling through the selenium. Using a 170 μm-thick DKDP crystal, the contrast ratio exceeds 70:1, and the limiting resolution exceeds 40 line pairs/mm. To erase the image, the whole photoconductor area is flooded with a 10 μs flash from a Xenon source while the two electrodes are short-circuited.

As in the Titus tube, DKDP addressed by a photoconductor performs better when cooled to near its Curie temperature. At room temperature, the half-wave voltage of DKDP is 3600 V while at T_c the half-wave voltage is approximately 200 V. Thus, one can obtain the same induced birefringence at T_c with a voltage many times lower than that at room temperature. However, the operation near $-52\,°C$ restrains the choice of the photoconductor. At this temperature, most photoconductors exhibit a low efficiency and trapping phenomena. In the case of amorphous selenium, however, the quantum efficiency can exceed 80 % at $-50\,°C$ [4.30] when light with a wavelength shorter than 450 nm is used and the electric field exceeds 10^5 V/cm. In the case of cooled DKDP, the voltage applied between the electrodes is about 200 V and the written potential difference between the two sides of the DKDP crystal varies from 0 to 100 V. Thus, a 10 μm-thick selenium film will have an electric field across it varying from 10^5 to 2×10^5 V/cm which results in almost constant quantum efficiency during writing. Operation near $-52\,°C$ presents the additional advantage that the dark conductivity is practically negligible; image decay exceeds one hour in the dark.

In optical data processing applications, image subtraction is possible by reversing the voltage polarity during the writing of a second image (Fig. 4.11). This allows an image to be reversed in contrast by subtracting a uniform image or to differentiate an image by the subtraction of a blurred image from the initial one. When the DKDP crystal is operated just above the Curie point in the paraelectric region, a linear relationship exists between the applied electric field and the induced optical phase shift. At the output of the polarizing beam splitter, the relationship between the modulation voltage and the light amplitude is linear up to 70 % of the modulation depth. Images can be written or erased optically with light flashes as short as 10 μs.

Electrically bistable ferroelectrics, such as bismuth titanate ($Bi_4Ti_3O_{12}$) single crystals, can also be used in a Photo-Titus configuration [4.30, 33]. The Curie temperature of $Bi_4Ti_3O_{12}$ is 675 °C so that room temperature operation is well below the Curie point placing the crystal in the bistable region. Optical addressing of bismuth-titanate devices has been demonstrated using CdS, ZnSe, and polyvinyl carbazole as the photoconductor [4.30]. The measured sensitivity of devices using CdS and ZnSe is near 1 mJ/cm^2 at a wavelength, λ, of 515 nm which is approximately two orders of magnitude lower than that obtained at $\lambda = 420$ nm using DKDP and a selenium photoconductor. This difference is due primarily to the large change in electrical polarization required with bismuth titanate. However, since this crystal is bistable, the storage of the information is permanent when the electric field is removed regardless of the intensity of the reading light.

The electro-optic effect in bistable crystals is based upon the switching of the spontaneous polarization of the crystal. Bismuth titanate exhibits a strong spontaneous polarization of about $50\,\mu C/cm^2$ along a direction lying in the ac plane and making an angle of about 4.5° with the a-axis. The crystal is optically biaxial. There have been several methods proposed for reading out the ferroelectric domains in bismuth titanate [4.33]. However, the "differential-retardation method" provides high efficiency and contrast and is insensitive to an optical degradation problem called depoling. This readout method is based upon the difference in birefringence in the written ferroelectric domains. The readout light must be monochromatic but not necessarily coherent. The crystal is placed between crossed polarizers, positioned 45° from extinction, rotated about the b axis, and illuminated with monochromatic light. The OFF or dark state is obtained by making the angle and crystal thickness such that for one of the polarization domains, the total retardation is equal to $N\lambda$ (N being an integer) which gives a "thickness extinction" for this domain. The other domain, having a different birefringence, will ordinarily have a different total retardation thus giving the ON or light-transmitting state. For useful crystal thicknesses and tilt angles, the birefringences are such that optimum $\lambda/2$ differences in total retardation are possible. This readout method is relatively insensitive to depoled surface layers which form on the crystal. The birefringence of the thin depoled layer is only slightly different from that of the bulk crystal and has only a small effect upon the total retardation. Optical contrasts of greater than 100:1 and a resolution of 90 line pairs/mm have been obtained with $Bi_4Ti_3O_{12}$ using this readout technique.

Because of the binary nature of the polarization domains in a bistable crystal, the grey-scale capability of these crystals is somewhat limited. The 90° polarization rotation involves a large optical change in the material which is in general rather slow and difficult to control electrically. The switching time is normally around 50 ms which is a little slow for TV rates. It is possible, however, to reduce the domain reversal time with high applied voltages. The activation time can be reduced to as little as 3 μs with an electric field of 10 kV/cm.

4.1.5 Ceramic Ferroelectrics

Several electro-optic effects have been reported in electrically poled, thin polished plates of hot-pressed lead zirconate-lead titanate ceramics [4.34–38]. Although many compositions of this class of ceramics have been studied, the most successful to date is a lead-lanthanum-zirconate-titanate of the composition $Pb_{0.99}La_{0.02}(Zr_{0.65}Ti_{0.35})_{0.98}O_3$, commonly called PLZT-7/65/35, where 7 denotes the atomic percent of La used and 65/35 is the $PbZrO_3/PbTiO_3$ ratio used in the mixture. This material is amenable to many

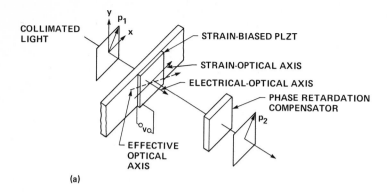

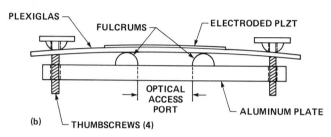

Fig. 4.12. (a) Strain-biased PLZT with transparent electrodes in one possible optical modulation arrangement. (b) Strain-biasing PLZT by bending plexiglas in a special jig [4.35]

fabrication procedures and is inexpensive. The electro-optic effects in PLZT are related to the electrical polarizability of the ceramics. Changes in the polarization fields are accompanied by changes in birefringence. The combination PLZT-7/65/35 exhibits electrical hysteresis and can be operated with or without memory.

Four operating modes have been reported for this material [4.37]. In general, this material must be electrically addressed through matrix electrodes, however, one particular implementation does allow for optical addressing through a ceramic-photoconductor sandwich. Only the strain-biased mode and the scattering mode of operation will be discussed here. The edge-effect mode and the differential phase mode do not provide any additional performance advantages in optical processing applications over the two operation modes discussed here.

The strain-biased mode of operation is based upon the stress-optic property of PLZT ceramics. A stress which induces a strain along a particular direction also induces a birefringence. The optical modulation operation in the strain-biased mode is illustrated in Fig. 4.12 along with a means of varying the strain-bias. The stress establishes an optical axis in the PLZT which is parallel to the stress vector in the plane of the thin PLZT plate. An electric field applied between transparent electrodes along the thickness dimension (z axis) of an unstrained PLZT plate will electrically pole the ceramic and thereby establish

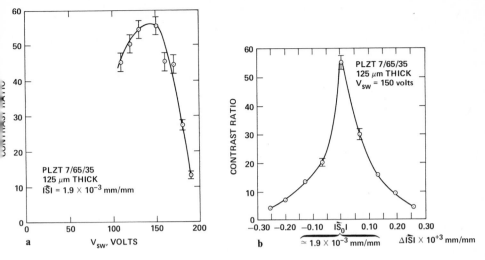

Fig. 4.13. (**a**) Contrast ratio of one BDC matrix element as a function of the switching voltage, V_{sw} [4.37]. (**b**) Contrast ratio of one BDC matrix element as a function of applied strain-bias, $|S|$ [4.37]

an optical axis. If the PLZT plate is strained along the x axis prior to the application of the electric field, the net effect is to establish an optical axis which is directed at some angle to both the x and z axis in the $x-z$ plane. Reversing the voltage and electric field across the PLZT plate electrically depoles the ceramic and returns the optical axis to the x axis. The birefringence experienced by a linearly polarized light beam propagating along the z axis through the PLZT plate can therefore be controlled by an electric field (a voltage) applied between transparent electrodes on the PLZT. The electrically controlled difference in birefringence is used to electro-optically alter the optical transmissivity of a PLZT device. In the absence of a strain in the plane of the PLZT plate, electric fields along the z axis cannot induce a net change in birefringence which can be detected by a polarized light beam propagating along the z axis. Without the strain, electric fields must be applied in the $x-y$ plane to induce detectable changes.

The procedure discussed above describes the operation of a single PLZT polarization modulating element. To operate on an image, a two-dimensional array of these elements must be fabricated with the transmitting electrodes deposited in a matrix format. The variation of the contrast ratio of a typical matrix element with switching voltage and with strain-bias are shown in Fig. 4.13. The graph of contrast ratio versus strain illustrates an advantage (high maximum contrast ratio, $\sim 55:1$) and a major disadvantage of this mode (rapid variation in contrast with strain). It has been very difficult in practice to maintain a uniform strain-bias over the whole matrix array. Uniformity in strain-bias can be achieved by reducing the strain-bias to the point where the contrast ratio is reduced to $20:1$.

The grain size of the PLZT used in the strain biased mode is less than 2.5 μm. There is very little scattering so that this type of device can be used for coherent optical processing. Fabrication procedures have been developed which allow matrix elements fine enough to resolve 50 line pairs/mm. Operating voltages from 50 to 300 V allow complete cycling of the material in 10 to 50 μs. Greater than 10^8 cycles have been achieved with no further degradation after the initial half-select disturbance signals in the matrix have partially switched the array positions to a lower contrast level. Net disturbed contrast ratios of 25:1 are considered possible with large matrix arrays. The PLZT platelets can be fabricated to >3.25 cm in diameter so that if the proper driving and addressing circuitry are developed, 1000 to 1500 elements are possible in one dimension. Uniformity in strain-bias remains a major problem, however, in fabricating such a large array.

The second basic mode of operation of PLZT light valves is the scattering mode [4.36, 37]. The grain size for this mode is larger than 4 μm. Light passing through the ceramic is scattered into a solid angle that depends on the state of electric polarization of the material. Light valves operating in the scattering mode can be electroded in a matrix array, as in the strain-biased mode, or they can be made into a continuous sandwich structure consisting of a coarse-grained PLZT ferroelectric ceramic plate and a photoconductor film between two transparent electrodes. The scattering mode of operation has the advantages that polarizers and strain-bias are not required and thickness variations are not critical.

The optical method for detecting the state of polarization employs a pinhole collector since the areas of the ceramic poled into a 1-state scatter the light into a much smaller solid angle than elements that have been depoled into the 0-state. In the image plane beyond the pinhole, the poled element areas appear bright and the depoled areas appear dark. The increased light scattering which produces minimum transmittance in the zero remnance or depoled state is throught to be related to the number of strain-relieving domain reorientations in this state [4.36]. The domains act as effective scattering centers, since there are significant refractive-index discontinuities at the doman walls.

The ratio of maximum to minimum transmittance (contrast ratio) depends on the ceramic preparation technique, on plate thickness, and on grain size and composition. The contrast ratio increases by more than two orders of magnitude as the average grain size is increased from 1 to 4.5 μm. With an average of 4.5 μm grains in a 600 μm-thick plate, contrast ratios as high as 1000:1 have been observed. The resolution obtainable with the ferroelectric-photoconductor plates has been as high as 40 line pairs/mm in 250 μm thick plates; however, this resolution decreases with plate thickness so that high contrast and high resolution cannot be obtained together.

The major problem with using the scattering mode PLZT in either configuration is its unsuitability for coherent optical processors. The scattering centers cause severe coherent speckle noise in coherent light which causes very poor signal-to-noise performance in the coherent processor.

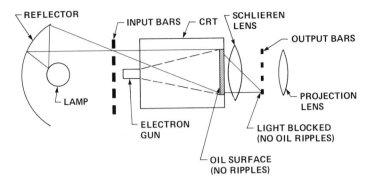

Fig. 4.14. GE light valve schematic [4.39]

4.1.6 Deformable Surface Devices

Surface deformation light valves have been made with both electron-beam and optical methods for generating a charge pattern on an elastomer surface. The elastomer may be an oil film or a thermoplastic, however, the effect of the charge pattern is the same; the surface of the elastomer deforms in an imagewise fashion as a function of the charge deposited or photogenerated. This surface relief pattern may then be read out in transmission or reflection by a coherent light source and a Schlieren optical readout system.

All of the surface-deformation devices have a band-limited, spatial frequency response. Primarily this is because the elastomer is an incompressible solid with an appreciable shear modulus. If depressed in one place, the same volume of elastomer must protrude in another place. This squeezing of the elastomer is opposed by internal shear forces. Low spatial frequency deformations of thin elastomer layers having only one free surface require more squeezing force than higher spatial frequencies, hence the dropoff in response to low spatial frequencies. At high spatial frequencies, the radius of curvature of the surface deformation becomes smaller, increasing the importance of the effective surface tension. Also the resolution of the electric field structure associated with the charge distribution on the elastomer falls off quickly at high spatial frequencies diminishing the contrast of the deforming force at the free surface of the elastomer.

a) Deformable Surface Tubes

The four devices described in this section all use an electron beam to write a charge pattern onto an elastomer-type material. The tubes have been developed for use as large-screen television projection devices and two of them have also found uses in coherent optical processing systems.

A schematic diagram of the General Electric (GE) light valve used in noncoherent light is shown in Fig. 4.14 [4.39]. An electron beam is scanned in a television raster format onto a liquid thermoplastic covering a transparent electrode. The recorded information appears as a charge pattern on the target which depth-modulates the thermoplastic by the generated electrostatic forces. The charge pattern then leaks through the liquid and the target again decays to a flat surface. The readout is performed in transmission with Schlieren optics. The deformation time is on the order of one msec and the decay time can be altered from 4 ms to 300 ms by controlling the beam current and the tube temperature.

The device is in a sealed vacuum tube which does not need to be pumped. The substrate disk rotates at three revolutions per hour and smooths and refreshes the liquid thermoplastic.

The major coherent processing applications of this light valve have been in wideband spectral analysis [4.40, 41] in which the two-dimensional optical Fourier transform of a one-dimensional signal (recorded in raster format) is produced. The folded spectrum [4.42] output is a more compact utilization of the available space than is the conventional optical transform.

A hybrid feedback system with the GE light valve is used [4.43] to correct geometric distortions due to the scan nonlinearity and phase errors due to thickness variations in the oil film target. This hybrid system is useful in correcting similar aberrations in other spatial light modulators. Holographic correction and compensation schemes also exist [4.27]. The vertical scan linearity is first sensed by imaging the raster scan lines onto a diffraction grating located in the image plane. The moire pattern produced describes the vertical scan distortions. Bright areas in this pattern correspond to regions where the raster and grating are aligned. The horizontal scan linearity can similarly be sensed by imaging a sine wave recorded on the oil film onto a second diffraction grating at right angles to the first. Detection of the horizontal linearity is achieved by imaging the different frequencies between two signals with the same ω_y coordinate and with ω_x coordinates that differ by the spatial frequency of the reference grating. Phase errors are similarly sensed by the phase modulating a carrier with the two input signals (whose difference frequency equals the spatial frequency of a reference grating). As in the detection of horizontal scan linearity, a pinhole placed in the transform plane can be used to filter this input and pass only one signal. These moire patterns are imaged onto a vidicon and subsequently interpreted and electronically measured.

In a calibration cycle, the appropriate signals are applied to the light valve by the calibration signal generator under control of the computer. The oil film target is divided into 32×32 cells and an individual error correction signal applied to each cell. The correction values are held in a 1024 word memory and synchronized such that the proper correction for a cell is fed to the D/A converter as that cell is being scanned in the light valve. The error in a given cell is determined by digitizing and averaging the brightness of the moire pattern in that cell and smoothing the correction values from cell to cell. The correction signals for the scan linearity error functions are applied to the light valve deflection circuitry. The phase error correction signals are used to spatially phase modulate the carrier on which the input data are present. The correction signals are integrated horizontally to produce a linear rather than a step correction.

The goal of the software correction algorithm is to drive the entire moire pattern to peak white. The first estimate of the correction voltage for a given cell is the correction applied to the prior cell. Values on each side of this initial estimate are then applied, the cell brightness obtained with each correction is stored, a parabolic fit calculated and the correction voltage corresponding to the peak of this parabola used as the correction value. This process may be repeated up to five times thus preventing the software routine from executing an endless loop. In practice, the least correction is required in the center of the target. Corrections then proceed to the left edge; this initializes the first column at which point corrections move up and down each column and left to right. One pass of corrections typically requires 20 s. 100 s are required for the correction cycle to converge. The dynamic nature of the oil thickness (the oil target rotates slowly) requires continuous correction. It is also often necessary to electronically preemphasize the signal to be recorded because of the dynamic and spatially varying decay mechanism. These techniques can also be used with other similar light valve devices.

The Eidophor tube was the first surface deformation tube developed for TV projection. The concept of the Eidophor tube for use as a theater projection system was developed in Switzerland by *Fischer* in the early 1940's. The device has since then been developed as both black and white and color television projection systems [4.44–47] and could find applications in optical processing systems. The Eidophor tube is similar to the GE light valve in operation. A thin layer of viscous oil acts as the deformable elastomer. The Eidophor tube was designed to operate at TV frame rates. The resolution obtainable on the oil film is limited to around 32 line pairs/mm, and contrast ratios have been obtained between 200:1 and 400:1.

Another deformable surface tube was developed at IBM and is called the Deformographic Storage Display Tube (DSDT) [4.48]. The target of the DSDT is a dielectric membrane which consists of an electronically controllable elastomer material layer and a reflective layer. The target is mounted in the tube so that the storage substrate faces the electron gun chamber of the tube. The deformable material with its conformal reflective layer is isolated in the separate front chamber of the tube. The addition of a thin reflective layer on the surface of the elastomer allows the use of a reflective optical system which eliminates the stringent optical requirements for the substrate and elastomer layer which are necessary for a transmissive optical system. The DSDT is limited to 15 line pairs/mm resolution. This tube was developed primarily for display purposes and the optical quality of the target is not adequate for coherent optical processing applications.

The Lumatron tube is another transmissive, deformable surface, electron-beam tube which was developed at CBS Laboratories [4.49, 50]. The electron

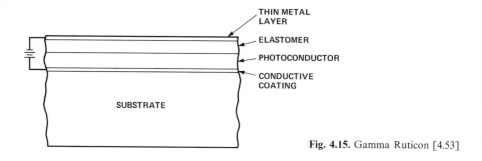

THIN METAL
LAYER

ELASTOMER

PHOTOCONDUCTOR

CONDUCTIVE
COATING

SUBSTRATE

Fig. 4.15. Gamma Ruticon [4.53]

beam writes onto a thin layer of reversible thermoplastic coated onto a transparent conductor on a glass substrate. The imagewise charge pattern depth-modulates the thermoplastic when it is heated above its melting point. The conductor acts both as a heater and as a reference-potential, faceplate electrode. Since the thermoplastic material must be heated to its melting temperature for writing, it may also be cooled after the image is scanned in, effectively freezing the image into the surface for days if desired. The heating and cooling cycle does limit the total write-erase cycle to slightly less than one second which is much too slow for TV rates. The electron beam and readout light paths in the Lumatron are coaxial. The resolution obtainable on the transmissive thermoplastic has been measured out to be 71 line pairs/mm. The thermoplastic material has been cycled for more than 30,000 times (full frames) before replacement was required.

b) Ruticon

Ruticons are optically addressed surface deformation light valves. The basic principles of operation are the same as the electron-beam deformation light valves. The Ruticons, however, are a layered structure consisting of a photoconductor layer coated on a conductive substrate and in turn coated with an elastomer layer. There have been several types of Ruticons developed [4.51–53] differing primarily in their means of placing an electric field across the photoconductor and elastomer layer. The method used to apply this field defines the type of Ruticon and its imaging properties.

Figure 4.15 shows the layered structure of a gamma Ruticon (γ-Ruticon), the structure most useful for optical processing applications. In this device, a thin, flexible, conductive metal layer is deposited on the surface of the elastomer layer. When image light exposes the light valve, charge carriers are photo-generated in the photoconductor and move in the electric field established between the two electrodes. The carriers move to the elastomer-photoconductor interface partly collapsing the field onto the elastomer. The resulting distribution of electric fields causes the elastomer layer to deform in an imagewise fashion. Since extended image storage is possible with some photoconductor elastomer combinations, it is thought that shallow trapping layers at this interface prevent lateral conduction of charge and loss of image resolution.

In the gamma Ruticon, the thin, conductive metal layer must conform to the elastomer to preserve the image information. The metal layer is generally opaque, so optical isolation of the input image and the readout light is possible. This device is read out with reflective Schlieren optics.

The optical quality of the metallized surface is quite good and has been used for coherent optical processing applications [4.53]. With recent photoconductors, an exposure of 10–50 ergs/cm^2 is required with approximately a 200 V bias for full modulation. The device can be written in less than 17 ms and can store the image for more than 5 min or be erased in less than 10 ms; thus, TV rates are quite feasible.

Another real-time, reusable material that has been used as an optically activated light valve is photothermoplastic [4.54]. This material requires a rather complex and well-controlled exposure development and erase cycle if reproducible results are to be obtained. In several instances [4.8, 55], the device has been electronically and computer controlled. Detailed recording procedures are discussed in Sect. 4.2.1d. The device consists of a thin thermoplastic and photoconductor on a substrate with a transparent conducting electrode. The front surface of the structure is initially uniformly charged. This charging is continued while the thermoplastic is exposed to a spatial light distribution corresponding to the data to be stored. After this exposure step, field variations proportional to the input data exist across the thermoplastic. If the material is heated to its development temperature, the structure will deform in a spatial pattern proportional to the recorded data. Upon cooling to room temperature, these deformations remain and can be used to phase modulate light.

Precise control of the exposure and temperature is needed to obtain a reproducible growth of deformations. Constant monitoring of the deformations and temperature is required or the deformations produced by two identical exposures will be different. Since erasure is also by heating, the development temperature must be adequate to produce strong deformations without erasing the stored charge. To further complicate operations, the duration of each step appears to depend on the prior history of the device. This is associated with the low (1000 cycle) lifetime of the device. The charging and reverse charging cycles must also be carefully controlled to insure adequate electrostatic force across the thermoplastic and large deformations and diffraction efficiencies without damage to the device.

Photothermoplastic has been used in both the input and spatial filter plane of an optical processor. Computer control of the device's operating cycle and the synchronization of the write, read and erase modes of the device with the input and output data and the remaining components in the system are essential operations in any real-time hybrid processor.

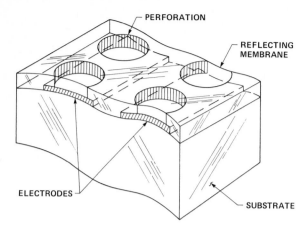

Fig. 4.16. Direct-wired membrane light modulator [4.60]

4.1.7 Membrane Light Modulator

The membrane light modulator (MLM) is a spatial optical phase modulator which uses electrostatic surface deformation of discrete, phase-modulating surface elements to modulate light reflected from its surface. Two types of membrane light modulators have been designed [4.56–61]: 1) the wired MLM in which surface deformations are produced by signals applied to stripe electrodes underlying the mirror surface, and 2) the photosensitive MLM where surface deformations occur on the mirror side of the device in response to the local light intensity present on the opposite side of the device.

Figure 4.16 shows a sectional view of a portion of a wired MLM. Stripe electrodes are deposited on an optically flat glass substrate with a thin ($\sim 1\,\mu$m) dielectric layer deposited over the electrodes. An array of small holes 5 to $100\,\mu$m in diameter is formed in the dielectric layer using photolithographic techniques. A thin membrane mirror, made out of a polymer such as $0.1\,\mu$m-thick collodion, is formed over the hole array. The membrane is metallized to enhance the reflectivity of the mirror surface and to make the mirror sufficiently conductive to hold a fixed electric potential.

When a voltage is impressed between the stripe electrodes and the metallized membrane, the unsupported regions of the membrane covering the small holes deform towards the substrate. Voltages of a few tens of volts are sufficient to deflect the membrane sufficiently to cause a half-wavelength of retardation in the illumination at the center of each deformation. The time response of these deformations is in the 0.1 to 1 μs range.

A 100×100 array wired MLM has been constructed on a 5×5 [mm] square substrate [4.59]. Forty volts were required to deflect the mirrors $\lambda/2$ in less than 1 μs. Another wired MLM was made into a 32×74 array on a 4×4 [mm] substrate [4.58] to optically calculate 32 point Fourier transforms of one-dimensional signals. This device was cycled for over 10^{12} cycles without failure.

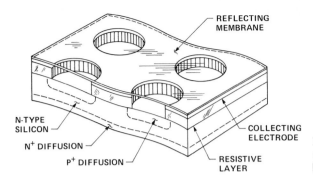

N-TYPE SILICON

N⁺ DIFFUSION

P⁺ DIFFUSION

REFLECTING MEMBRANE

COLLECTING ELECTRODE

RESISTIVE LAYER

Fig. 4.17. Photosensitive-membrane light modulator [4.60]

The wired MLM has a resonant frequency of the membrane elements in the 1 to 10 MHz range. In using a wired MLM, problems arise because of oscillatory vibrations of the membrane elements when excited at resonance. Mechanical or electrical damping must be used to control this resonance. The major problem with the wired MLM, however, is that it can handle only one-dimensional signals. This occurs because the phase modulation is identical along the entire length of each stripe electrode and varies only from electrode to electrode.

A two-dimensional, optical spatial phase modulator has been devised and is called the photosensitive-membrane light modulator [4.60]. A sectional view of this device is shown in Fig. 4.17. It is constructed from a semiconductor crystal, such as n-type silicon, sliced and polished to a thickness of 50 to 100 μm. An array of $p-n$ junction diodes is diffused into the mirror side of the device, with one diode for each membrane element. Next, a perforated layer of a highly resistive material, such as a semi-insulating glass, is deposited, The resistivity of this layer is chosen to be intermediate between the effective dark resistivity of the $p-n$ junction diodes and their effective resistivity when fully illuminated. A perforated collecting electrode is then deposited which makes ohmic contact with the resistive layer. The metallized polymer membrane is then applied over the entire structure. Finally, the side of the photosensitive MLM opposite to the membrane mirror surface is heavily doped in a shallow diffusion. This diffusion forms an electrode which is transparent in the long wavelength region of the visible and in the near infrared.

In operation, a voltage is applied between the collecting electrode and the transparent electrode such that all of the $p-n$-junction diodes are back-biased. The metallization on the mirror is held at the same potential as the collecting electrode. With no illumination onto the photo-MLM, most of the applied voltage appears across the $p-n$ junctions. When light falls on the side of the photo-MLM opposite to the mirror surface, hole-electron pairs are created. Holes diffuse toward the p regions, are collected by the nearest $p-n$ junction, and swept through its depletion layer. This current flows through the associated portion of the resistive layer and produces a potential difference between the

p region of the $p - n$ junction diode and the metallization in the membrane. This causes the associated membrane element to deflect.

There are several potential noise sources in either type of membrane light modulator. Since the individual membrane mirrors deflect the light out of the main optical path, some light falls out of the field-of-view of the processing optics. This is called truncation noise. Other sources of noise are due to surface irregularities, undamped mechanical oscillations, mirror position hystereses during cycling, and intermodulation noise among the neighboring mirror elements. MLM's have been constructed with a high quality $\lambda/10$ mirror surface across the aperture of the device. It has been found that MLM's operate without position hysteresis and have essentially unlimited lifetime when operated in a vacuum. When an MLM is operated in an air ambient, the deflection of a membrane element compresses the air molecules which are trapped in the perforation between the membrane and the underlying electrode. The thin membrane (~ 0.1 µm thick) is porous so that some of the air molecules diffuse through it tending to equalize the pressure on either side of the deflected membrane. When the input signal is removed, the membrane element does not return to its initial undeformed state until a number of air molecules equal to those which originally diffused out of the perforation have returned. Diffusion times are typically several seconds.

4.1.8 Summary of Interface Devices

Most of the interface devices discussed in Sect. 4.1 are still in the research and development stage, and, in some cases, not all of the performance parameters are available. Progress on these devices is, in general, quite slow since 1) they are usually considered proprietary by the companies which developed them and therefore have limited or no distribution to other potential users, and 2) the application of hybrid optical-digital processing has been a very slow maturing field, again limiting the applications of the devices. Several of the devices are still receiving active government and internal company support and are coming very close to providing the optical processing goal of a "real-time transparency".

A review of Table 4.1 and Sects. 4.1.1–7, however, will show that each device still needs some improvement to reach the goal of a fast, efficient, high resolution optical interface device. In general, the light controlled devices are simpler in construction and provide higher spatial resolution than their electrical accessed counterparts. The liquid crystal light valves have progressed significantly in the last few years with the major problem areas of poor lifetime due to electrochemical deterioration, slow speed, and low contrast being solved by the latest devices. The problems of ease of fabrication, reproducibility, and useful operating temperature range still remain with these devices. The electron-beam accessed DKDP tube has optical uniformity fabrication problems on adequately large crystals. The two-stage Peltier cooling makes the device bulky and slightly more difficult to use. The PROM device has required extensive work in fabricating high quality bismuth silicon-oxide crystals although recent devices have demonstrated that this can be achieved. The requirement for writing with blue light and reading out with red light could present problems in some applications. The DKDP Photo-Titus light valve has the same problems associated with cooling to near the Curie point as does the Titus tube. The bismuth titanate device has too slow a response for TV rates and also suffers from poor sensitivity. The strain-biased ceramic light valves suffer from rapid variation in contrast with local strain variations, sensitivity to thickness variations, and initial electrical and optical fatique due to half select disturbances. Another problem, common to all matrix-addressed devices, is the interconnection and addressing of a 1000×1000 matrix array. The scattering mode ceramic light valves are not useful in coherent light and also require large amounts of charge in the photo-conductors to alter the field in the ceramic. The present models of the Eidophor and GE light valve are of adequate optical quality for use in noncoherent light. Only with elaborate electro-optic feedback methods can these devices be used in a coherent optical processor [4.43]. The IBM deformographic tube has rather low spatial resolution and with some of the elastomers tested suffers from incomplete erasure, slow time response, and very poor lifetime. The Lumatron tube is too slow for use at TV rates due to the long erase time required for the thermoplastic. The optical quality and thickness variations in the thermoplastic cause a high noise level. The thermoplastic is also quite limited in lifetime. The Ruticon has a bandlimited response and, when screened, suffers from a lower overall spatial frequency response. The elastomer in this device is also limited in lifetime. The membrane-light modulator requires fabrication with more elements to be useful and also suffers from low optical efficiency.

In spite of the above limitations, most of the devices are finding some limited use in coherent optical processing applications. As the field of optical processing matures, the economics will develop for solving the remaining engineering problems listed above for one or more of the devices. The development of applications of optical data processing should also insure the commercial availability of more of these devices in the future.

4.2 Memory Materials

There is some overlap between the material requirements for interface devices and optical storage media. Some of the recording materials discussed in this section are, in fact, used in several of the devices of the previous section. Although similar materials may be used, the requirements of optical interface devices and optical storage media are quite different. This section reviews the present status of optical recording materials and discusses their suitability for optical storage.

Optical storage can take several forms: direct image storage, holographic storage (two and three-dimensional), and digital storage in the form of small, well-resolved data bits. Each of these forms of optical storage has a wide range of applications, and it is the application which will determine the requirements on the optical recording media. This cannot be stressed strongly enough. Although attempts should be made to develop the ideal recording material with very high sensitivity and readout efficiency, low noise, high resolution, instant readout with no development and infinitely cyclable, few applications really demand this perfect material.

There are some general requirements which most applications would require of a storage material. The material integrity must be good, i.e., it must be physically and chemically stable over a reasonable time span. In some optical processing applications where the material will be rewritten often, this time-span might be minutes, whereas in large scale digital storage the time-span must be years. The operating environment for the recording material must be suitable for ease of use. This normally implies room temperature operation in ambient air. Several of the storage materials to be discussed, however, exhibit much better performance at cryogenic temperatures and in a vacuum. This is a serious limitation of these materials. However, some optical processing applications could probably withstand these inconveniences. Wet chemical development processes are regarded as undesirable both with respect to the messy processing and also the time delay encountered between recording and readout. Although wet chemical processing has these drawbacks, photographic emulsions still are the most heavily used optical recording material for almost all applications. In spite of the above limitations, the extremely high recording sensitivity, high resolution, and familiarity and availability of photographic materials make them hard to displace. Another advantage of most photographic materials is low cost, an advantage gained strictly from batch fabrication procedures and those materials amenable to batch fabrication will have this advantage.

Write-erase cyclability (or reversibility) is normally regarded as an optical recording material requirement in most optical storage applications. The reversibility of the materials discussed will vary from none for the permanent recording materials of Sect. 4.2.2a to tens of reversals for some organic photochromics, to seemingly infinite reversibility of magneto-optic materials. In a hybrid optical computing device, it would be highly desirable to use reversible materials for "real-time" operation. In this application, a capability of write-read-erase cyclability in the thousands might be more than adequate. If an optical material is to be used as a general purpose digital store for a computer system, however, then semi-infinite reversibility is highly desirable. If the optical store is very large ($> 10^{12}$ bits) and is considered tertiary, then reversibility no longer becomes important and, in fact, can be considered a disadvantage. The material for this application should have an add-on or "postable" capability with instant readout such as the permanent recording materials of Sect. 4.2.2a.

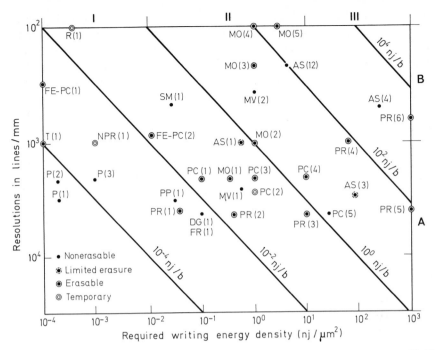

Fig. 4.18. Writing energy density and resolution of various optical memory materials [4.63]. The abbreviations refer to:

AS   Amorphous semiconductor: 1) $Te_{88}Ge_7As_5$; 2) $Te_{81}Ge_{15}Sb_2S_2$; 3) As_2S_3; 4) Au–Se–SnO$_2$ composite

DG   Dichromated gelatin

FR   Free radical organic dye in polymer

FE–PC   Ferroelectric-photoconductive: 1) $Bi_{12}SiO_{20}$; 2) $Bi_4Ti_3O_{12}$-ZnSe composite

MV   Metal vaporization: 1) Si engraving; 2) Rh film

PR   Photorefractive: 1) $Sr_{0.75}Ba_{0.25}Nb_2O_6$; 2) LiNbO$_3$ @ $\eta = 1.5\%$; 3) LiNbO$_3$:Fe–Mo; 4) $Ba_2NaNb_5O_{15}$; 5) LiNbO$_3$; 6) PLZT

MO   Magnetooptic: 1) MnBi films; 2) EuO; 3) GdIG; 4) YIG:Ga; 5) CrO$_2$

NPR   Nonlinear photorefractive: 1) KTN

P   Photographic: 1) Agfa 8E70; 2) Kodak SO285; 3) Kodak 649F

PC   Photochromic: 1) KCl:Na; F_A-centers; 2) Salicylideneaniline; 3) NaFM-centers

PP   Photopolymer

SM   Semimetal-metal transition: 1) VO$_2$

T   Thermoplastic-photoconductor composite

   High resolving power is a general requirement of all optical recording materials, however, this can vary from hundreds of lines per millimeter for optical processing to thousands of lines per millimeter for holographic storage. It is desirable to have the resolution of the storage process limited by the optics or wavelength rather than the material.

   Low recording sensitivity is probably the most serious single limitation of the alterable recording materials. Optical storage materials which employ photon-induced latent recording with development gain, such as photographic film, are usually 3 to 6 orders of magnitude more sensitive than direct recording materials. Significant progress is being made, however, in our understanding of and our ability to increase the recording sensitivity especially in photochromic and electro-optic crystals. The recording sensitivity required also strongly depends upon the application. If micron-sized bits are to be recorded for a digital optical store, then sensitivity in the range of 1 to $10\,mJ/cm^2$ is perfectly acceptable for low power, low cost laser write-sources. If one millimeter holograms are to be recorded, however, then this sensitivity range is rather low, especially when very short recording pulses must be used to decrease the effects of thermal diffusion. It is also desirable to have a high material damage threshold with respect to the typical write energy range so that the material is not permanently damaged.

High readout efficiency combined with low noise is very important for good signal-to-noise performance. The combination of high efficiency with low noise (optical scatter, optical and electromagnetic material nonuniformities, etc.) is the requirement, not high efficiency or low noise alone. A low efficiency material such as that found in a magneto-optic readout is perfectly acceptable if the readout noise level is very low. Recording dynamic range usually enters into the absolute signal-level readout considerations. A large induced change in material absorption and/or refractive index usually leads to a strong output signal. This, combined with low scattering and a good optical quality material, lead to high readout signal-to-noise ratios.

There are several excellent review articles in the literature which compare various optical recording materials [4.62–65]. Figure 4.18 from *Chen* and *Zook* [4.63] shows the required writing energy density versus resolution of erasable and nonerasable recording media. The diagonal lines in Fig. 4.18 indicate the required energy per bit or per square resolution cell. For holographic recording, only materials in region IA of Fig. 4.18 have the required light sensitivity and resolution. For bit-oriented memories, the required energy density can be much higher since typically micron sized bits are written, one at a time. For this application, materials in regions IA or IB are both suitable. An excellent review of holographic and bit-oriented optical memory applications and systems can be found in [4.64].

4.2.1 Photon-Induced Effects

a) Photochromics

Photochromic materials may exist as pure solid crystals, a solid solution of the compound in a host crystal, a polycrystalline layer of the compound on a substrate, a suspension of solid particles in a solid matrix, homogeneous solutions in plastics, liquid solutions, or suspensions of liquid droplets in a solid matrix.

A photochromic material reversibly changes its color (i.e., its absorption spectrum) on exposure to visible or ultraviolet radiation. This color change can revert to its original condition in the dark by thermal processes or a reversal may be stimulated by photons of a wavelength different to those causing the forward reaction. Many different mechanisms and materials which cause and exhibit photochromic changes have been reported. An excellent review of the potential mechanisms is available in [4.66].

Color changes in organic compounds have been observed to be caused by the following mechanisms: 1) trans-cis isomerism caused by rotation about a double bond, 2) bond ruptures, 3) transfer of an H atom to a different position in the molecule, 4) photoionization, and 5) oxidation-reduction reactions. Metastable electronically excited states of molecules provide another group of materials. Inorganic compounds exhibit photochromism with electron transfer reactions in the crystal lattice. These reactions can be due to ions (which may be present as a main constituent or as an impurity) with the ability to exist in two relatively stable oxidation states, and the trapping of photoreleased electrons in crystal lattice defects. The latter effect has been extensively studied recently as color center formation in alkali halide crystals. Photochromic glasses are another class of inorganic photochromic. Particles of silver halide, 50–300 Å in diameter are uniformly dispersed in the glass several hundred angstroms apart. Upon illumination, photolysis yields elemental silver which can later recombine with the halogen.

None of the above reactions involve gain or a chain reaction. Thus, the maximum recording sensitivity of a photochromic material would involve unit quantum efficiency where one photon would create or destroy the optical absorption of one color center. Competing side reactions reduce the quantum efficiency to less than one. In an ideal photochromic material with unit quantum efficiency, an exposure of approximately $0.4 \, mJ/cm^2$ would be required for an optical density change of 0.1. This limitation on the quantum yield is the most serious limitation of photochromic materials, a sensitivity which is orders of magnitude lower than that of photographic emulsions.

After irradiation, photochromic compounds recover naturally in the dark. This recovery time is strongly dependent upon temperature, but at room temperature the recovery may vary from microseconds to months. With some photochromics, the rate of the reverse reaction can be increased by photons of a wavelength absorbed by the activated form. Repeated forward and reverse reactions in most photochromic materials eventually cause fatigue due to the formation of irreversible photo-products. Organic materials suffer most in this respect; the inorganic materials are more stable. Most organic films last only 200–500 cycles before total fatigue. Some photochromic glasses have been cycled for more than 10^6 times without any observable fatigue.

The resolution of photochromic materials is potentially the resolution of the molecular size of the active compound, typically 5–10 Å. The actual resolution, then, in optical recording will be limited by the wavelengths of the recording and readout light and the quality of the optics.

No attempt will be made to give examples of all photochromic mechanisms and materials, instead only recent investigations of photochromic materials for optical recording and holography will be discussed. A theoretical treatment of holographic recording in thick absorption media is given in [4.67, 68], where, in addition to specifics of photochromic recording, the difficulties of limited dynamic range and low diffraction efficiency are discussed.

Photochromic films and plastics [4.69–72] are potentially useful substitutes for photographic film when enough recording and erasure energy is available. Holographic recordings have been made in salicylideneaniline, a polycrystalline thin film organic material [4.70, 72]. This material has a yellow color which changes to reddish orange when illuminated near UV (380 nm). The orange state fades back to yellow in about 30 h at room temperature. Cycling between the two states was performed 50,000 times with no observed fatigue which is extremely good for an organic photochromic. There are two crystalline forms, α_1, and α_2, of salicylideneanaline with the α_1 form requiring about 40 mJ/cm^2 exposure and the α_2 form, 200 mJ/cm^2. The α_2 state retains its color centers for about 30 h whereas the α_1 form decays 100 times faster. Contrast ratios of between 10:1 and 20:1 have been obtained in 10 μm thick films. Holographic gratings were written at resolutions greater than 3,300 line pairs/mm and with readout diffraction efficiencies of up to 1%.

There have been extensive studies recently of color center formation in alkali halide crystals [4.73–80]. The interest is centered primarily around the photodichroic properties of these crystals; that is, the property of some of the absorption bands to be strongly dependent on the polarization of the exciting light.

There have been many color centers identified in alkali halides. Four simple excess electron centers which can exist in alkali halides are the F, F_A, M, and M_A centers. The F-center is an isolated negative-ion vacancy that has trapped an electron. The M-center consists of two F-centers on nearest neighbor sites. The F_A and M_A-centers are F and M-centers associated with a substitutional ion impurity as their neighbors.

M and M_A-centers have been used to store bit patterns and images in NaF and KCl:NaCl [4.73, 76]. Imaging with grey scale has also been demonstrated in alkali halide crystals [4.81–84]. In an unaligned crystal containing M_A-centers, the centers will occupy six equally probable orientations within the crystal lattice. A particular M_A-center may shift from one orientation to another should a nearest neighbor anion make a vacancy jump. These vacancy jumps can be optically induced with F-band light and all M_A-centers can be aligned into one orientation. Figure 4.19 shows a typical absorption spectrum for a KCl:NaCl crystal containing F and M_A-centers. In Fig. 4.19, the centers are unaligned and read out with unpolarized light. Figure 4.19 is the corresponding plot of dichroic absorption after alignment with polarized F-band light at 531 nm. The large contrast change between the two polarization states can be read out destructively at 531 nm or nondestructively at 825 nm.

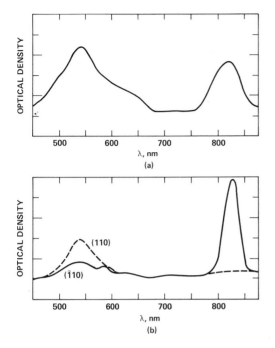

Fig. 4.19a, b. Absorption spectra of a KCl:NaCl crystal containing F and M_A centers at 77 K. (a) The absorption spectrum of an unaligned F and M_A center system measured with unpolarized light. (b) The absorption spectrum measured with ($\bar{1}$10) and (110) polarized light after alignment with ($\bar{1}$10) polarized 531 nm light [4.76]

A differential detection scheme has been proposed and tested for reading out the two possible M_A-center orientation states [4.76]. Circularly polarized 531-nm light is used to read out a bit state. If the bit state is written so as to pass (110) polarized light, the ($\bar{1}$10) component of a circularly polarized reading pulse will reorient the M_A centers and these centers will begin to absorb (110) polarized light. Simultaneously the ($\bar{1}$10) component will become more transmitting. Thus, if polarization filters are placed behind the crystal and oriented to pass (110) and ($\bar{1}$10) light, respectively, the bit state can be determined from the polarity of the difference between the (110) and ($\bar{1}$10) transmissions. The technique uses differential absorption, not absolute absorption, to detect a bit state and can therefore be used with thinner photodichroics having lower absorption. This readout scheme does rely on destructive readout.

A readout technique employing a write suppression beam has been proposed by *Schneider* [4.77] for use in KCl and by *Caimi* et al. [4.82] for use in NaF at room temperature. Suppressively written images and matched filters have also been recently demonstrated by *Casasent* and *Caimi* [4.83, 84]. In this technique, the M_A-centers are aligned and then bits written with write wavelengths λ_w and λ_R. The λ_w can be noncoherent since it just acts as a write catalyst. Readout is done at wavelength λ_R alone. Without λ_w present, the readout is nondestructive. This is very important to holographic storage where writing and readout at the same wavelength is desirable.

Most color centers in alkali halide crystals will thermally decay back to an unwritten state with time, this time being strongly dependent upon temperature. The optical memory tested with KCl:NaCl crystals [4.76] used liquid nitrogen cooling of the crystal to stabilize the centers. *Tubbs* and *Scivener* [4.80] have used very low temperatures (80 K) in conjunction with an applied electric field to increase the quantum efficiency of KCl crystals. Optical conversion of F-centers becomes efficient at much lower temperatures when a strong electric field is applied across the crystal. At the lower temperatures, the F-centers can be converted with approximately unit quantum efficiency. The sensitivity obtained is close to the theoretical maximum for photochromic materials. Crystals of KCl with 100 to 500 μm thickness were recorded on with exposures of 0.3 to 1.0 mJ/cm^2 for a 0.1 change in optical density at 80 K. Contrast ratios up to 10:1 with image resolution of a few microns were obtained with 4 mJ/cm^2 exposure.

Holographic recording has been successfully performed in several alkali halide crystals. Volume holograms were recorded with a 15 mJ/cm^2 exposure onto Na-doped KCl crystals using dichroic absorption of F_A-centers [4.74]. Permanent holographic recording has been done in KCl and KBr crystals using F to X-center conversion [4.78]. This type of recording in an F-center crystal is done at a high temperature. The F-centers are destroyed and X-centers are produced at the elevated temperature. The recorded X-center pattern can be fixed by quenching the sample to room temperature. The X-centers are then very stable and unlimited optical readout is possible. Holographic subtraction has also been demonstrated using selective erasure in volume holograms stored in alkali halide crystals [4.85]. In another coherent optical processing application, photodichroic crystals have been used as adaptive spatial filters in realtime optical spectral analysis [4.86]. NaF has been used at room temperature as a realtime matched spatial filter for pattern recognition [4.84].

A class of organic photochromic materials has been reported in which reversible photodimerization (bond rupture) is used with both states of the material being very stable at room temperature [4.87]. In most organic photochromics which use bond rupture as the color-change mechanism (such as spiropyrans) there are photon-induced side effects which limit the stability of the written state and also limit reversibility. Photodimers can be optically written into polycyclic aromatic hydrocarbons, such as anthracene, which are thermally stable at room temperature. The dimers can then be cleaved to the original monomers by irradiation with light of a suitable wavelength. In the materials tested, the dimer can be remade efficiently because the two monomers are held in a rigid matrix in the proper relationship for photodimerization. The absorption spectra of monomer and photodimer differ substantially and associated with this is a difference in refractive index. Typical experimentally observed index changes are 10^{-3} for single crystals and 10^{-5} for dimers in polymer matrices. Thus, this class of materials is attractive for reversible-phase holographic recording with nondestructive readout. The nondestructive readout is obtained at the expense of reading and writing at different wavelengths.

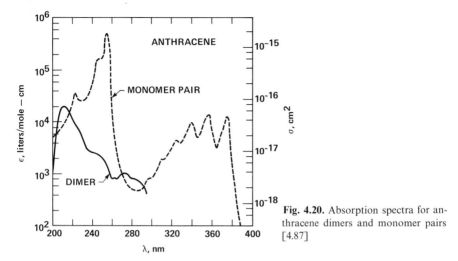

Fig. 4.20. Absorption spectra for anthracene dimers and monomer pairs [4.87]

The absorption spectra for anthracene dimers and monomer pairs are shown in Fig. 4.20. The anthracene dimer can be cleaved by absorption of light anywhere below 310 nm. Reformation of the anthracene dimer is caused by absorption in the longer wavelength band. The absorption of this light by one of the monomers leads to singlet-excited anthracene. This excited molecule then forms an excited complex (excimer) with the unexcited anthracene and this short-lived species can undergo bond formation resulting in the formation of the dimer. Typical photosensitivities are around $0.1 \, J/cm^2$ for this process.

If the dimers are held in a rigid matrix such as a polymer or an organic glass, then the pairs of monomers resulting from photon breaking of the bonds are not free to move about and tend to be in a favorable orientation for dimer formation. The form that generally provides the highest quantum efficiency is a single crystal of the dimer. Crystal dimers, however, are difficult to grow and some show signs of mechanical failure under cycling between the dimer and broken dimer states. The most successful experiments were on a nitrogen-substituted derivative of anthracene called acridizinium.

Another photochromic mechanism in organic materials is trans-cis isomerism caused by rotation about a double bond. Large refractive index changes have been observed in cis-trans isomers of stilbene [4.87–89] which make the mechanism attractive for phase holographic storage. Cis and trans isomers are molecules which have the same chemical constituents but different molecular geometry due to hindered rotation about a carbon-carbon double bond. The recording material is a composition of the cis and trans forms. A hologram is written by the absorption of one of the isomer forms which causes the cis-trans ratio to change. The cis-compounds are nonplanar and the trans-compounds are planar; therefore, changing the cis-trans ratio causes a marked change in refractive index. The hologram can be erased by a light wavelength absorbed by the new compound restoring the film to the original composition. Refractive

ELECTRO-OPTIC MEDIUM

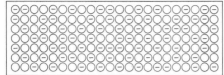

BEFORE EXPOSURE TO LIGHT

Fig. 4.21. Pictorial representation of holo-gram storage by diffusion of photogenerated free electrons [4.95]

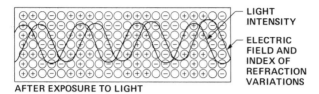

AFTER EXPOSURE TO LIGHT

LIGHT INTENSITY

ELECTRIC FIELD AND INDEX OF REFRACTION VARIATIONS

index differences as high as 10^{-2} have been measured between cis-trans pairs. The hologram can be read out with radiation absorbed by neither isomer form, so nondestructive readout is possible.

b) Electro-Optic Crystals

Electro-optic crystals are attractive materials for volume-phase holographic storage. Multiple hologram storage with very high diffraction efficiency and reversibility is possible in these materials, the only serious limitation being their relatively low sensitivity for recording and erasure. Several electro-optic crystals have been investigated over the past ten years with dramatic gains being made in material sensitivity, diffraction efficiency, and optical and electrical control of the recording and erasure properties [4.90–122].

The storage mechanism in electro-optic crystals is based upon photon-generated differences in the local index of refraction. Exposure of such a crystal to a light interference pattern excites electrons from traps to the conduction band. The electrons migrate to low light intensity areas, get retrapped and set up local charge patterns which cause changes in the index of refraction through the first-order electro-optic effect. A pictorial representation of a crystal containing localized traps (e.g., impurity centers, vacancies, or other defects) is shown in Fig. 4.21. Some of the localized states are assumed to contain optically excitable electrons, the remaining ones should be empty so that they may act as traps to allow for redistribution of the charge. All the traps are assumed to be thermally stable, and charge neutrality exists throughout the volume before it is exposed to light. When the light interference pattern is applied as shown in Fig. 4.21b, its effect is to generate a free electron concentration of the same shape into the conduction band. This concentration of free carriers diffuse thermally, or drift under applied or internal electric fields and become retrapped preferentially in regions of low intensity light. The end result is a net space-charge pattern that is positive in regions of low intensity as shown schematically in Fig. 4.21b for a sinusoidal interference pattern. The space-charge buildup continues until the field completely cancels the effect of diffusion and drift and makes the current zero throughout. The equilibrium space-charge generates a field that modulates the index of refraction and gives rise to a phase hologram.

Lithium niobate ($LiNbO_3$) and strontium-barium niobate ($Sr_{0.75}Ba_{0.25}Nb_2O_6$ or SBN) have been the most heavily investigated of the electro-optic crystals. Holographic storage was first demonstrated in lithium niobate in 1968 [4.90] and many significant advances in this material have been reported since that time [4.93, 95–100, 103–105, 107–110, 112–117, 122]. A very dramatic gain in the sensitivity of $LiNbO_3$ was made by doping the crystal with transition metal impurities during the crystal growth [4.95]. It was found that the sensitivity (i.e., the incident light energy needed to reach a certain diffraction efficiency) is determined by the concentration and absorption cross section of the electrons in the impurity centers and by the quantum efficiency of the process. Relatively pure crystals of lithium niobate exhibit low sensitivity due to the low concentration and optical absorption of the intrinsic defects or residual impurities. Crystals doped with iron were found to have dramatically increased recording sensitivity and diffraction efficiency over undoped crystals. In crystals containing 0.1 mole% Fe, it was possible to achieve diffraction efficiencies of 60% in a 0.25 cm thick sample, and the exposure required to reach 40% diffraction efficiency was only 1 J/cm^2, a factor of 500 improvement over previous results. Besides increasing the sensitivity for recording, heavy iron doping was found to lower the sensitivity to optical erasure which is useful when many holograms are to be recorded within the same volume.

The electro-optic storage process described above produces a hologram which erases optically unless longer wavelength light is used during readout. However, readout of thick holograms at a different wavelength causes distortion in the readout pattern. Thermal techniques were discovered to fix the refractive index pattern in the crystal [4.95]. The $LiNbO_3$ crystal was heated for 20 to 30 min to 100 °C during or after recording. The recording process creates a space-charge pattern due to the optically excited electrons. Heating the crystal allows ionic migration to take place without erasing the electronic hologram pattern since the activation energy for ionic or vacancy motion is lower than that for exciting trapped electrons. The increased ionic mobility allows them to drift and neutralize the electronic space charge. At this point, the hologram is nearly erased. However, upon cooling and reexposure to readout light, the electrons are redistributed evenly over the volume. The ions are not excited by the light and leave an ionic space-charge pattern which is resistant to optical erasure. A thermally fixed hologram can be erased by either heating the crystal to 300 °C which allows the ions to redistribute, or by heating to 100 °C while exposing it to uniform light of the proper wavelength to excite the trapped electrons and the storage process can then be repeated.

The diffraction efficiency of a fixed hologram is higher than that of the initial hologram due to a self-enhancement effect. This effect is brought about by the interference of the readout beam and the diffracted beam creating an additional electronic pattern that increases the field. When the readout beam is applied and part of it is diffracted by the fixed hologram, the readout beam and diffracted beam intersect within the crystal to form an interference pattern that writes a new hologram by the normal mechanism. The new hologram is identical to the fixed one and can enhance the net diffraction efficiency as long as the two holograms constructively interfere.

The doping of $LiNbO_3$ with iron has caused some difficulties. The diffraction efficiency of a thermally fixed and heavily doped crystal is typically much smaller than that of the original hologram [4.97]. Reductions of one to two orders of magnitude are common for samples doped with 0.01 % or more of iron. When the crystal is cooled after thermal fixing and then exposed to uniform illumination, the electrons tend to diffuse away from regions of higher concentration leaving behind a portion of the ionic pattern unneutralized. The fraction of ionic charge that actually becomes unneutralized depends upon the total concentration of Fe donors. As the doping level increases, a smaller percentage of the initial space charge is recoverable after fixing.

Because of the above problem, a new fixing technique was developed [4.99]. If the hologram is initially recorded in a crystal held at a high temperature where the ionic conductivity is very high, the electronic charge pattern which forms is continuously neutralized by the moving ions and no net fields are generated. Under these conditions, there is no limit on the magnitude of the space charge and one can establish sufficient index modulation to achieve good diffraction efficiency in the fixed hologram in an Fe-doped $LiNbO_3$ crystal.

Another problem encountered in transition-metal doped crystals of $LiNbO_3$ exhibiting high sensitivity is the appearance of optical scattering induced by exposure to coherent light [4.97, 105]. The scattering appears only during exposure to coherent light and shows pronounced angular selectivity. The diffraction cones of scattered light result from the internally recorded interference pattern resulting from the original incident laser beam interfering with light scattered from material inhomogeneities. The scattering interference patterns may be erased by illumination with uniform incoherent light or by writing additional superposed holograms at new angles.

A significant increase in writing sensitivity was found for Fe-doped $LiNbO_3$ when short-duration laser pulses were used [4.100]. When 30 ns laser pulses were used for recording, a sensitivity of $2 \, mJ/cm^2$ at 476 nm and $2.5 \, mJ/cm^2$ at 488 nm were required to record a hologram of 1 % diffraction efficiency. The erasure sensitivity of Fe-doped $LiNbO_3$ was also dramatically increased by heat-treatment reduction techniques on the crystal [4.103]. Only $12 \, mJ/cm^2$ of 488 nm radiation was required to erase a hologram. This investigation showed that the parameters of Fe-doped $LiNbO_3$ can be adjusted by control of both the concentration and valence state of the impurity ions. Figure 4.22 shows the write and erasure behavior of a low Fe-doped $LiNbO_3$ crystal (0.005 %) and a heavily doped crystal (0.1 %). The low-doped crystal was heat-reduced in an

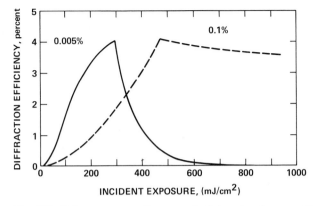

Fig. 4.22. Hologram storage (increase of efficiency) and erasure (decrease of efficiency) for two different LiNbO$_3$ samples: 0.1 % Fe-doped, 0.8 mm thick, partially reduced (argon at 850 °C); 0.005 % Fe-doped, 1.8 mm thick, almost fully reduced (argon at 1050 °C) [4.103]

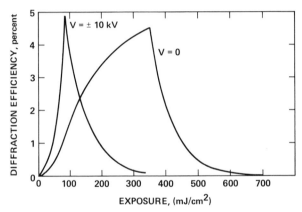

Fig. 4.23. Enhancement of storage, but not erasure, upon application of c-axis electric field [4.103]

argon atmosphere so that more than 90 % of the traps were filled and the heavily doped crystal was partially reduced so that only 15 % of the traps were occupied. Both crystals had equal absorption strength. The heavily doped crystal shows the typical behavior characterized by an erase sensitivity much lower than the write sensitivity. The lightly doped, heavily reduced crystal has higher write and erase sensitivities, both of which are nearly equal. In addition to the higher sensitivity, the 0.005 % sample did not exhibit self-enhancement during readout or erasure, nor did it exhibit optical scattering. A reduction in optical scattering has also been found when holograms are recorded and simultaneously fixed at elevated temperatures [4.108].

The storage sensitivity can also be enhanced by the application of an external electric field [4.103, 113, 115]. Figure 4.23 shows the storage and erasure sensitivity of a 0.005 % Fe-doped crystal with and without a c-axis field applied to the sample. The applied field does not appreciably change the erasure sensitivity, but it did increase the maximum diffraction efficiency from 5% to 20% in the crystal tested for Fig. 4.23.

The doping-ratio and heat-reduction treatment experiments also showed that diffraction efficiency is sacrificed for an improvement in erase sensitivity. The dynamic range of a crystal depends on the maximum space-charge density that can be produced which in turn depends on the density of traps. In a heavily reduced crystal in which a large fraction of the traps are occupied, the modulation is limited by the number of empty traps. Thus, there is a progressive lowering of dynamic range and diffraction efficiency with increasing erasure sensitivity.

Recent investigations have unveiled a new recording process in $LiNbO_3$, two-photon photorefractive recording [4.107, 110, 114, 122]. Reversible changes of the refractive index of pure and doped $LiNbO_3$ have been obtained by multiphoton absorption of intense, ultrashort light pulses. Recording at 530 nm was performed with a train of 10 picosecond pulses from a frequency-doubled, mode-locked Nd-glass laser. The recording sensitivity in high-purity $LiNbO_3$ was less than $0.4\,J/cm^2$ for a 25% diffraction efficiency. The change in index was 6×10^{-5} which is 130 times more birefringence change than obtained with the same energy but low intensity. The intensity of the short pulses was approximately $500\,MW/cm^2$. A further fourfold enhancement of the birefringence was obtained when the crystal was simultaneously exposed to both 533 nm and $2\,J/cm^2$ of 1.06 μm radiation.

This recording mechanism has been identified as intrinsic two-photon absorption of the $LiNbO_3$ host and is not related to the impurity contents as the impurities are not essential as electron traps in this process. The recorded holograms can be nondestructively read out without any fixing process with low-intensity 533 nm radiation. The holograms can be erased with a high-intensity uniform beam at 533 nm and following each erasure, holograms can be rewritten with the same sensitivity.

Strontium-barium niobate (SBN) has also proved to be an excellent material for volume-phase holographic storage [4.91, 101, 102, 118]. The material is ferroelectric at room temperature and exhibits a very large electro-optic effect. In initial recording tests, an exposure of $14\,J/cm^2$ was required to obtain an induced refractive index change of 5×10^{-4}. It was found, however, that use of an external electric field to provide a drift field would increase the recording sensitivity [4.91]. In an applied electric field of 10 kV/cm, images were written in SBN with $1.2\,J/cm^2$ exposure. Once the hologram was recorded, with the index change determined by the electric field and exposure, the reconstructed image oculd be made to disappear and reappear repeatedly by cycling the potential applied to the crystal. The reconstructed image increases in intensity with the field until the original efficiency is reached.

The recorded image in SBN will decay with optical readout at the same wavelength. The recorded hologram in SBN, however, can be electrically fixed with an applied voltage pulse larger than the coercive voltage of SBN [4.101]. This allows nondestructive readout at the recording wavelength. Figure 4.24

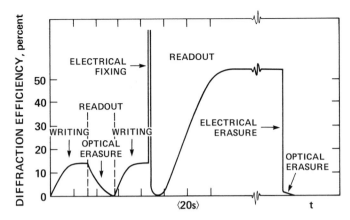

Fig. 4.24. Holographic diffraction efficiency during recording, fixation, readout, and erasure processes in a cube shaped SBN crystal; edges = 8 mm [4.101]

shows typical results of the recording, fixation, and erasure processes in an 8 mm, cube-shaped SBN crystal. During recording at 488 nm, the diffraction efficiency peaks at 14 %. Readout at this wavelength erases the hologram. Upon rewriting the hologram, a voltage pulse of 1000 V is applied for 0.5 s to fix the hologram. A diffraction efficiency enhancement occurs during the polarization switching time and is attributed to the increase of the electro-optic coefficient near the coercive field. The electrical pulse redistributes the ions to cancel the electron pattern which causes the readout efficiency to start at zero; however, during readout the electrons migrate until a saturation value of 52 % efficiency is reached. The hologram can now be read out nondestructively. The electrically fixed hologram can be erased by applying a voltage pulse with amplitude high enough to saturate the polarization in the crystal.

An effect unique to SBN is the ability to voltage-control a latent-to-active image reconstruction. The latency condition is obtained from a recorded image by applying a bias field antiparallel to the spontaneous polarization of the ferroelectric and large enough to start polarization reversal. Hysteresis in the polarization-reversal characteristic, which is normal in a ferroelectric, allows switching between the active and latent state by applying voltage pulses to the crystal. Suppression of the readout is done with a polarity opposite to the field used for enhancing the recording sensitivity during recording. A change in diffraction efficiency of 10^3 is produced between maximum (active) and minimum (latent) diffraction efficiency.

Two-photon photorefractive recording mechanisms have been recently reported in $KTa_{0.65}Nb_{0.35}O_3(KTN)$ as well as $LiNbO_3$ [4.110]. KTN is cubic at room temperature and exhibits a quadratic electro-optic effect. Because of the absence of a polar axis, there can be no spontaneous charge transport mechanisms and an electric field is required for recording in KTN. The sensitivity varies as the square of the bias field. With 10 picosecond pulses at

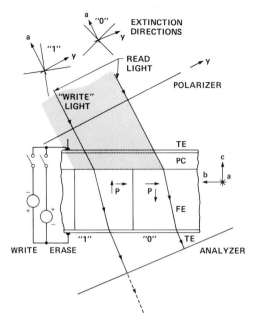

Fig. 4.25. $Bi_4Ti_3O_{12}$ photoconductor device write and read processes [4.124]

530 nm and a bias field of 6 kV/cm, holograms with several percent diffraction efficiency have been recorded in a 0.2 cm thick KTN crystal. The total exposure was approximately 0.1 mJ/cm², which is about 5000 times more sensitive than that of two-photon recording on Kodak 649 F emulsion. Holograms in KTN can be erased by turning off the bias field and illuminating the crystal uniformly with an intense light pulse with approximately the same intensity required for recording. Very interesting reversible holographic recording experiments have also been performed with materials such as $Bi_{12}SiO_{20}$ and $Bi_{12}GeO_{20}$ [4.119, 4.121].

c) Ferroelectric – Photoconductor Media

Erasable storage devices have been fabricated from ferroelectric-photoconductor sandwich configurations and also from ferroelectric crystals alone which demonstrate photoconductivity in addition to their ferroelectric properties. An example of a sandwich configuration is the combination of the ferroelectric bismuth titanate ($Bi_4Ti_3O_{12}$) and the photoconductor zinc selenide (ZnSe) between two transparent electrodes for use as an optical storage media [4.123–125]. Ferroelectric $Bi_4Ti_3O_{12}$ has an a-axis component as well as a c-axis component of remanent polarization and exhibits different orientations of the optical indicatrix corresponding to the two states of c-axis remanent polarization. Optical readout is based upon the birefringence induced by the internal polarization states.

Figure 4.25 shows schematically the write and readout process for a typical device configuration. A thin photoconductive layer (approximately 2 μm) of

ZnSe is evaporated onto a $50\,\mu m$ thick crystal of $Bi_4Ti_3O_{12}$ and then transparent electrodes are added. An optical pattern imaged onto the ferroelectric-photoconductor (FE-PC) device modulates the conductivity of the photoconductor. A voltage applied between the transparent electrodes switches these regions of ferroelectric beneath the high conductivity PC to a specific remanent polarization state. To erase the image, the ferroelectric is returned uniformly to the opposite remanent polarization state by applying a negative voltage with or without coincident light. The birefringence properties of the FE are used for readout. In Fig. 4.25, if the polarizer-analyzer pair is set for extinction of the zero state, then light will be transmitted only in one-state regions, allowing nondestructive readout of the stored polarization pattern.

There are two possible readout modes which work for this device. In the extinction-direction mode shown in Fig. 4.25, the crystal is tilted about the a-axis to obtain different extinction directions for the two states. In the refractive index readout mode, the crystal is tilted about the b-axis to obtain a difference in the relative refractive indices of the two polarization states for light polarized perpendicular to the b-axis.

Holograms have been stored in these devices with grating structure as fine as $1.26\,\mu m$. This resolution of the FE-PC device exceeds what would normally be expected for polarization domains going through the $50\,\mu m$ crystal thickness parallel to the c-axis. A surface storage mechanism has been discovered which explains the high resolution [4.125]. The high resolution patterns written into the FE-PC device are stored as c-axis remanent polarization domains of triangular cross section which have grown from the *ab* surface of the $Bi_4Ti_3O_{12}$ crystal adjacent to the photoconductor layer into the bulk. For a $1.5\,\mu m$ grating, the domains penetrate about $7\,\mu m$ into the crystal. Thus, a FE-PC grating acts more like a thin rather than a thick hologram and may be read out over a wide range of angles.

The recording sensitivity of the device is dependent upon the gain of the photoconductive layer. With ZnSe as the PC, a recording sensitivity of $1\,mJ/cm^2$ at $515\,nm$ was measured. Readout diffraction efficiencies are typically 0.01% at $633\,nm$ using the refractive index or phase readout mode. The birefringence-mode readout technique produced diffraction efficiencies in the range of $10^{-4}\%$ to $10^{-3}\%$ at $633\,nm$. The readout efficiency of this mode is low because of the low efficiency of converting light in the switched grating regions from linear polarization in one direction to linear polarization in an orthogonal direction. The refractive-index mode of readout also eliminates image degradation due to optical nonuniformities caused by a depoling of the a-axis remanent polarization. No deterioration in performance has been observed in this device after more than 10^5 repeated erase, write, and read cycles.

An optical storage device has also been constructed from single active layer photosensitive, electro-optic crystals. The typical device configuration is the Pockels readout optical modulator (PROM) described in Sect. 4.1.3 and illustrated in Fig. 4.6. Since the construction and operation of that device have been covered, only performance parameters pertinent to optical storage will be discussed here. Three photosensitive electro-optic crystals have been used in PROM storage devices: zinc sulfide (ZnS), zinc selenide (ZnSe), and bismuth silicon oxide ($Bi_{12}SiO_{20}$) [4.21, 126–128].

Recent PROM devices have used $Bi_{12}SiO_{20}$ exclusively [4.21]. $Bi_{12}SiO_{20}$ has a much larger electro-optic effect than ZnS or ZnSe with a Pockels effect approaching that of DKDP. Furthermore, bismuth silicon oxide can be grown by the Czochralski method into large crystals with over $6\,cm^2$ useful area.

The required half-wave voltage of this material is about a factor of 3 lower than ZnS or ZnSe. Peak recording sensitivity in $Bi_{12}SiO_{20}$ PROM devices has been measured at $25\,ergs/cm^2$ and observable optical modulations still exist at a spatial frequency of 300 line pairs/mm. The resolution of the PROM device is limited during write-in by the cone angle and absorption of the exposing radiation.

d) Thermoplastic-Photoconductor Media

Hologram formation in thermoplastic film was first demonstrated by *Urbach* and *Meir* in 1966 [4.129]. Because of the advantages of high sensitivity and resolution, dry and nearly instantaneous *in situ* development, erasability, and high readout efficiency, development of thermoplastic-photoconductor media has continued to draw support and show improvements in the recording media parameters [4.54, 130–135].

A hologram is recorded in a transparent thermoplastic film as a spatial variation in film thickness which corresponds to the input light fringe pattern. Thermoplastics are usually resin-type materials, are not sensitive to light and must, therefore, be combined with a photoconductor. The photoconductor can be dissolved or finely dispersed in a layer of thermoplastic or the thermoplastic may be coated as a layer over a photoconductor film. The substrate for either structure is a glass plate in which a thin film of transparent conductor is deposited. Since thin, thermoplastic film thicknesses are required for high-resolution recording, the sandwich structure has been preferred.

There are a couple of operating modes for a thermoplastic-photoconductor device. In the sequential mode [4.130], the first step in the recording operation is to sensitize the material with an electrostatic charging device, such as a corona device (a thin wire at a high voltage). The corona device is scanned across the thermoplastic to establish a uniform potential on the surface with respect to the transparent conductor. The corona device ionizes the air near the surface of the thermoplastic. The positive ions are attracted toward the grounded conductor and are deposited on the surface on the thermoplastic where they are held by negative charges induced on the transparent conductor. The second step is the holographic exposure. The light discharges the photoconductor in the illuminated areas and reduces the potential at the surface of the thermoplastic in the same area. The electric field across the thickness of the thermoplastic, however, remains unchanged as it is proportional to the surface charge density which is unaffected by the exposure. In the third step, the corona device is used to charge the surface once more, the potential everywhere on the surface reaching its preexposure value. The electric field across the thermoplastic increases in the illuminated areas as a result of the second charging, although the surface potential is now uniform. The thermoplastic is developed with a pulse of heat in the fourth step. Either a hot air stream, or current in the transparent conductor is used to raise the temperature of the thermoplastic to the softening or melting point, typically between 60° and 100 °C. A this temperature, the thermoplastic film deforms according to the imagewise local electric field and becomes thinner at high-field (illuminated) regions and thicker elsewhere. Cooling quickly to room temperature, the deformation is frozen into the thermoplastic as a thickness variation and is stable at room temperature. The last step before starting a new recording is thermal erasure. The thermoplastic is heated so that the surface tension of the softened or molten thermoplastic evens out the thickness variation and erases the hologram. The electrostatic charges are also neutralized due to the increased conductivities of the photoconductor and thermoplastic at the elevated temperature.

Recording on a thermoplastic-photoconductor device can be reduced to a two-step process by a "simultaneous" recording process [4.131]. Simultaneous recording can be done in two ways: 1) exposure to incident radiation during the corona charging before heating, and 2) simultaneous charging and exposure after or during heating. In the first case, the charge pattern is recorded at room temperature and the film is then heat-developed. In the second case, the charge pattern is formed as the thermoplastic cools, and charging and exposure continue while the surface of the thermoplastic is deforming. Since the potential surface is maintained as deformation occurs, the charge density in the grooves will be increased. In this recording mode, the deformation continues until either the surface tension and viscous forces balance the increased electrostatic force or until the thermoplastic film is torn apart along the bottom of the grooves.

The simultaneous mode with thin thermoplastic layers is preferred to the sequential mode for high efficiency holograms. First-order diffraction efficiencies of up to 40% have been achieved, greater than the theoretical maximum of 34% for thin sinusoidal-phase grating [4.131]. These high efficiency gratings are actually "blazed" into a nonsinusoidal pattern due to the strong electrostatic forces. The high efficiency gratings were recorded in a sample of 1.2 µm Staybelite Ester 10 thermoplastic at a spatial frequency of 1170 cycles/mm. The diffraction efficiency was measured at 442 nm. The write temperature was in the 40°–50 °C range, and the erase temperature was approximately 70 °C.

In the absence of a light pattern, i.e., with a uniform surface charge on the thermoplastic, a "frost" pattern can be developed. The frosting is a random deformation of the thermoplastic which has a dominant spatial frequency of approximately twice the thermoplastic layer thickness. The simultaneous recording mode gives virtually frost-free images. Evidently an electric field threshold for frost formation exists [4.131] and so during simultaneous recording the small starting signal deformations are driven by an increasing field as corona charging proceeds and by the time the frost threshold is reached, the signal deformation is strong enough to suppress frost.

The thermoplastic recordings exhibit a bandpass spatial frequency response just as in the elastomer recordings on the Ruticon (see Sect. 4.1.6b). The bandpass characteristic serves to reduce intermodulation distortion in the thin-phase recordings on thermoplastic. Intermodulation distortion is an intrinsic problem of most thin-phase holograms and shows up as ghost images due to beats between the object spectrum and low-frequency autocorrelation terms. The attenuated low-frequency response of the thermoplastic serves to attenuate the intermodulation. The lack of response to uniform or low spatial-frequency illumination also allows holographic recording at high ambient-light levels.

If the sequential mode of recording is used, the resulting diffraction efficiency is usually quite low. A post-development recharging and reheating (either simultaneously or sequentially) step has been found to dramatically enhance the diffraction efficiency [4.131]. The recharging establishes a uniform potential on the deformed surface; however, the charge distribution is nonuniform with more charge located in the valleys. Thus, electrostatic forces are formed which tend to deepen the corrugations, hence increase the diffraction efficiency of the recording.

Thermoplastic-photoconductor devices eventually fatigue after repeated cycling. Thermoplastic damage under prolonged corona charging with the existence of repeated heating has been shown to be one of the major causes of fatigue [4.54]. After repeated write-read-erase cycling, the thermoplastic film begins to look nonsmooth and sticky as a result of the combined effects of corona charging and heat.

Some thermoplastics also suffer from residual image problems. The residual images are a result of residual deformation rather than residual charges. Double or triple erasure heat pulses are useful for removing the residual images. The speed of cycling a thermoplastic is limited mainly by the cooling rate of the substrate after each thermal erasure. Short heat pulses help speed up this process by greatly reducing the heat energy dumped into the substrate.

The best performance reported to date for thermoplastic-photoconductor devices was for a thermoplastic of a terpolymer of styrene, octyl methacrylate, and decyl methacrylate, and a PVK/TNF photoconductor [4.133]. Diffraction efficiencies of 10% with $60 \, \mu J/cm^2$ exposure were obtained, which is comparable to the sensitivity of high-resolution photographic emulsions. These thermoplastics were capable of over 5000 record-erase cycles before fatigue set in. Sensitivities as high as $5 \, \mu J/cm^2$ and recording resolutions higher than 4000 lines pairs/mm have been reported for other thermoplastic-photoconductor combinations [4.131]. Thermoplastics, as well as elastomers, also hold the promise of replication since the information is recorded as a surface relief pattern.

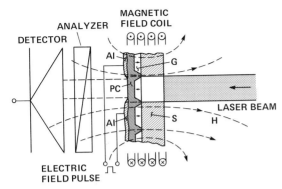

Fig. 4.26. Experimental setup for studying the feasibility of MOPS for reversible high-sensitivity optical recording. The composite device is kept at its compensation temperature where not heated via photocurrent in the illuminated photoconductor. Al: aluminum electrodes, PC: photoconducting Cu–CdS film, G: magnetic garnet memory cell, S: nonmagnetic garnet substrate, and H: applied magnetic field [4.136]

e) Other Effects

A novel optical recording mechanism has been recently reported which employs photoelectric access and magneto-optic readout [4.136–140]. The composite optical memory structure is called MOPS (magneto-optic photoconductor sandwich). The memory cell is triggered by a low-power laser beam and a separate electric field supplies the energy for warming up the addressed ferrimagnetic storage cell. A cross-sectional view of an experimental magneto-optic photoconductor sandwich is shown is Fig. 4.26. A mosaic pattern of 4 μm thick single-crystal ferrimagnetic garnet islands, confining the bit sites, is covered by a 4 μm photoconducting CdS layer and two 0.5 μm thick Al electrodes with 106 μm spacing and 5 mm length. Each island is a single magnetic domain with the magnetization, M, oriented either up or down to the large surface of the garnet crystal. The magnetic state of the garnet film is read out in transmission via the magneto-optic Faraday effect. Recording and erasure of bits correspond to the reversal of M between the two polar domain states, just as in MnBi recording.

When light strikes, the photoconductor charge carriers are generated in the environment of the addressed magnetic island. An electric pulse is applied to the electrodes around the island just before termination of the light pulse and causes a current to flow in the illuminated region of the photoconductor. The ohmic heat produced raises the temperature of the adjacent magnetic garnet island allowing an applied magnetic field to switch the magnetic domain into the direction of the field. Thus, in the switching of a domain in a MOPS array, the bit position is selected with a laser beam, and an applied electric field performs the actual heating allowing a magnetic field to switch the bit state.

The single crystal ferrimagnetic storage medium used in experiments of the MOPS device is film of the composition $Gd_{2.5}Yb_{0.5}Fe_{4.8}Al_{0.2}O_{12}$ grown on a nonmagnetic garnet substrate and has a compensation temperature at 46 °C. The whole device is kept at this temperature so that a temperature rise of tens of degrees centigrade will drop the film coercivity to the point where a small applied magnetic field (typically 50–150 Oe) will switch the domain.

Recording experiments have shown an exposure sensitivity of 10^{-4} J/cm^2 for 1 % read-out efficiency and a magnetic switching field of < 100 Oe. Typical cell size is 10×10 μm^2. To optically switch the cell, about 1 erg/bit would be required for heating the magnetic garnet film via optical absorption. In the MOPS device, less than 1 millierg/bit of optical energy is required, a sensitivity gain of greater than 1000:1. The sandwich has been switched at a rate above 400 Hz and more than 10^6 switching cycles have shown no degradation in performance. Applications of the MOPS device to large scale optical memories are given in [4.138, 139].

Changes in the physical properties of magnetic crystals have also been induced by irradiation with polarized light [4.141]. Photoinduced magnetic anisotropy and photoinduced optical dichroism have been studied in silicon-doped YIG at cryogenic temperatures. The photoinduced effects are associated with motion of electrons from iron site to iron site. With liquid helium cooling to 4.2 K, optically induced magnetic torques up to 1.6×10^4 dyn-cm/cm^3 have been measured. The magnetic field required to produce such a torque would be 80 Oe. The mechanisms for such effects are still not completely understood. Any application to optical storage would at the present be severely limited by the extremely low temperatures required.

4.2.2 Thermal Effects

a) Permanent Recording Media

Laser thermal recording has been used to record "real-time" transparencies for display purposes and large scale read-only optical memories. Thermal recording uses the temperature rise in a thin film due to the absorption of energy from a laser beam to cause a physical or chemical change in the film. Such recordings are "real time" since the data are immediately available for readout and no subsequent chemical or heat processing is required to develop a latent image. Thermal film recording is presently one of the most active optical media development areas, especially with respect to optical disk recorders.

One of the earliest reports of thermal recording [4.142] demonstrated spot by spot image recording on metallic and organic thin films. The three films used in this investigation were a 500 Å evaporated coating of lead, a 500 Å evaporated coating of tantalum, and a 1.0 μm coating of a triphenylmethane dye in a plastic binder. The thin films were all on glass substrates. A 50 mW beam at 633 nm was used to record line widths of the order of 2 μm.

Maydan [4.143] has given a rather complete analysis of laser machining of thin metallic films. A model of the machining process is developed which predicts the temperature rise of the thin film as a function of time. Both the theory and the experimental results show that machining of thin films on substrates of very low thermal conductivity is most efficiently obtained with short laser pulses in the 10 ns range. For a long illumination duration t, the temperature rise of the thin film at the end of the light pulse is proportional to $\sqrt{t}$. For very short light pulses, however, the temperature rise approaches a linear dependence on time. For long light pulses, a large portion of the laser energy is lost to the substrate by heat conduction during the pulse.

The experiments were performed primarily on 600 Å thick bismuth film on glass and Mylar substrates. The write sensitivity was 60 mJ/cm² for 25 ns pulses from an intracavity, acousto-optic modulated argon laser. The intensity of the laser pulse was modulated to vary the size of the holes and to write halftone transparencies. Each laser pulse machines a single, nearly circular hole in the metal film by displacing and removing metal from the transparent substrate. Near the write threshold, almost no bismuth film is evaporated; the metal is only melted and the surface tension in the film pulls the metal back from the center, forming a crater. In larger holes (∼ 10 μm) up to 50 % of the bismuth film is evaporated, the rest rolls back under surface tension to the melting point diameter. The machining of metal films involves a sharp recording threshold so that it is easy to write spots smaller than the beam waist diameter. The machining could be done with front or back illumination with approximately equal efficiency. Back illumination through the substrate has the advantage that the bismuth vapor from the surface is directed away from the focusing and deflecting optics.

One of the earliest laser mass memory systems was based upon laser machining of a thin metallic film [4.144]; in this case the metal film was rhodium. A 10^{12} bit memory was developed by using an argon laser to ablate data bits on approximately 12 cm by 75 cm data strips. The plastic-based strips are coated with a thin film of rhodium. The data strips are accessed from a carrousel and loaded on a drum. A carriage drives the focus optics across the strip with fine tracking done with a galvanometer. Although this type of memory is not erasable, it has the advantage that it is postable, i.e., new data can be written at any time into unused portions of the memory material. Metallic recording produces holes in a highly reflective media and thus produces very high contrast and the possibility of very high signal-to-noise ratios.

The ablative type bit-oriented memory materials have recently been extensively applied to optical disk recorders based upon features of the consumer video disk players such as the Philips VLP [4.145–150]. Extremely sensitive thin film (few hundred angstrom) metallic layers, such as tellurium, and amorphous metallic compounds, such as AsTeSe, have been found to be sensitive to a few milliwatts of laser power for micron sized bit recordings. These metal layers also possess the desirable features of high contrast, high signal-to-noise readout, and apparently stable and archival qualities. With the recent advent of single mode diode-lasers, there has been substantial work on ablative materials for the near infrared (around 8400 Å). An antireflective structure employing thin metallic films has been developed which can be absorption and reflection tuned for writing and reading at any wavelength

[4.145, 148]. An example of a potential low cost optical disk memory based upon an ablative metal layer disk and a diode laser write source is described in [4.150].

b) Thermomagnetic Materials

Thermomagnetic recording materials for optical readout, or more concisely, magneto-optic materials, have received more attention in the past fifteen years than any other class of alterable optical storage materials. These materials offer most of the advantages of magnetic storage which is so prevalent today with computer tape and disk systems, and yet offer potential advantages in storage density, noncontact recording and readout, and addressing speed. This section will not cover the systems implications of these materials for optical storage since they are covered in excellent review articles [4.151, 152]. Suffice it to say that the primary motivation for research on these materials has been for bit by bit recording on magneto-optic disk-type systems for potential use in the computer industry. Holographic recording has been demonstrated on magneto-optic materials, but as will be discussed below, these materials are rather poor candidates for holographic storage.

In bit by bit thermomagnetic recording, a focused laser beam is used to locally heat the thin film material so that the magnetic state of that particular location may be switched. There are several thermomagnetic effects which can allow this magnetic switching to take place. The major advantage of using a laser source to control the switching is the small size (down to $1\,\mu m^2$) of the switched area when compared to typical ferrite head magnetic-switching techniques. Readout of the written bits is also performed with a laser beam and typically employs one of the two first-order magneto-optic effects.

The Faraday effect employs a transmitted linearly polarized beam through the medium of thickness, d. The polarization of the beam is rotated by an angle ϕ_F given by $F\hat{m}\cdot d$, where F is the specific Faraday rotation in degrees per centimeter (typically $10^5\,deg/cm$) and $\hat{m}$ is the unit vector of saturation magnetization. In reflective readout, there is a rotation of the polarization, ϕ_k, through the Kerr effect. This Kerr rotation is a sensitive function of the incidence angle. For uncoated ferromagnetic materials, ϕ_k seldom exceeds 1 degree although ϕ_k can be increased with proper antireflection coating.

A figure of merit commonly used to compare magneto-optic materials is the amount of magneto-optic rotation per unit optical loss. For Faraday rotation, the figure of merit is $2F/\alpha$ in degrees per neper optical loss where α is the optical absorption coefficient in cm^{-1}; for Kerr rotation it is $R\phi_k$, where R is the surface reflectivity. The transmitted or reflected signal is analyzed by an analyzer and then imaged onto a photodetector. The difference in signal-current levels from region of positive and negative magnetization is directly proportional to the respective figure of merit, hence the signal-to-noise ratio is also proportional to this figure. Optimizing the medium thickness for maximum signal in a transmissive system gives $d_{opt} = 1/\alpha$.

In holographic recordings on magneto-optic materials, the read-out signal is proportional to the square of the figure of merit, and the optimum medium thickness is $2/\alpha$. The first-order diffraction from the magnetic grating is derived from the component of the magnetic field perpendicular to the original polarization direction so that no analyzer is required upon reconstruction. The major problem with using magneto-optic materials for holography is the very low diffraction efficiency, typically 10^{-4}. The thin materials do not diffract light well and increasing the material thickness to increase the diffraction efficiency only serves to lower the resolution possible on the medium. The resolution in a magnetic thin-film material is proportional to the smallest magnetic domain size, and this, in turn, is proportional to medium thickness.

A second problem with holographic recording on magneto-optic materials is the medium sensitivity. Typical write sensitivities for magneto-optic materials fall in the range 0.01 to 0.1 J/cm^2. In the writing of individual, micron-sized bits at 10 Mbits/s rates, this implies a 1 mW to 10 mW laser source. For 1 mm diameter holograms, however, the laser must provide a peak power of greater than 1000 W, and the pulse width cannot exceed much more than 10 ns or thermal diffusion will destroy the high-resolution magnetic grating. Lasers with this power and pulse duration requirement with a suitable pulse repetition rate are not presently available.

There are four identified phenomena used for thermomagnetic writing in thin film magneto-optic materials [4.152] and the particular phenomenon depends upon the recording medium to be used. Curie-point writing is the method most commonly used. The material is heated with the laser beam until the temperature of the heated spot exceeds the Curie temperature of the medium. During cooling from above the Curie temperature, the magnetic closure flux as well as an applied external magnetic field can effectively determine the direction of magnetization of the cooling bit. Once cooled, the bit will not switch because of either magnetic field, and thus only the area heated above the Curie temperature is affected. Thus, a large-area magnetic field may be used in conjunction with the laser beam to control the magnetic state of the stored information bit. Thin films of manganese bismuth (MnBi) and europium oxide (EuO) have been most extensively studied using Curie point writing.

A second thermomagnetic writing technique is called compensation temperature writing. In some ferrimagnetic materials such as gadolinium iron garnet (GdIG), the two sublattice magnetizations are in opposite directions. At the compensation temperature, the magnitudes of these sublattice magnetizations cancel out, and the medium attains extremely high coercivity. A few degrees away from this compensation temperature, the coercivity drops and magnetization switching becomes easy. The laser pulse is used to raise the spot above a temperature at which the coercivity drops below the applied magnetic switching field.

A third writing technique also involves the temperature dependence of the material coercivity. In materials like phosphorous-doped cobalt [Co(P)] there is a strong temperature dependence of the coercivity. At a temperature of about

150 °C, the coercivity is decreased by a factor of three from that of the room temperature material.

The fourth technique is called thermo-remanent writing. In materials such as chromium dioxide (CrO_2), the magnetization can be thermally induced at a temperature below the Curie temperature. The remanent magnetization is strongly dependent on the peak temperature and the applied magnetic field during cooling. The remanent magnetization is equal to the statistical average of an ensemble of positive and negative magnetic domains within the heated spot so that analog recording is possible with this technique.

The major noise source in magneto-optic readout is fixed pattern noise within the medium. Possible sources for this noise are optical scattering by the magnetic domain walls and grain boundaries, birefringence induced by strain, and noise fluctuations caused by magnetic, metallurgical, and structural inhomogeneities in the material.

Examples can be found in literature [4.153–182] of all four classes of thermomagnetic writing materials. The discussion to follow gives emphasis to MnBi, one of the most extensively studied magneto-optic materials. Curie-point writing on MnBi has received particularly intensive study [4.153–155, 159, 160, 162–166, 168–176] because of its very high specific Faraday rotation, its good write sensitivity, high resolution and room temperature operation.

Curie-point writing on MnBi requires that a laser pulse heat a spot from approximately 20 °C to 360 °C. The write sensitivity of the various forms of MnBi can vary from 0.01 to 0.1 J/cm², which implies a laser power in the milliwatt range with a pulse width of 1 μs for 1 μm diameter spots. The demagnetizing field of this material will cause an isolated spot heated above the Curie temperature to switch magnetic polarity without an applied field; however, for erasure, an applied field of approximately 750 Oe is required. The stability of the magnetic domain walls and the square magnetic hysteresis loop and high coercivity ensure that the data stored are stable. The material can be operated at room temperature in air if the film is overcoated. Repeated write and erase cycles up to 10^7 cycles have been successfully demonstrated with no write or readout degradation [4.168]. The minimum magnetic domain size in MnBi is below 1 μm, and the grain size is of the order of 0.1 μm so that the thin film material can support an optics-limited recording density. The specific Faraday rotation and the figure of merit, $2F/\alpha$, for MnBi are both very high and among the best found to date in any magneto-optic material.

Chen et al. [4.166] have outlined a method of MnBi thin film preparation which consistently yields high quality film. The key to his process lies in depositing the bismuth layer onto the substrate before the manganese layer and then annealing the two to form MnBi. The easy axis of magnetization of the thin film is oriented perpendicular to the substrate surface. With the MnBi normally prepared on a glass substrate, a silicon monoxide protective coating is then evaporated. The coating thickness is carefully adjusted to provide an optical refractive index match which allows a more efficient use of laser power during writing and improved read-out signal-to-noise ratio. The second

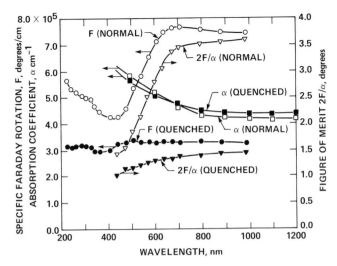

Fig. 4.27. Specific magneto-optic Faraday rotation and absorption coefficient as function of wavelength for both low temperature phase and quenched low temperature phase MnBi films [4.166]

objective of the coating is to protect the MnBi from contamination by moisture and oxidation.

Dynamic monitoring of Curie-point writing on MnBi indicates that it is essentially a Curie-point process but that it is somewhat complicated by the existence of two crystallographic phases within the operating temperature range of the material [4.168]. The normal low temperature phase (LTP) is ferromagnetic with a Curie temperature around 360 °C. During repeated write cycles above 360 °C, a first-order phase transition takes place. The crystallographic phase change converts the material at the particular location into a high-temperature phase (HTP) which when it is quenched from above 360 °C to room temperature goes to the quenched state (QHTP). The quenched, high-temperature phase film has a Curie temperature of 180 °C. *Chen* and *Stutius* have shown [4.172] that the LTP and HTP films are separate compounds with the chemical formula MnBi and $Mn_{1.08}Bi$, respectively. Above 360 °C, the LTP material MnBi decomposes into QHTP $Mn_{1.08}Bi$ plus Bi. Thus, thermomagnetic writing on normal LTP MnBi will lend to the formation of areas consisting of QHTP and a nonmagnetic Bi-rich phase. The quenched phase transforms back to the normal phase by what appears to be a nucleation and growth process. This process takes about 30 days at room temperature [4.168]. Both phases of the material will dissociate at 446 °C, at which point the material is destroyed.

The absorption coefficient, Faraday rotation, and the figure of merit, $2F/\alpha$, of the two phases of MnBi are shown in Fig. 4.27. The rotation, hence the figure of merit for QHTP films is about a factor of 2 lower than that for the LTP film

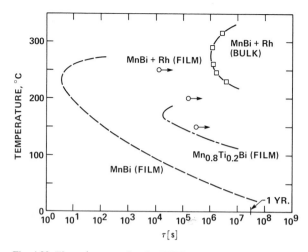

Fig. 4.28. Time constant τ for the QHTP→LTP phase transformation as a function of temperature. The three data points (O) shown for a film with the nominal composition $Mn_{0.84}Rh_{0.16}Bi$ represent annealing times during which no transformation was observed and the arrows indicate the corresponding τ's are longer in time [4.173]

at 633 nm. The figure of merit of LTP MnBi of $>3°$ at 633 nm represents the highest value of any room temperature magneto-optical material.

Because of the reduced Curie temperature of 180° of the QHTP film, it requires about one-fourth as much laser power to write as the normal phase [4.155]. The lower Curie point also gives a much larger margin between the write threshold at 180 °C and the damage threshold at 466 °C. For optical memory applications, the higher write sensitivity and the increase in write-to-damage power margin of the QHTP film make it more attractive than the LTP film, in spite of the reduction in $2F/\alpha$ and the readout signal by a factor of two. However, these advantages of the QHTP film cannot be fully utilized if it reverts back to normal LTP during the life of the memory.

Several investigations have been made to dope MnBi with other elements to stabilize the material in the high temperature phase. The QHTP films have been successfully stabilized by doping with titanium [4.162, 164] and with rhodium and ruthenium [4.173]. The compounds $Mn_{0.8}Ti_{0.2}Bi$ and $Mn_{0.84}Rh_{0.16}Bi$ are the most successful and result in films with specific Faraday rotation of approximately 2×10^5 deg/cm which is about a factor of two lower than undoped QHTP MnBi. Thus, one must suffer an additional reduction in signal output to stabilize the material which is not a completely satisfactory solution. Figure 4.28 shows the time constant for the QHTP to LTP transformation for undoped MnBi film, Ti-doped MnBi, and Rh-doped MnBi. No transformation at all was observed for the MnBi + Rh thin film so that the time constants shown by the three data points on Fig. 4.28 lie considerably to the right of the points shown. A doping of MnBi which stabilizes the film in either phase

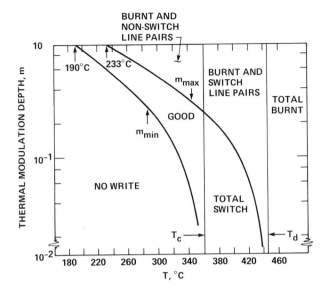

Fig. 4.29. Characteristics of holographic recording on MnBi films [4.166]

without a major reduction in figure of merit could lead to an extremely useful material for bit-oriented optical memories.

Holograms have also been written onto MnBi [4.159, 163, 166, 174, 175]. The holographic pattern generated from the interference of an object beam and a reference beam produces a pattern of temperature rise in the film due to the absorption of light energy. The transverse heat flow in the material, however, rapidly diffuses the local temperature differences. In order to capture the temperature profile before considerable diffusion takes place, a laser pulse in the region of 10 to 15 ns is required. As the thermal fringes cool below the Curie temperature, magnetic flux reversal occurs producing a magnetic fringe pattern. Thus, the thermal modulation must exceed the Curie temperature but not the damage temperature, and yet the average temperature must be below the Curie temperature to maintain fringe contrast. These requirements define a very specific writing region for holographic recording on MnBi films and this region is shown in Fig. 4.29. As stated earlier, the laser power required to attain the writing temperature of MnBi in 10–15 ns over a 1 mm diameter hologram is in the kilowatt range. The magnetic grating on MnBi will diffract a linearly polarized beam with a typical efficiency of 10^{-4}.

A second thermomagnetic writing technique, described earlier, is compensation temperature writing. Gadolinium iron garnet (GdIG) is a typical material that has been studied using this thermomagnetic process [4.161]. GdIG is a ferrimagnet that exhibits compensation at 13 °C. At this temperature, the magnetization becomes zero and the coercivity rises sharply. Thin polycrystalline wafers (14 to 40 µm thick) are prepared by hot pressing. Typical grain diameters of 2 to 20 µm are obtained in the wafers. During writing, the material is kept at a temperature slightly above the compensation temperature.

The pulse of light is used to heat the spot by only 5 to 10 °C which causes the coercivity to drop dramatically, allowing a 100 to 300 Oe field to switch the magnetic domain. The material has a high figure of merit, $2F/\alpha = 2°$; however, the minimum wafer preparation thickness of around 10 μm means that the minimum switchable magnetic domains are of the order of 3 μm which seriously limits the data packing density of this material.

Another compensation point magneto-optic material has been recently examined in optical memory experiments at IBM [4.167]. This material is GdCo, an amorphous ferrimagnet which does not suffer from the grain noise associated with polycrystalline materials. GdCo films have perpendicular magnetic anisotropy and a compensation temperature which can be adjusted in the vicinity of room temperature by changes in the composition. The writing power for GdCo covers a broad range, 0.02 to 0.08 J/cm², depending upon the compensation temperature. At room temperature, the slope of the coercivity versus temperature curve is very steep so that only a small change in temperature is required to reduce the coercivity below the magnetic bias field level. By varying the composition ratio of GdCo so that the compensation temperature is −100 °C, the coercivity changes less rapidly and considerably more write power is required. However, a problem has been uncovered with the more sensitive room-temperature samples – small domains or bits are unstable. Domains less than 4 μm in diameter tend to collapse. GdCo samples with $T_c = -100$ °C will retain 2 μm diameter domains. These samples, however, have low coercivity, ~50 Oe, which requires a tight control of the bias field. The figure of merit of this material is fairly high, ranging from 0.5° to 1°, so that if methods are developed to stabilize small domains near room temperature, this material would be quite useful for optical storage.

In the past few years solutions to the polycrystalline and temperature phase problems of MnBi have been found in a new class of amorphous magneto-optic materials [4.177–182]. These materials exhibit excellent record sensitivity, good readout rotation and unusually uniform quality thin films. Although the basic Faraday and Kerr rotation of the amorphous magneto-optic thin films, such as TbFe, is not as high as MnBi, it has been found that overcoating with dielectric thin films can significantly enhance the readout signal [4.180].

c) Amorphous-Crystalline Transition

Reversible changes in the optical properties of amorphous semiconductors can be induced by application of light energy. Focused beams of laser light have been used to write and erase micron-sized areas in these materials, and holographic storage has also been successfully used in these materials [4.183–199]. The materials used as recording media have been alloys of various combinations of Te, Ge, Sb, S, Se, and As deposited on glass and metal substrates as thin films a few microns thick. The change in the thin film induced by the light energy causes a variation in optical transmissivity of the film which can be detected with optical readout.

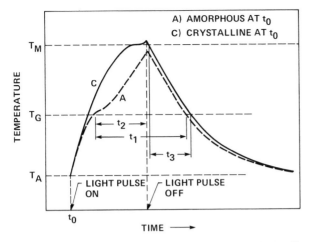

Fig. 4.30. Temperature-time profile of the active material when illuminated by a square light pulse of energy above both the amorphous and crystalline band gaps. Curve *A* refers to material initially in the amorphous phase, curve *C* to material initially in the crystalline phase [4.189]

Two distinct writing effects have been reported in amorphous semiconductors [4.193]. The first was a change from an amorphous (glassy) state, in which the film is transparent, to a crystalline state in which it is opaque. This effect is reversible. The second effect involves the creation of scattering centers induced by material flow in the amorphous phase. The spots, or voids created can be erased by annealing with less intense laser radiation.

A model for the first effect, photo and thermal-crystallization, is shown in Fig. 4.30. With the material initially in the amorphous phase, curve *A* shows the condition for crystallizing the amorphous material initially at ambient temperature, T_A. The laser-pulse intensity and duration are such as to bring the material to a temperature well above the glass-transition temperature, T_G, but just below the melting temperature, T_M; in this range, the crystallization growth rate is maximum. After the pulse ends, the material rapidly cools from this temperature. During a time, τ_1, the material is in the temperature range between T_G and T_M in which crystallization can proceed. During a fraction of this time, τ_2, the light is on and the absorption of photons creates a large density of broken bonds creating an accelerated crystallization growth rate; this is called photocrystallization. Once crystallized, the material will remain so after cooling to T_A. The reverse process is shown in curve *C*. The crystallized material to be revitrified is heated to a temperature above T_M. In this case, a glass will be reformed provided that after the light pulse ends the material is cooled through the region between T_M and T_G sufficiently rapidly to prevent crystallization. During the time the material is in this critical regime, the light is off so no photocrystallization occurs and the material is quenched into the amorphous phase.

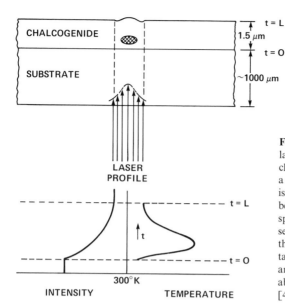

Fig. 4.31. Schematic diagram of the laser interaction with a long chain chalcogenide film illuminated through a transparent substrate where the film is partially absorbing to the laser beam. Included is an approximate spatial profile of the beam, a cross section of the written spot including the (cross-hatched) bubble and qualitative profiles of both laser intensity and temperature through the film at about the peak in the laser pulse [4.186]

A model for the second reversible writing phenomenon observed in amorphous semiconductors is shown in Fig. 4.31. A focused beam of laser light is absorbed throughout the thickness of the film. In the amorphous, or chalcogenide film, the glass transition temperature is reached at only tens of degrees above ambient at which point there a marked drop in viscosity of the glassy material. Thus, the exposed material quickly becomes very fluid with a vapor pressure which increases with the continued absorption of light until either a vapor bubble nucleates or thermal expansion produces a bulge. If the cooling, or quenching of the material is rapid enough after the light pulse goes off, the material around the bubble deformation will remain frozen in and the interior of the bubble or void will remain more or less empty. The light intensity and temperature profiles responsible for formation of a bubble or void in the interior of the thin film are shown in Fig. 4.31. The observed optical effect of this bubble upon optical readout is a dark spot due principally to the refraction and scattering of the light. To reverse this effect, the material surrounding the bubble must absorb enough light to reach a temperature well above the glass transition temperature but not as hot as in the formation light pulse. At this point, the bubble collapses due to surface tension and the material continues to flow and reform a more or less smooth film. Write-read-erase reversibility up to thousands of cycles has been demonstrated using the above two effects [4.192].

d) Semiconductor-Metal Transition

Vanadium dioxide (VO_2) exhibits a semiconductor-to-metal transition at 68 °C which is accompanied by large changes in its electrical and optical properties

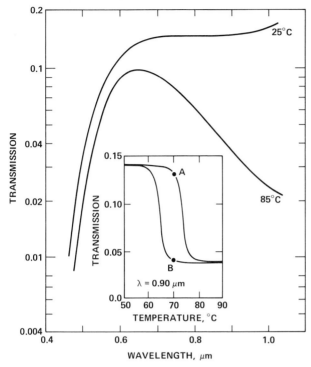

Fig. 4.32. Optical transmission of a 0.15 μm VO$_2$ film vs. wavelength below (25 °C) and above (85 °C) the semiconductor-to-metal transition. Inset: Hysteresis in the transmission at wavelength 0.90 μm [4.201]

[4.200–202]. The changes in optical transmission and reflection are small in the visible spectrum but increase considerably in the near infrared as shown in Fig. 4.32. A hysteresis in transmission versus temperature occurs when the material changes state (see inset of Fig. 4.32) so that the material can be used for optical storage. During a typical write cycle, the film is thermally biased at point A in Fig. 4.32, and a pulse of laser light will thermally switch a spot to the lower transmission state, point B.

Writing and reading out VO$_2$ films is best done at near infrared with a GaAs injection laser (0.84–0.9 μm) or a Nd–YAG laser (1.06 μm). The thermal response of the films of VO$_2$ appears to be less than 10 ns. The write sensitivity for a 0.15 μm thick film is about 3 mJ/cm^2 to switch a spot. The written spots can be erased by lowering the temperature of the film approximately 10 °C.

Holographic storage has also been demonstrated on VO$_2$ films [4.200]. High-resolution holographic storage is feasible since in the vicinity of the transition temperature of VO$_2$, the semiconducting and metallic phases can coexist with sharp phase boundaries. A resolution of 10^3 line pairs/mm has been achieved.

The permanence of the storage in VO_2 is associated with the width of the hysteresis and in practice allows the ambient to vary by 7 °C without affecting the stored data. The films can be recycled through the phase transition indefinitely without deterioration.

4.2.3 Recording Material Application and Summary

A broad range of optical recording materials has been described with widely varying performance characteristics. The suitability of any one of the materials depends almost completely upon the intended application. Optical processing applications demand recyclable materials to be used as interface devices and to act as temporary image or data buffers. For optical storage applications, however, reversibility becomes less important, especially for very large memories of 10^{12} bits or more.

The recording materials still most widely used today are the permanent recording materials such as photographic films, dichromated gelatin, photopolymers and photoresist. These materials combine high recording sensitivity, high readout efficiency and a low noise output. Thermal recording materials such as thin films of tellurium, however, have been receiving extensive investigations for applications to large scale optical disk memories where the lack of erasability can be considered an advantage. The integrity of large data banks is important and these films possess good sensitivity, a "postable" or add-on recording capability, and instant readout.

Among the alterable recording materials, magneto-optic materials seem destined for bit-oriented optical stores, such as optical disk devices. The magneto-optic materials are quite unsuitable for holographic storage, yet they appear as the single most attractive alterable media for high-density digital optical recording. Recent advances in amorphous magneto-optic materials are leading to more practical formulations for system applications. The marriage of solid state lasers with magneto-optic materials should eventually lead to large capacity, low cost-per-bit optical stores, probably in a disk format.

The amorphous-crystalline transition materials exhibit a much higher contrast than magneto-optic materials but eventually fatigue with cycling. This, of course, is not acceptable to the digital optical stores for which these materials seem most suited. The major limitations of the semiconductor-metal transition material (VO_2) are the requirements for thermal biasing and, more important, the requirements for cooling to erase a bit location.

Thick, photochromic crystals and electro-optic crystals offer tremendous potential for excellent three-dimensional holographic storage media. For economic as well as technical reasons, it is unlikely that a volume holographic memory will be built to compete with available magnetic storage technology. The crystal storage devices do, however, offer multiple hologram storage capability for optical processing applications such as complex spatial filters, image buffers, optical logic devices, etc. The primary limitations of these crystals is their relatively low sensitivity, a much more serious limitation for holographic storage than for bit-oriented storage. Dramatic improvements continue to be made in all aspects of these materials and these combined with improvements in laser sources might make them very useful for future optical-processing applications.

Most work of ferroelectric-photoconductor media has concentrated on display applications. The sandwich materials are not suited for volume storage and cannot compete with magneto-optics for digital storage. The thermoplastic-photoconductive media has proved to be very useful for optical memory investigations. Although this sandwich material is obviously not suited for high speed, high density optical storage, time and time again it has been this class of media which has been chosen as the most acceptable for demonstrating prototype erasable holographic optical memory systems.

References

4.1   D. Casasent: Proc. IEEE, **64** (1976)
4.2   D. Casasent: Proc. S.I.D. **15**/3, 131 (1974)
4.3   J. Flannery, Jr.: IEEE Trans. ED-**20**, 941 (1973)
4.4   A. D. Jacobson, T. D. Beard, W. P. Bleha, J. D. Margerum, S.-Y. Wong: Pattern Recognition **5**, 13 (1973)
4.5   A. D. Jacobson, W. P. Bleha, L. Miller, J. Grinberg, L. Fraas, D. Margerum: "Optical-to-Optical Interface Device" Techn. Rept. NASA CR-144700, Hughes Research Laboratories (1975)
4.6   J. Grinberg, A. D. Jacobson, W. P. Bleha, L. Miller, L. Fraas, D. Boswell, G. Myer: Opt. Eng. **14**, 217 (1975)
4.7   A. D. Jacobson, J. Grinberg, W. P. Bleha, L. Miller, L. Fraas, G. Myer, D. Boswell: Information Display **12**, 17 (1975)
4.8   R. Alrich, F. Krol, W. Simmons: IEEE Trans. ED-**20**, 1015 (1973)
4.9   W. J. Poppelbaum: Proc. Symposium Automatic Photointerpretation, Washington, D.C. **387** (1967)
4.10   W. J. Poppelbaum, M. Faiman, D. Casasent, D. S. Sand: Proc. IEEE **56**, 1744 (1968)
4.11   G. Groh, G. Marie: Opt. Commun. **2**, 133 (1970)
4.12   D. Casasent, W. Keicher: Proc. Elec.-Opt. Sys. Des. Conf., Washington, D.C. **99** (1972)
4.13   D. Casasent: IEEE Trans. ED-**20**, 1109 (1973)
4.14   D. Casasent: IEEE Trans. C-**22**, 852 (1973)
4.15   D. Casasent: Proc. Intern. Opt. Comp. Conf., Zürich **18** (1974)
4.16   D. Casasent: Opt. Eng. **13**, 228 (1974)
4.17   D. Casasent, F. Casasayas: IEEE Trans. AES-**11**, 65 (1975)
4.18   D. Casasent, W. M. Sterling: IEEE Trans. C-**24**, 348 (1975)
4.19   D. Casasent: Proc. Intern. Opt. Comp. Conf., Washington, D.C., 5 (1975)
4.20   D. Casasent, F. Casasayas: App. Opt. **14**, 1364 (1975)
4.21   J. Feinleib, D. S. Oliver: App. Opt. **11**, 2752 (1972)
4.22   P. Nisenson, S. Iwasa: App. Opt. **11**, 2760 (1972)
4.23   P. Vohl, P. Nisenson, D. S. Oliver: IEEE Trans. ED-**20**, 1032 (1973)
4.24   P. Nisenson, J. Feinleib, S. Iwasa: Proc. SPIE Conf. Opt. Mapping, Rochester **241** (1974)
4.25   J. Feinleib, S. Iwasa: Proc. 2nd Eur. E–O Markets & Tech. Conf., Montreaux **365** (1974)
4.26   S. Iwasa, J. Feinleib: Opt. Eng. **13**, 235 (1974)
4.27   S. G. Lipson, P. Nisenson: Appl. Opt. **13**, 2052 (1974)
4.28   P. Nisenson, R. A. Sprague: App. Opt. **14**, 2602 (1975)

4.29 S.Iwasa: App. Opt. **15**, 1418 (1976)
4.30 G.Marie, J.Donjon: Proc. IEEE **61**, 942 (1973)
4.31 M.Grenot, J.Pergrale, J.Donjon, G.Marie: Appl. Phys. Lett. **21**, 83 (1972)
4.32 J.Donjon, F.Dumont, M.Grenot, J.-P.Hazan, G.Marie, J.Pergrale: IEEE Trans. ED-**20**, 1037 (1973)
4.33 S.E.Cummins, T.E.Luke: IEEE Trans. ED-**18**, 761 (1971)
4.34 W.C.Stewart, L.S.Cosentino: Ferroelectrics **1**, 149 (1970)
4.35 H.N.Roberts: App. Opt. **11**, 397 (1972)
4.36 W.D.Smith, C.E.Land: Appl. Phys. Lett. **20**, 169 (1972)
4.37 M.D.Drake: App. Opt. **13**, 347 (1974)
4.38 J.W.Burgess, R.J.Hurditch, C.J.Kirkby, G.E.Scrivener: App. Opt. **15**, 1550 (1976)
4.39 W.E.Good: Proc. Natl. Elect. Conf. **24**, 771 (1968)
4.40 R.Markevitch, D.Rodal: Proc. E–O Syst. Design. Conf., Anaheim (1975)
4.41 D.Casasent (ed.): *Applications of Optical Signal Processing*, Topics in Applied Physics, Vol. 23 (Springer, Berlin, Heidelberg, New York 1979)
4.42 C.Thomas: App. Opt. **5**, 1782 (1966)
4.43 T.M.Turpin: Proc. Intern. Opt. Comp. Conf., Zurich **34** (1974)
4.44 E.Labin: J. SMPTE **54**, 393 (1950)
4.45 E.I.Sponable: J. SMPTE **60**, 337 (1953)
4.46 E.Baumann: J. SMPTE **60**, 344 (1953)
4.47 W.E.Glenn: J. SMPTE **79**, 788 (1970)
4.48 B.J.Ross, E.T.Kozol: Proc. IEEE Intercon **5**, 2613 (1973)
4.49 R.J.Doyle, W.E.Glenn: IEEE Trans. ED-**18** (1971)
4.50 R.J.Doyle, W.E.Glenn: Appl. Opt. **11**, 1261 (1972)
4.51 N.K.Sheridon: IEEE Trans. ED-**19**, 1003 (1972)
4.52 A.I.Lakatos: Digest SID Intern. Symp. **28** (1973)
4.53 N.K.Sheridon, M.A.Berkovitz: Proc. 2nd Eur. E–O Markets & Tech. Conf., Montreaux (1974)
4.54 T.C.Lee: App. Opt. **13**, 888 (1974)
4.55 A.Vander Lugt, A.A.Friesem, G.E.Hoffmann, H.N.Roberts, R.G.Zech, P.J.Peters, E.N.Tompkins: "Optical Read/Write Memory System Design", NASA Tech. Rept. CR-103058, Elect. Opt. Center, Radiation Inc., (1971) pp. 60–68
4.56 K.Preston, Jr.: Proc. IEEE Intern. Solid-State Ckt. Conf. **100** (1968)
4.57 K.Preston, Jr.: Opt. Acta **16**, 579 (1969)
4.58 K.Preston, Jr.: IEEE Trans. AES-**6**, 458 (1970)
4.59 D.G.Grant, R.A.Meyer, D.N.Qualkinbush: IEEE J. Quant. Elect. **322** (1971)
4.60 K.Preston, Jr.: *Coherent Optical Computers* (1972) pp. 139–148
4.61 R.A.Meyer, D.G.Grant: Proc. Elect. Opt. Syst. Des. Conf. **107** (1972)
4.62 M.R.Tubbs: Opt. & Laser Tech. **5**, 155 (1973)
4.63 D.Chen, J.D.Zook: Proc. IEEE **63**, 1207 (1975)
4.64a J.P.Huignard, F.Micheron, E.Spitz: "Optical Systems and Photosensitive Materials for Information Storage"; in *Optical Properties of Solids-New Developments* (North-Holland, Amsterdam 1976) pp. 847–925
4.64b H.M.Smith: *Holographic Recording Materials*, Topics in Applied Physics, Vol. 20 (Springer, Berlin, Heidelberg, New York 1977)
4.65 R.A.Bartolini, H.A.Weakliem, B.F.Williams: Opt. Eng. **15**, 99 (1976)
4.66 G.Jackson: Optica Acta **16**, 1 (1969)
4.67 A.A.Friesem, J.L.Walker: Appl. Opt. **9**, 201 (1970)
4.68 W.J.Tomlinson: Appl. Opt. **14**, 2456 (1975)
4.69 G.D.Baldwin: Appl. Opt. **8**, 1439 (1969)
4.70 D.S.Lo, D.M.Manikowski, M.M.Hanson: Appl. Opt. **10**, 978 (1971)
4.71 M.Lescinsky: Opt. Commun. **5**, 104 (1972)
4.72 D.S.Lo: Appl. Opt. **13**, 861 (1974)
4.73 I.Schneider, M.Marone, M.N.Kabler: Appl. Opt **9**, 1163 (1970)

4.74 F.Lanzl, V.Roder, W.Waidelich: Appl. Phys. Lett. **18**, 56 (1971)
4.75 I.Schneider: Appl. Opt. **10**, 980 (1971)
4.76 J.V.Burt: "An Experimental High Density Optical Memory", Techn. Rept. R-557, Coor. Sci. Lab., Univ. Illinois (1972)
4.77 I.Schneider: Appl. Opt. **11**, 1426 (1972)
4.78 B.Stadnik, Z.Tronner: Opt. Commun. **6**, 199 (1972)
4.79 G.E.Scivener, M.R.Tubbs: Opt. Commun. **6**, 242 (1972)
4.80 M.R.Tubbs, G.E.Scivener: J. Photo. Sci. **22**, 8 (1974)
4.81 D.Casasent, F.Caimi: Appl. Opt. **15**, 815 (1976)
4.82 F.Caimi, D.Casasent, I.Schneider: Proc. SPIE Conf., San Diego **83** (1976)
4.83 D.Casasent, F.Caimi: Appl. Phys. Lett. **29** (1976)
4.84 D.Casasent, F.Caimi: Appl. Opt. **15**, 2631 (1976)
4.85 I.Schneider, W.C.Collins, M.J.Marrone, M.E.Gingerich: Appl. Phys. Lett. **27**, 348 (1975)
4.86 W.C.Collins, M.J.Marrone: Appl. Phys. Lett. **28**, 260 (1976)
4.87 W.J.Tomlinson, E.A.Chandross, R.L.Fork, C.A.Pryde, A.A.Lamola: Appl. Opt. **11**, 533 (1972)
4.88 D.E.Klingler, A.A.Friesem, R.R.Basson, W.S.Colburn, G.E.Hoffmann, W.-H.Lee, H.N.Roberts, F.B.Rotz, E.N.Tomkins, P.J.Peters, R.G.Zech: "Optical Read/Write Memory Components," NASA Tech. Rept. CR-122559, Elect. Opt. Center, Radiation, Inc. (1972) pp. 5-36 to 5-44
4.89 C.M.Verber, R.A.Nathan, A.H.Adelman, D.R.Grieser, V.E.Wood: Investigative Study of Holographic Materials Development, NASA Tech. Rept. CR-112248, Battelle Columbus Labs (1972)
4.90 F.S.Chen, J.T.Lamacchia, D.B.Fraser: Appl. Phys. Lett. **13**, 223 (1968)
4.92 J.B.Thaxter: Appl. Phys. Lett. **15**, 210 (1969)
4.92 J.J.Amodei, D.L.Staebler, A.W.Stephens: Appl. Phys. Lett. **18**, 507 (1971)
4.93 T.K.Gaylord, T.A.Rabson, F.K.Tittel: Appl. Phys. Lett. **20**, 47 (1972)
4.94 F.Micheron, G.Bismuth: Appl. Phys. Lett. **20**, 79 (1972)
4.95 J.J.Amodei, W.Phillips, D.L.Staebler: Appl. Opt. **11**, 390 (1972)
4.96 J.J.Amodei, D.L.Staebler: RCA Rev. **33**, 71 (1972)
4.97 W.Phillips, J.J.Amodei, D.L.Staebler: RCA Rev. **33**, 94 (1972)
4.98 A.Ishida, O.Mikami, S.Miyazawa, M.Sumi: Appl. Phys. Lett. **21**, 192 (1972)
4.99 D.L.Staebler, W.Phillips, B.W.Faughnan: "Materials for Phase Holographic Storage", Tech. Rept. PRRL-73-CR-17 RCA Laboratories (1973)
4.100 P.Shah, T.A.Rabson, F.K.Tittel, T.K.Gaylord: Appl. Phys. Lett. **24**, 130 (1974)
4.101 F.Micheron, C.Mayeux, J.C.Trotier: Appl. Opt. **13**, 784 (1974)
4.102 J.B.Thaxter, M.Kestigian: Appl. Opt. **13**, 913 (1974)
4.103 D.L.Staebler, W.Phillips: Appl. Opt. **13**, 788 (1974)
4.104 O.Mikami: Opt. Commun. **11**, 30 (1974)
4.105 R.Magnusson, T.K.Gaylord: Appl. Opt. **13**, 1545 (1974)
4.106 J.M.Spinhirne, T.L.Estle: Appl. Phys. Lett. **25**, 38 (1974)
4.107 D. Von Der Linde, A.M.Glass, K.F.Rodgers: Appl. Phys. Lett. **25**, 155 (1974)
4.108 D.L.Staebler, W.J.Burke, W.Phillips, J.J.Amodei: Appl. Phys. Lett. **26**, 182 (1975)
4.109 G.A.Alphonse, R.C.Alig, D.L.Staebler, W.Phillips: RCA Rev. **36**, 213 (1975)
4.110 D.Von Der Linde, A.M.Glass: Appl. Phys. **8**, 85 (1975)
4.111 S.F.Su, T.K.Gaylord: J. Appl. Phys. **46**, 5208 (1975)
4.112 E.Okamoto, H.Ikeo, K.Muto: Appl. Opt. **14**, 2453 (1975)
4.113 J.P.Huignard, F.Micheron: Opt. Commun. **16**, 80 (1976)
4.114 D.Von Der Linde, A.M.Glass, K.F.Rodgers: J. Appl. Phys. **47**, 217 (1976)
4.115 W.D.Cornish, M.G.Moharam, L.Young: J. Appl. Phys. **47**, 1479 (1976)
4.116 G.A.Alphonse, W.Phillips: RCA Rev. **37**, 184 (1976)
4.117 H.Kurtz: Philips Tech. Rev. **37**, 109 (1977)
4.118 K.Megumi, H.Kozuka, M.Kobayashi, Y.Furuhata: Appl. Phys. Lett. **30**, 631 (1977)
4.119 M.Peltier, F.Micheron: J. Appl. Phys. **48**, 3683 (1977)

4.120 E. Kratzig, R. Orlowski: Appl. Phys. **15**, 133 (1978)
4.121 J. Herrian, J. P. Huignard, P. Aubourg: Appl. Opt. **17**, 185 (1978)
4.122 C. Chen, C. Kim, D. Von Der Linde: Appl. Phys. Lett. **34**, 321 (1979)
4.123 S. A. Keneman, G. W. Taylor, A. Miller, W. H. Fonger: Appl. Phys. Lett. **17**, 173 (1970)
4.124 S. A. Keneman, G. W. Taylor, A. Miller: Ferroelectrics **1**, 227 (1970)
4.125 S. A. Keneman, A. Miller, G. W. Taylor: Ferroelectrics **3**, 131 (1972)
5.126 D. S. Oliver, P. Vohl, R. E. Aldrich, M. E. Behrndt, W. R. Buchan, R. C. Ellis, J. E. Genthe, J. R. Goff, S. L. Hou, G. McDaniel: Appl. Phys. Lett. **17**, 416 (1970)
4.127 S. L. Hou, D. S. Oliver: Appl. Phys. Lett. **18**, 325 (1971)
4.128 D. S. Oliver, W. R. Buchan: IEEE Trans. ED-**18**, 769 (1971)
4.129 J. C. Urbach, R. W. Meir: Appl. Opt. **5**, 666 (1966)
4.130 L. H. Lin, H. L. Beauchamp: Appl. Opt. **9**, 2088 (1970)
4.131 T. L. Credelle, F. W. Spong: RCA Rev. **33**, 206 (1972)
4.132 W. S. Colburn, L. M. Ralston, J. C. Dwyer: Appl. Phys. Lett. **23**, 45 (1973)
4.133 W. S. Colburn, E. N. Tompkins: Appl. Opt. **13**, 2934 (1974)
4.134 T. Saito, S. Oshima, T. Honda, J. Tsujiuchi: Opt. Commun. **16**, 90 (1976)
4.135 T. C. Lee: Appl. Phys. Lett. **29**, 190 (1976)
4.136 J.-P. Krumme, B. Hill, J. Krüger, K. Witter: J. Appl. Phys. **46**, 2733 (1975)
4.137 J.-P. Krumme, H. J. Schmitt: IEEE Trans. MAG-**11**, 1097 (1975)
4.138 B. Hill, J.-P. Krumme, G. Much, R. Pepperl, J. Schmidt, K. P. Schmidt, K. Witter, H. Heitmann: Appl. Opt. **14**, 1607 (1975)
4.139 B. Hill, J.-P. Krumme, G. Much, D. Reikmann, J. Schmidt: J. Appl. Phys. **47**, 3697 (1976)
4.140 H. Heitmann, B. Hill, J.-P. Krumme, K. Witter: Philips Tech. Rev. **37**, 197 (1977)
4.141 J. F. Dillon, Jr., E. M. Gyorgy, J. P. Remeika: J. Appl. Phys. **41**, 1211 (1970)
4.142 C. O. Carlson, E. Stone, H. O. Bernstein, W. K. Tomita, W. C. Meyers: Science **154**, 1550 (1966)
4.143 D. Mayden: Bell. Syst. Tech. J. **50**, 1761 (1971)
4.144 E. E. Gray: IEEE Trans. MAG-**8**, 416 (1972)
4.145 R. Bartolini: Proc. SPIE **123**, 2 (1977)
4.146 J. Corcoran, H. Ferrier: Proc. SPIE **123**, 17 (1977)
4.147 A. Bell, F. Spong: IEEE J. QE-**14**, 487 (1978)
4.148 R. Bartolini, A. Bell, R. Flory, M. Lurie, F. Spong: IEEE Spectrum **15**, 20 (1978)
4.149 G. Blom: Appl. Phys. Lett. **35**, 81 (1979)
4.150 F. Bulthuis, M. Carasso, J. Heemskerk, P. Kivits, W. Kleuters, P. Zalm: IEEE Spectrum **16**, 26 (1979)
4.151 R. L. Aagard, T. C. Lee, D. Chen: Appl. Opt. **11**, 2133 (1972)
4.152 D. Chen: Appl. Opt. **13**, 767 (1974)
4.153 D. Chen, R. L. Aagard, T. S. Liu: J. Appl. Phys. **41**, 1395 (1970)
4.154 G. Lewicki, J. E. Guisinger: Appl. Phys. Lett. **16**, 240 (1970)
4.155 D. Chen, R. L. Aagard: J. Appl. Phys. **41**, 2530 (1970)
4.156 K. Y. Ahn: Appl. Phys. Lett. **17**, 347 (1970)
4.157 R. K. Waring, Jr.: J. Appl. Phys. **42**, 1763 (1971)
4.158 J. C. Suits, K. Lee, H. F. Winters, P. B. Phipps, D. F. Kyser: J. Appl. Phys. **42**, 3458 (1971)
4.159 T. C. Lee, D. Chen: Appl. Phys. Lett. **19**, 62 (1971)
4.160 R. L. Aagard, F. M. Schmidt, W. Walters, D. Chen: IEEE Trans. MAG-**7**, 380 (1971)
4.161 P. Coeure, J.-C. Gay, J. Carcey: IEEE Trans. MAG-**7**, 397 (1971)
4.162 W. K. Unger, R. Rath: IEEE Trans. MAG-**7**, 885 (1971)
4.163 T. C. Lee: Appl. Opt. **11**, 384 (1972)
4.164 E. Feldtkeller: IEEE Trans. MAG-**8**, 481 (1972)
4.165 G. Lewicki, J. E. Guisinger: J. Appl. Phys. **44**, 2361 (1973)
4.166 D. Chen, G. N. Otto, F. M. Schmidt: IEEE Trans. MAG-**9**, 66 (1973)
4.167 C. F. Shelton: IEEE Trans. MAG-**9**, 398 (1973)
4.168 R. L. Aagard, F. M. Schmidt, T. S. Liu, D. Chen: IEEE Trans. MAG-**9**, 463 (1973)
4.169 Y. Ono, M. Nagao: Jpn. J. Appl. Phys. **12**, 1907 (1973)
4.170 W. Streifer, B. A. Huberman: Appl. Phys. Lett. **24**, 147 (1974)

4.171 T.Chen: J. Appl. Phys. **45**, 2358 (1974)
4.172 T.Chen, W.E.Stutius: IEEE Trans. MAG-**10**, 581 (1974)
4.173 K.Lee, J.C.Suits, G.B.Street: Appl. Phys. Let. **26**, 27 (1975)
4.174 M.Tanaka, Y.Nishimura: Trans. Inst. Elec. Comm. Eng. Jpn. **58**, 36 (1975)
4.175 H.Rull, K.Kempter: Opt. Commun. **16**, 83 (1976)
4.176 P.Dekker: IEEE Trans. MAG-**12**, 311 (1976)
4.177 Y.Mimura, N.Imamura, T.Kobayashi: Jpn. J. Appl. Phys. **15**, 933 (1976)
4.178 Y.Mimura, N.Imamura, T.Kobayashi: IEEE Trans. MAG-**12**, 779 (1976)
4.179 K.Sunago, S.Matsushita, Y.Sakurai: IEEE Trans. MAG-**12**, 776 (1976)
4.180 M.Urner-Wille, T.Te Velde, P.Van Engen: Phys. Stat. Sol. (a) **50**, K 29 (1978)
4.181 Y.Mimura, N.Imamura, T.Kobayashi: Jpn. J. Appl. Phys. **17**, 2007 (1978)
4.182 P.Hansen, M.Urner-Wille: Proc. Joint InterMag-MMM Conf., N.Y. (1979)
4.183 S.A.Keneman: Appl. Phys. Lett. **19**, 205 (1971)
4.184 M.Terao, H.Yamamoto, S.Asai, E.Maruyama: Proc. 3rd Conf. Solid State Dev., Tokyo (1971)
4.185 A.Hamada, T.Kurosu, M.Saito, M.Kikuchi: Appl. Phys. Lett. **20**, 9 (1972)
4.186 J.Feinleib, S.Iwasa, S.C.Moss, J.P.De Neufville, S.R.Ovshinsky: J. Noncryst. Solids 8–10, 909 (1972)
4.187 Y.Ohmachi, T.Igo: Appl. Phys. Lett. **20**, 506 (1972)
4.188 T.Igo, Y.Toyoshima: Jpn. J. Appl. Phys. **11**, 117 (1972)
4.189 D.Adler, S.C.Moss: J. Vac. Sci. Technol. **9**, 1182 (1972)
4.190 R.J. von Gutfield, P.Chaudhari: J. Appl. Phys. **43**, 4688 (1972)
4.191 K.Weiser, R.J.Gambino, J.A.Reinhold: Appl. Phys. Lett. **22**, 48 (1973)
4.192 S.R.Ovshinsky, H.Fritzsche: IEEE Trans. ED-**20**, 91 (1973)
4.193 R.G.Neale, J.A.Aseltine: IEEE Trans. ED-**20**, 195 (1973)
4.194 V.I.Mandrosov, E.I.Pik, G.A.Sobolev: Opt. Spectrosc. **34**, 695 (1973)
4.195 V.I.Mandrosov, E.I.Pik, G.A.Sobolev: Opt. Spectrosc. **35**, 75 (1973)
4.196 W.Tokuda, T.Katoh, A.Yasumori, K.Nakamura: J. Appl. Phys. **45**, 5098 (1974)
4.197 S.B.Gurevich, N.N.Ilyashenko, B.T.Kolomiets, V.M.Lyubin, V.P.Shilo: Phys. Stat. Sol. (a) **26**, K 127 (1974)
4.198 S.Zembutsu, Y.Toyoshima, T.Igo, H.Nagai: Appl. Opt. **14**, 3073 (1975)
4.199 I.Shimizu, H.Fritzsche: J. Appl. Phys. **47**, 2969 (1976)
4.200 W.R.Roach: Appl. Phys. Lett. **19**, 454 (1971)
4.201 A.W.Smith: Appl. Phys. Lett. **23**, 437 (1973)
4.202 I.Balberg, S.Trokman: J. Appl. Phys. **46**, 2111 (1975)

5. Hybrid Processors

D. P. Casasent

With 24 Figures

At the 1974 International Optical Computing Conference [5.1], *Preston* surveyed the highlights in the history of optical computing and noted the emergence of the optical/digital processor as one of the major accomplishments of the 1970s. In a recent survey of optical processing [5.2], *Thompson* noted that such hybrid processors may finally breach the gap that has separated laboratory systems from practical ones. These sentiments were again echoed in a recent panel discussion [5.3] in which leading optical and digital researchers participated. An entire session of an EOSD Conference [5.4] was devoted to the major hybrid optical/digital processors in existence. In the special issue of the Proceedings IEEE on "Optical Computing", hybrid processors was again the subject of an invited paper [5.5]. Every year, new hybrid optical/digital processor architectures and systems are reported upon in the literature. These hybrid systems are thus of major current interest and importance.

In this chapter, the system philosophy of these hybrid processors will be discussed with emphasis on the optical/digital interface and applications of these novel systems. Because the architecture of these systems depends on the specific application, one cannot divorce the system from the application. The application also determines the specifications and performance requirements of the interface. Since some of the more salient applications of optical processing have already been discussed in [5.6], the discussion of the applications of these processors included in this chapter will thus be intentionally brief. The reader is referred to [5.6] and the references in this chapter for further details.

5.1 Overview

One feature of a hybrid processor is the ability to control the format, etc., of the input data with a digitally-controlled, decision-making feedback loop. The full embodiment of such a system and any optical processor requires the use of real-time and reusable input transducers and often spatial filter materials, too. Several surveys of these devices and materials exist elsewhere [5.6–14] and also in Chap. 4. However, several such devices used in the hybrid systems to be described are not extensively discussed elsewhere and thus a few general remarks on these devices are included in this chapter together with adequate references to provide a more complete understanding.

One of the first published suggestions to combine the advantages of optical and digital processing in a hybrid system was made by *Huang* and *Kasnitz* [5.15] in 1967. Several of the first descriptions of the architecture for a hybrid system were presented by *Casasent* [5.16–18] from 1972–1974. Since 1974, several systems have been assembled and demonstrated. They are in various stages of research and are intended for rather diverse applications. However, they can be divided into three basic classifications.

In the first system to be discussed (Sect. 5.2), an array of 64 specially shaped detectors is placed in the Fourier transform plane of an optical processor. The outputs of all 64 photodetector elements are available simultaneously and provide measures of the intensity in various portions of the Fourier transform plane. The transform plane is sampled and the data fed to a digital computer for subsequent analysis. This hybrid system classification is termed "diffraction pattern sampling" and is discussed in Sect. 5.2.

The output from the optical portion of the second class of hybrid processor to be discussed (Sect. 5.3) is either a texture-variance image, a power spectrum, a monochromatic or binary or enhanced image, or some other similar light distribution. This distribution is then detected by a closed-circuit camera and subsequently analyzed by a minicomputer. These systems are best classified as optical pre-processors.

The last class of hybrid processor to be discussed (Sect. 5.4) is the most powerful since the optical system is a correlator. Two of the hybrid systems included in this section utilize real-time and reusable devices in the input and/or spatial filter planes. These latter systems represent some of the most powerful processors in existence.

After describing these hybrid processors and their various applications, the operations required in the analysis of the various output optical patterns will have been well defined. At this point a description of several of the optical/digital interfaces that have been employed will be presented (Sect. 5.5).

5.1.1 Advantages of Optical Processing

The advantages of optical processing have been expounded upon on numerous occasions. They can briefly be summarized as follows:
1) Parallel Processing: An optical processor is inherently a two-dimensional system. When illuminated by collimated light, all points in the transparency placed in the input plane are operated on in parallel.
2) High Throughput: The rate at which data is transmitted through an optical system is essentially limited only by the rate at which data can be placed in the system and detected at the output.
3) Powerful Operations: A spherical lens inherently forms the complex two-dimensional Fourier transform of the light distribution in its input plane. With a joint transform [5.19] or matched spatial filter [5.20] system, two complex two-dimensional functions can be correlated.

All of these powerful operations can be performed in parallel on two-dimensional data at data rates that can approach the speed of light.

5.1.2 Advantages of Digital Processing

Optical processors are not a panacea. Their limitations are mainly a lack of flexibility. By comparison with digital computers, these optical systems lack the programability, control, analysis and decision-making features that are the hallmark of digital computers. Digital systems are at a far more advanced state and far more hardened and commercially available than their optical counterparts. Digital systems are also more powerful since they can perform all operations; however, not with equal efficiency.

By comparison, optical processors are somewhat like discrete transistors before the dawn of LSI; they are truly in their infancy. More commercially available optical systems, such as the point-of-sale checkout systems, fingerprint analysis systems, currency sorters, credit card verifiers, etc., appear each year, however.

The comparisons made between these powerful optical and digital processing technologies are often inappropriate. Besides the differences in the state of development of both technologies, the shortcomings of one system are the advantages of the other. The two technologies are actually complementary rather than similar.

5.1.3 Hybrid Processing

From these prior remarks, it should be apparent that direct comparisons of optical and digital processing technologies are not appropriate. The philosophy of a hybrid optical/digital processor is simply to combine an optical system and a digital processor such that the resultant system has the best features of each. In its simplest embodiment, the optical system is used to form the transform of two-dimensional data (as a transparency). This transform pattern or some similar distribution of light is then analyzed by digital algorithms.

Far more elaborate and powerful hybrid systems result when an optical correlator is used. A simplified diagram of such a system is shown in Fig. 5.1. The top portion of the diagram is one version of a conventional optical correlator while the lower portion represents the digital section of the hybrid system. The input plane P_1 contains the data to be processed. Its Fourier transform is formed in plane P_2 by lens L_2. If spatial filters or holographic matched filters are placed in plane P_2, plane P_3 will contain either a filtered version of the input or the correlation of the input data and the source, from which the holographic filter was formed. As shown in Fig. 5.1, the contents of planes P_2 and/or P_3 are analyzed by the digital processor. The digital section also has control over the contents of the input plane P_1 and the spatial filter or holographic matched filter(s) placed in plane P_2.

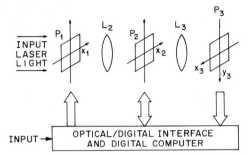

Fig. 5.1. General schematic diagram of a hybrid optical/digital processor

The potential of such a hybrid processor and the manner in which this architecture produces a system with the throughput and parallel processing operations of an optical processor and the flexibility, control and decision-making properties of a digital computer are quite apparent. In the following sections, various applications of several versions of this basic configuration will be presented. Several ways by which the requisite interface functions can be implemented are then discussed.

5.2 Diffraction Pattern Sampling

This is the simplest as well as the most developed and commercially available of the hybrid systems to be discussed. It is commercially available and has been used in numerous research and factory applications [5.21]. The hybrid system consists of a special detector array placed in the Fourier transform plane of an optical processor. This detector array enables one to sample the contents of the Fourier transform plane and subsequently transmit this data to a digital computer for analysis. This sampling of the Wiener spectrum of an object is used to extract data or information about the input. The principles of this operation are based on fundamental properties of diffraction patterns which are reviewed in Chapts. 1, 2 of this volume and elsewhere [5.22], and are thus not extensively discussed here.

The diffraction pattern properties used depend on the specific application. One of the most useful features of the Fourier transform is its translation invariance by which input features of similar size contribute energy in the same region of the transform, regardless of their positions in the input. Another aspect of importance is the inverse relationship which exists between the size of the objects and the size of the spectrum; this facilitates investigation of small objects.

5.2.1 System Description

The RSI (Recognition Systems, Inc.) diffraction pattern sampling detector array [5.21] which is the basis for this hybrid system, consists of 32 wedge-shaped and

32 annular ring-shaped photodetectors arranged on the two semicircular regions of a circular detector. The outputs of the 32 ring-shaped detectors provide information on the distribution of input data with spatial frequency. The wedge-shaped detector outputs provide information on the orientation of the objects and data in the input. The detector array is fabricated from silicon by planar diffusion on a common substrate. The spectral response of the detector extends from the visible to 1 μm; its rise time is less then 0.1 μs and cross-talk between elements is less than 0.1%.

The outputs of all 64 detector elements are available in parallel and may be fed to separate auto-ranging amplifiers with 60 dB dynamic range. These outputs are coupled to data loggers or mini-computers through various electronic packages such that the analog signals can be converted to digital formats, including floating point. Peak detectors and sample-and-hold circuits can also be incorporated. The outputs of all 64 detectors can be instantaneously sampled with a window as short as 10 μs and over 500 of these instantaneous snapshots per second can be read into the minicomputer with a channel rate of two megabits/s. Data rate depends greatly on the application and digital computer used. If a 10 bit dynamic range (1000:1) is acceptable, measurements can be made in 5 μs. The data can also be recorded on eight punched cards using a system with a special operator entry and control unit.

The general software approach uses nonparametric pattern recognition, in which no assumptions need be made about the statistics of the data. Preliminary data analysis usually includes statistical tests to determine if the data are separable and if recognition algorithms can be developed. Once this preliminary analysis has been completed and algorithms determined, future data processing is automatic and fast.

This hybrid system is well-suited for automatic data analysis in applications requiring coarse sorting or analysis. It is especially useful in applications requiring the analysis of large amounts of data. Many applications will be briefly noted in which decisions on the input data can be made based on the sampled diffraction pattern. When such an analysis is possible, this system is faster, cheaper and requires the analysis of far less data than the entire input. This results in a reduction in data bandwidth and data storage requirements. If 12 bit data from only the 32 ring measurements are adequate (usually the outputs of only a few of the ring elements are required), a given input can be characterized by $12 \times 32 = 384$ bits. If the input is 25 mm in diameter and contains a limiting resolution of 50 lines/mm and 64 shades of gray, the conventional all-digital analysis of the input would require the analysis and storage of 6×10^7 bits. In such a case, diffraction pattern sampling represents a data compression ratio in excess of 150,000:1.

Several criteria, upon which to base decisions on whether the input or transform space should be sampled and whether a diffraction pattern analysis or a holographic matched filter should be used, follow. If all N points in the input plane are required to sort the data, matched filtering should be used. If the resolution needed is "too fine" and if the input area is large and uninteresting, diffraction pattern sampling should be used. The use of this wedge/ring detector array and the entire concept of diffraction pattern sampling can best be seen by example. The ten or so examples discussed in the next two sections provide a rather complete description of this hybrid processor and a good introduction to the hybrid systems to be discussed later. A brief discussion of the algorithm analysis used will be presented for several of these examples.

5.2.2 Laboratory Applications

The more salient laboratory applications of this hybrid detector system are summarized in Fig. 5.2. Typical inputs, diffraction patterns and intensities vs wedge or ring elements are provided in the three columns in Fig. 5.2 for most of the applications. A schematic diagram of the basic system is shown in Fig. 5.2a. The input is illuminated with laser light and the diffraction pattern focused on the detector array whose output is subsequently fed to a digital computer for analysis and decision-making.

The initial applications [5.22] of such hybrid systems involved the analysis of aerial imagery. As shown in Fig. 5.2b, urban imagery can be distinguished from nonurban imagery by diffraction pattern analysis. Urban imagery contains high resolution detail and man made regular structures. These image features appear as high-frequency information and cause significant peaks of light to occur at selected angular directions in the diffraction pattern. The intensity in all 32 wedge elements for a typical urban input image is plotted in Fig. 5.2b. A discriminant analysis of the outputs of the wedge-shaped detector elements is usually used. The total variation of the wedge elements, the maximum difference between adjacent wedge elements and similar data are usually employed in making the final discrimination.

Spatial information from ERTS-1 (Earth Resource Technology Satellite) has also been found to provide sufficient data from which different physiographic regions can be distinguished [5.23]. For this specific study, ERTS-1 imagery of eight physiographic regions of Kansas were analyzed (e.g. Red Hills, Osage Plains, etc.). Each region had a unique geological character reflected in characteristic and unique topography. In the preliminary analysis of this imagery, the ring measurements were divided into two bands (0–0.9 cycles/km and 1.1–2.8 cycles/km). This essentially separated the high-frequency information due to stream patterns in rough terrain from low-frequency information due to field patterns. The wedge data was divided into four sectors of approximately $40°$ each from $25°$ to $185°$ CW from north. The sample areas analyzed were 37 km in diameter. Data in the 1.1–2.8 cycles/km band were found to contain most of the useful information on which to distinguish the various physiographic categories. Wedge or orientational data were not found to be so useful in discriminating between categories.

The use of diffraction pattern sampling has also been shown to be useful in screening imagery prior to processing for the occurence of cloud cover [5.24].

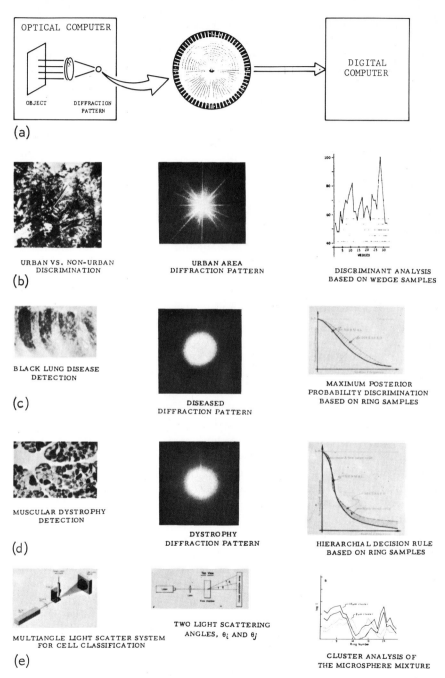

Fig. 5.2a–e. Representative laboratory applications of the hybrid diffraction-pattern sampling system [5.21]. (**a**) Schematic diagram; (**b**) aerial image analysis; (**c**) black lung disease detection; (**d**) muscular dystrophy detection; (**e**) cell classification

The technique involves monitoring the energy in certain frequency bands in which the signal level exceeds that associated with film grain noise. This same analysis can provide an estimate of image quality and the limiting resolution of imagery. When applied to large quantities of imagery, average estimates of image quality and resolution result that are less dependent on image contrast. By monitoring the energy in selected spatial frequency band(s), estimates of the amount of defocus in imagery can also be obtained.

This hybrid detector system has also been used in the analysis of x-ray images [5.25] for the presence of black-lung disease or pneumoconiosis (Fig. 5.2c). This represents an example of image textural discrimination. A brief summary of various pattern classification methods and probalistic difference measures of tone and texture applied to black-lung disease discrimination exists [5.26]. Texture refers to the visual sensation associated with the structural arrangement of an image while tone refers to the global gray level perceived. An image with small dynamic range fluctuations and a smooth consistency is dominated by tonal quality. If the image structure consists of predominately rapid, semire-petitive spatial variations, the image is best characterized by its texture. Fine texture corresponds to large dynamic range differences with little spatial pattern, while coarse texture is associated with more definite spatial patterns.

Image texture is both structured and statistical. Both digital and optical diffraction pattern sampling measures were used to develop an accurate classification basis using a test set of 141 chest radiographs. The basic discrimination hypothesis used was that lesions in lung regions would appear as well-defined textured areas and produce more high-frequency diffraction for abnormal lung regions than for normal lung regions. The graph of intensity vs ring element shown in Fig. 5.2c verifies this hypothesis.

For this specific application, the lung region was located by digital recognition techniques and the transform of a 6.25 cm diameter region of the lung imaged onto the detector and spatial frequencies above 8.8 cycles/mm were not considered. A wedge filter was oriented normal to the direction of the rib structure to reduce its effects on the spectral data. Features by which lung classification could be made were obtained by the divergence measure. Diagnostic classifications were then made assuming a multivariate Gaussian distribution in which various Gaussian distributions are assumed and the variance and mean computed for each. Sixty-two features (wedge-ring measure-ments) for each of six lung zones were measured for each film. These measures included: maximum brightness, number of points, standard deviation, skew coefficient, coefficient of excess and various spatial moments and the five first and second-order moments. Feature extraction was achieved by a stepwise discri-minant analysis. A correct classification rate from 86–100 % was achieved using data from only two ring detectors for most cases.

In a prototype system being assembled, the input will be digitally scanned at low resolution by an image dissector to detect and measure gross anatomical features. The data from the sampled diffraction pattern will then be used to remove all obvious normal lung x rays from further analysis and to select potentially suspicious x rays for further digital analysis.

A related biomedical application of diffraction pattern analysis concerns a search of photomicrograph inputs for muscular dystrophy [5.27]. Normal muscle tissue is characterized by muscle fibers of similar size, while diseased tissue contains an uneven distribution of fibers. The smaller fibers in these diseased tissues produce an increased high spatial frequency content and a decreased low spatial frequency content. The intensity of the spectrum at specific spatial frequencies can thus be used to determine the presence of muscular dystrophy (Fig. 5.2d). These data were then analyzed to determine a sequence of ring samples that would produce a reliable decision on the presence of diseased muscle tissue.

This wedge/ring detector interface and subsequent digital computer analysis have also been used to screen cervical cytologic samples for malignancy [5.28]. The success of the technique depends on the ability to relate diffraction pattern data to cell and nuclear diameters, nuclear density and clumping of acid. All inputs were 200X magnifications of individual cells on 649F film. Annular ring data were taken in the 10–100 cycles/mm, and 100–1000 cycles/mm frequency bands plus 360° angular scans with a 130 cycle/mm wide aperture centered at 430, 485, 565, and 630 cycles/mm.

As in previous experiments, the transform of malignant cells exhibited a mottled pattern with more intense high spatial frequency components. The mottled pattern has been attributed to the clumping of deoxyribonucleic acid within the cell nucleus. Plots of intensity vs spatial frequency for the 10–100 cycles/mm band showed a concave shape for normal cells and a convex shape for malignant cells. Fifteen parameters were measured from the annulus data and eight parameters from the angular data. Classical statistical methods were then employed to determine those parameters and combinations of parameters that provide the best discrimination. Divergence calculations were made, scatter plots and quadratic and linear discriminatory functions were produced and similar statistical analyses performed. Although only 80 cells were examined, initial results were most encouraging.

A cell-flow analysis system [5.29] using this detector is shown in Fig. 5.2e. A laser beam is passed through the flow chamber and strikes the detector array. Each ring element detects the light scattered by a cell at a different polar angle. A pulse of scattered light occurs whenever a cell passes through the laser beam. A trigger signal activates the peak sample and hold circuits to capture and hold the peak pulse heights of all detector elements while the cell was present. A mathematical clustering algorithm is then used to locate and classify clusters. This flow system has been used to determine the differential count in white blood cell analysis, to determine invasive carcinoma in cervical cells and to discriminate between particles of different sizes. The discrimination is based on the relative intensity of the scattered energy at relatively wide scattering angles. The scatter pattern for 8–10 μm balls shown in Fig. 5.2e is seen to cluster at angles from 0.74 to 14.4°.

a   "SIMILARITY" MEASURE

<u>OBJECTIVE</u>: CREATE AN ALGORITHM TO MEASURE THE SIMILARITY OF
DIFFRACTION PATTERNS NOT TO BE INFLUENCED BY:

    1. INPUT ORIENTATION
    2. ILLUMINATION LEVEL SHIFT
    3. DATA ACQUISITION NOISE

<u>EXAMPLE</u>:   WEDGE RANK DIFFERENCE SUM

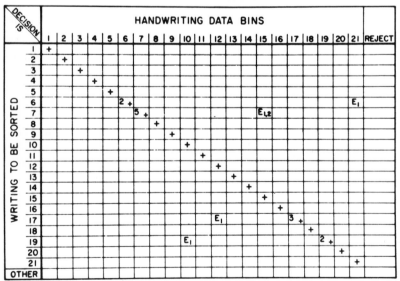

Magnitudes. $x_1 = (m_1^1, m_2^1, ..., m_n^1)$

$\Downarrow$ ← Rank Replacement

Ranks: $s_1 = (r_1^1, r_2^1, ..., r_n^1)$

Measure of Similarity: $S(s_1, s_2) = \sum\limits_{w=1}^{n} |r_w^1 - r_w^2|$

$\hat{S}(s_1, s_2) = \min S(s_1 \circlearrowright s_2)$

b   SORTING OF HANDWRITING
              SUMMARY OF RESULTS

DECISION IS → / WRITING TO BE SORTED ↓	HANDWRITING DATA BINS																					
	1	2	3	4	5	6	7	8	9	10	11	12	13	14	15	16	17	18	19	20	21	REJECT
1	+																					
2		+																				
3			+																			
4				+																		
5					+																	
6						2 +														E₁		
7							5 +								E₁,₂							
8								+														
9									+													
10										+												
11											+											
12												+										
13													+									
14														+								
15															+							
16																+						
17												E₁					3 +					
18																		+				
19								E₁											2 +			
20																				+		
21																					+	
OTHER																						

Fig. 5.3a, b. Algorithm (**a**) and results (**b**) of the use of diffraction pattern sampling in the automatic sorting of hand-writing [5.32]

A final laboratory application of this system is in the analysis of handwriting. These results and the analysis algorithm used are summarized in Fig. 5.3 [5.32]. This example also serves to demonstrate the possible simplicity of the computer processing used. The experiment involved the identification of handwritten pages from 21 individuals into 21 separate bins. One page was used to establish the feature vector for each bin. One to six components of a feature vector were found to be adequate for most cases. A feature vector consisted of the intensities measured in a number of wedges. A second page with different text was used in the data base to be sorted. In one version of this algorithm, sample I was used as the comparison base and the magnitudes of the detected energy in six wedge elements were tabulated and ranked for samples I and II. To achieve orientational insensitivity, the comparison is made under a cyclic wedge rotation. This is indicated in the middle section of Fig. 5.3. The minimum value of the sum of the wedge rank differences is used as the similarity measure. This is equivalent to an angular cross correlation between the two samples. The difference Δ between the six wedge readings for samples I and II are tabulated in the figure and total 14 for the orientational insensitive case. A rotation of the data from sample II by two wedge locations yields a sum of zero as shown. It should be noted that these rotations are implemented in the digital processing section and not by physically rotating the detector.

The results of various measures are summarized in the lower half of Fig. 5.3. With a three component feature vector, all samples were correctly sorted as indicated by the " $+$ " signs in the table. A reduction to a two-component vector caused an error in the identification of sample 7 (as sample 15). A reduction to a single component vector also caused samples 6, 17, and 19 to be misidentified. In general, a sorting accuracy of 95 % could be obtained with a three-component feature vector. The reduction in data storage and the required computation were quite large for the particular wedge photodetector configuration used in this application.

5.2.3 Factory Applications

The salient factory inspection applications of this hybrid system are summarized [5.21] in Fig. 5.4. The difference between the laboratory and factory systems can be seen from the schematic diagram in Fig. 5.4a, in which a material handling capability is required to position the data in the input of the optical computer. Additional obvious differences in the two systems are the high reliability of these factory systems and the ability to operate them with nontechnical personnel.

The input, diffraction pattern and system view for each application are shown in the three columns of Fig. 5.4. In the automatic inspection of disposable medical syringe needle points, the tip of the needle is illuminated with a laser beam. As shown in Fig. 5.4b, bent or broken needle tips produce a diffraction pattern with vertical components. The outputs of the wedge elements in the

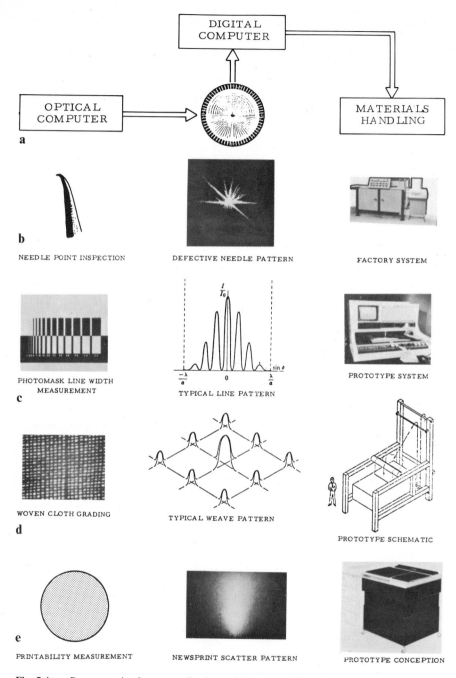

Fig. 5.4a–e. Representative factory applications of the hybrid diffraction-pattern sampling system [5.21]. (**a**) Schematic diagram; (**b**) needlepoint inspection; (**c**) photomask inspection; (**d**) weave defect inspection; (**e**) paper printability determination

detector indicate the presence of these needle point defects. As in most of these factory systems, the minicomputer controls the material handling subsystem as well as performs the data analysis. The throughput of this system is ten needles per second.

The measurement of linewidths in photomasks (Fig. 5.4c) and other applications is another important industrial application of this technique [5.30]. While the locations of the nulls in the diffraction pattern of a line in a transmissive aperture are related to the width of the line, the size of the aperture must also be known or accurate determination of the linewidth is not possible. However, by sampling the intensity of the optically produced diffraction pattern and subsequently performing a digital Fourier transform of these diffraction pattern samples, the autocorrelation functions of the input aperture result. The autocorrelation function is piecewise linear, but only one of these sections has a positive slope. If this segment is extended back toward the axis, its point of intersection with the axis will be proportional to the width of the line independent of the input aperture. Repeatable accuracies of 1 % for line widths as small as 1.5 μm have been achieved by this technique. Added operations performed by the digital computer in this application include control of the image focus and illumination intensity.

One of the more recent factory applications of this wedge/ring detector system is in the inspection of griege goods (undyed or unprinted cloth) as shown in Fig. 5.4d. The detection of weave defects and the grading of their size and severity is a tedious and time-consuming operation that is very susceptible to human error. The diffraction pattern of a weave sample (see Fig. 5.4d) is similar to that of a crossed grating. The occurence of defects such as a double thread causes deformations in the patterns of the first-order lobes that can easily be detected. The present system under development will be used to analyze webs up to 120 cm wide traveling past the light source at 100 m/min. Besides controlling the fabric flow, the minicomputer will analyze the diffraction pattern and, based on preassigned point counts for various defects, will print out a grade for each piece of cloth passing through the system.

The final application of this hybrid system is depicted in Fig. 5.4e. It is an application of diffraction pattern sampling for determining surface texture, in this case the printability of paper [5.31]. This application is more unique because it and the cell identification example in Sect. 5.2.2 represent extensions of diffraction pattern sampling from transmissive flat objects to reflective and forward scattering three-dimensional objects. As in the cell analysis case, only a few angular measurements are needed. For the present application, it is sufficient background to know that the printability of paper is related to its surface roughness or texture. The reflective diffraction pattern formed with laser light at oblique incidence would be an Airy pattern for a smooth surface. A diffused spread in the scatter would result upon reflection from a rough surface. By decreasing the grazing angle of the incident light, the diffraction pattern will begin to resemble the Airy disc pattern. The difference in path length for a surface with a step of height h is related to the grazing angle θ by $2h\sin\theta$. When this path difference is less than $\lambda/4$ where λ is the wavelength of the illuminating light, the surface is arbitrarily said to be smooth.

Twelve different types of paper were tested by reflecting a 4.7 mm diameter beam at a 6° angle from the surface of the paper. The output of the ring elements in the detector were analyzed for approximately 500 sample areas on these twelve paper types in an effort to determine a method of classifying the paper's printability. (The normal measure of printability is the Vandercook smoothness which is a measure of the number of dots that are missed when a standard halftone test pattern is printed on this paper.) The twelve paper types tested had known average smoothness values from 1 to 132.

The annular diffraction pattern data was analyzed by stepwise linear regression. This involves selecting multipliers and constants for various ring measurements to reduce the minimum standard error estimate. The best fit required the measurement of the intensity in six ring elements and summing the total intensity in five others. The final relationship between the Vandercook smoothness S and the measurement R_N of various ring elements N was found to be

$$\sqrt{S} = 3 + (61{,}000 - 14.1R_{12} - 257R_{19} + 5.6R_{14}$$
$$- 0.19R_{31} + 4.8R_6 + 0.118R_{32})/R_{8-12}, \qquad (5.1)$$

where R_{8-12} denotes the sum of the measurements for rings 8 to 12, R_{12} represents the measurement for ring 12, etc. Using this relationship, the standard error was only 1.23 missing dots.

In summary, the wedge/ring detector represents a well-engineered, developed and proven device for a hybrid system. However, its use is restricted to somewhat general pattern analysis applications. The recognition of specific image features such as orchards, corn fields, etc. requires the use of quite sophisticated digital recognition algorithms. This is due in part to the sampled nature of the diffraction pattern and in part to the loss of phase information that occurs in the recording of the transform by nonholographic methods. However, these very features are the reason why this system is so useful in the rapid, coarse analysis of data and why a large savings in storage and bandwidth results from the use of such a hybrid system.

5.3 Optical Preprocessing [5.33, 34]

The optical section of this second type of hybrid processor is primarily used to preprocess the input data, while the digital section performs analysis and makes decisions based on the output of the optical processor. A block diagram of a possible system is shown in Fig. 5.5. The optical configuration and even the source of illumination (coherent or noncoherent) vary with the application. The preprocessed output from the optical system is a texture-variance image, power

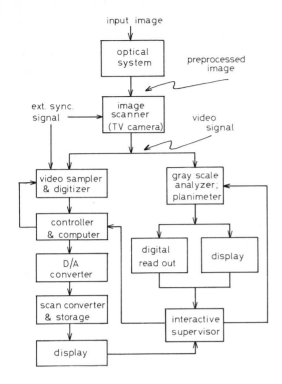

Fig. 5.5. Block diagram of hybrid optical preprocessor system [5.33]

spectrum, binary image, enhanced image, monochromatic image, or some similar distribution of light. The optical output is then scanned by a television camera. The video output is fed either to a gray-scale analyzer and planimeter or is sampled and digitized as indicated in separate branches in Fig. 5.5. Various displays complete the right-hand branch in the figure. The digital subsystem on the left-hand branch consists of a remote META-4 computer with 8000 16 bit words of high-speed core, a disc drive, seven track tape, card reader and teletype input.

A gray scale analyzer can be used to quantize the gray levels in the preprocessed image into ten different levels adjustable from a density of 0 to 3 with a resolution of 0.15 optical density units. These levels are then encoded with one of ten different colors (blue for the brightest tone and red for the darkest). This enhances subtle gray level variations in the original image. A planimeter is used to measure the area of the input covered by certain classes of objects. These data appear on a digital readout meter as a percent of the total area.

Alternatively, the optical output can be sampled and digitally encoded as a 512×512 matrix with 64 gray levels per point. The video digitizer encodes one point on each of the 256 horizontal scan lines per half frame at a maximum scan rate of 6.7 horizontal positions per second, and 262.5 vertical positions per half frame with a 2.25 k bit/s output rate. These data are then stored on tape, disc, or in core. Digital pattern recognition routines are initiated by the interactive supervisor. The main operations of the supervisor consist of subroutine calls, parameter assignments and execution commands.

5.3.1 Texture-Variance Analysis [5.35]

In the analysis of aerial imagery, it is useful to be able to distinguish between different types or classes of regions of the scene. This can often be done from an analysis of the texture of the scene. For example, it is possible to distinguish strongly textured areas (orchards, groves, etc.) from weakly textured ones (fields, etc.) by an analysis of a texture-variance version of the original image. This technique has proven to be quite appropriate for the analysis of agricultural scenes and is similar to the analysis technique used in determining the presence of pneumoconiosis (Sect. 5.2.2). In this present system, the formation of a texture-variance image of the scene and a special use of the image analyzer are required.

The texture variance is defined as

$$\sigma^2(x, y) = \overline{[g(x, y) - g_0]^2}, \tag{5.2}$$

where $g(x, y)$ is the transmittance of the input transparency, g_0 is its average transmittance, and the bar in (5.2) denotes the ensemble average over one transparency. This variance can be realized optically by coherently illuminating a transparency of the scene in question, forming its optical Fourier transform, blocking the dc portion of the transform and inverse transforming the result. The desired variance image is formed on a ground glass diffuser on which a TV camera is focused. The video output is then fed to the image analyzer, which is operated in the "probe mode". In this mode, only objects in certain texture classes are displayed. The texture-variance signature

$$T_L < \sigma^2 < T_U, \tag{5.3}$$

and the limits T_L and T_U on the range of σ^2 values are determined by the type of imagery. Various texture classes can be separated and individually highlighted by varying the width and position of the electronic tone-texture window defined by (5.3) in the analyzer.

A black and white print made from an infrared Ektachrome aerial photograph of an agricultural scene is shown in Fig. 5.6a. Its texture-variance image was formed as described above. With the tone-window set to display only the highly textured areas ($\sigma^2 \simeq 0.017$), the image in Fig. 5.6b results. Only certain orchards, trees and groves are highlighted. As the probe is adjusted to admit medium textured objects and then weakly textured regions ($\sigma^2 \simeq 0.0003$), new objects, fields, grassy areas and fallow fields are displayed and the objects previously displayed are no longer apparent.

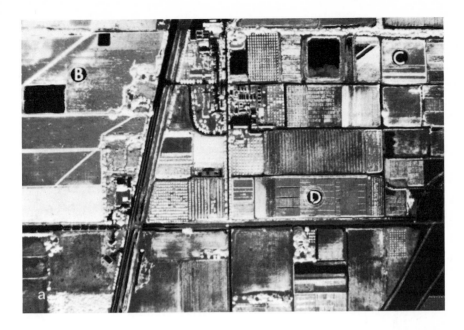

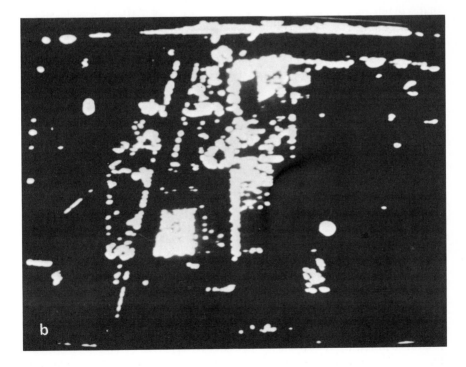

Fig. 5.6a, b. Texture-variance image analysis. (**a**) Input; (**b**) highly-textured areas [5.33]

The same technique used to produce the binary texture masks, which are used to discriminate various agricultural features, has also been used to locate and isolate the lung field in chest x rays. Subsequent analysis of these x rays for characteristics of pneumoconiosis has also been demonstrated on this system.

If the color-encoded image of a scene is scanned with a narrow tone window containing all ten quantizing and color levels, the boundaries of agricultural fields and other objects can be delineated. For any setting of the tone window, field tones (in the case of an input agricultural scene) will either be above or below the sampling window and no output will occur. When tonal gradients occur (e.g., at discontinuities such as field boundaries) that intersect the sampling window, an output is present.

5.3.2 Particle Size Analysis

A new and promising application of this system is particle size analysis [5.36]. In this case, the Fourier transform $S(\omega)$ of a transparency of the input scene containing particles of various sizes and populations is scanned by a TV, digitized, smoothed and stored in the digital computer. The signatures $G_i(\omega)$ (the smoothed, spatial-frequency spectrum) for particles of various classes or sizes i must also be produced and digitally stored. The digital system then uses a least-squares inversion technique to obtain the number N_i of particles of each class or size i in the input. The average diffraction pattern for a scene of N particles (N_1 of size d_1, N_2 of size d_2, etc., up to N_L particles of size d_L) at a spatial frequency ω is

$$S(\omega) = \sum_{i=1}^{L} N_i G_i(\omega). \tag{5.4}$$

$S(\omega)$ is digitized, smoothed and sampled at frequencies $\omega_1, ..., \omega_M$, from which a set of M equations result:

$$S(\omega_j) = \sum_{i=1}^{L} N_i G_i(\omega_j), \tag{5.5}$$

where $j = 1, ..., M$. In matrix notation, $[S] = [G][N]$. The computer furnishes an estimate of N from the pseudo-inversion of (5.5), i.e.,

$$[N] = ([\tilde{G}][G])^{-1}[\tilde{G}][S], \tag{5.6}$$

where $[\tilde{\ }]$ denotes the transposed matrix.

The results of five runs on a scene containing 90 particles in three classes (small, medium, and large particles) showed an accuracy of 96%. This approach has also been applied to the analysis of noncircular particles. In this instance, the areas rather than the diameters of the particles were used to define the various classes and the diffraction patterns of a large number of randomly oriented particles from the same class were averaged to form the stored representative signatures of each class or size of particle. Careful selection was needed of the spatial frequency band over which the diffraction pattern of the scene was sampled. From a scene containing 85 particles in three classes, the number of particles in each class was determined with a 97% accuracy.

5.3.3 Planimetric Analysis

The final application of this system to be considered demonstrates the use of the planimeter and is an example in which the light source used is noncoherent [5.37]. This configuration is useful in instances where planimetric data (e. g., the amount of acreage covered by water, forests, etc.) is desired, rather than topographic delineations or boundaries. A transparency of the input color image is illuminated with narrow-band filtered light at various wavelengths λ_j. The quasimonochromatic energy in the spectrally filtered image is integrated by a detector. This produces an integrated-radiance datum-vector component D_j.

   If the input scene is populated by an unknown subset of objects of a known set of N, the vector $[K]$ where K_i is the area in the input covered by the ith object can be found. The color signatures H, where H_{ij} is the irradiance of the ith object at λ_j, must be obtained and stored. The measured D_j must be repeated at each λ_j and are related to the unknown areas K_i by

$$D_j = \sum_{i=1}^{N} H_{ij} K_i. \tag{5.7}$$

The digital computer can then obtain the least-squares area-vector estimation $[K]$ from

$$[K] = ([\tilde{H}][H])^{-1}[\tilde{H}][D]. \tag{5.8}$$

In all of these instances, the parallel processing features of the optical system are utilized. Although the color signatures H_{ij} in this last example and the particle signatures $G_i(\omega)$ in the previous application must be known beforehand and stored, once this is done the system operates equally fast on an input containing 10 or 10,000 particles, etc.

5.4 Hybrid Correlators

In the previous hybrid systems, the Fourier transform property of a coherent optical system was used. Although considerable processing is possible using such systems, the optical correlator is a far more powerful system. The basic principles and methods of optical correlations have been discussed in Chap. 2 and elsewhere. The preliminary step in performing correlation is the synthesis of a

Fourier transform hologram (FTH) of the reference function(s), i.e., a matched spatial filter (MSF). This MSF is placed at P_2 of Fig. 5.1. The correlation of the input and reference function appears at P_3 as a peak of light if the reference function is present in the input. Other optical correlation techniques are possible (Sects. 5.4.6–10), but we will concentrate on this matched spatial filter system.

If any optical computer is to realize its full potential, real-time devices at P_1 and P_2 are essential and the MSF synthesis must be automated. In a hybrid system, the cyclic operation of these real-time devices, correction of their phase errors (see Chap. 4) and automated MSF synthesis and correlation are possible. Thus, hybrid correlators represent one of the most viable types of optical pattern recognition system and merit considerable attention.

An automated computer-controlled correlator and holographic MSF synthesis system is described in Sect. 5.4.1. In many instances a simple spatial filter rather than an MSF at P_2 is required. A digitally controlled real-time frequency plane filter for such applications is described in Sect. 5.4.2. In the remaining sections, eight hybrid correlators and applications are discussed. These systems differ primarily in their intended application. Thus, they are best described with reference to specific applications. For a more complete discussion of these applications, the reader is referred to [5.6].

A discussion of these various applications and hybrid systems serves to define many of the operations that the optical/digital interface (Sect. 5.5) must perform and to show, by example, how hybrid systems are configured. In most of these optical correlators, the digital section of the hybrid system is used to analyze the contents of the output correlation plane. However, in space-variant and signal processing applications, the digital processor is also of use in controlling the input data format as we will show. The proper input signal format facilitates design of the optical system and results in more powerful and flexible optical processors.

5.4.1 Automated Matched Filter Synthesis

The laboratory version of the optical correlator of Fig. 5.1 is shown in Fig. 5.7 in which the system is redrawn as a hybrid semiautomated system for generating and reading holographic Fourier transform matched spatial filters [5.38]. The half-wave plate ($\lambda/2$), placed at the output of the laser, produces a horizontally polarized beam. The electro-optic shutter (EOS) can rotate the plane of polarization by $90°$ when the half-wave voltage is applied. The Glan prism (GP) reflects the vertical polarization component of the input beam (this forms the reference beam) and passes the horizontal polarization component (this forms the signal beam). The ratio of intensities of the reference and signal beam can be varied by the voltage applied to the EOS. This control is necessary to optimize the Fourier hologram [5.39]. Shutters (S) are located in both beams together with conventional collimating optics. The half-wave plate ($\lambda/2$) in the signal beam insures that the polarizations of the reference and signal beams are matched as required for interference at the filter holder (FH). Upon reflection

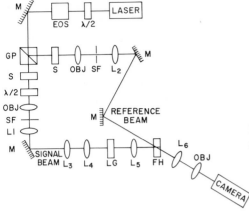

Fig. 5.7. Semi-automated hybrid holo-gram filter synthesis and reconstruction system

from the mirror (M), the signal beam is passed through a variable aperture used to isolate portions of the input image which is placed in the liquid gate (LG). The confocal lens configuration (L_3 and L_4) is used to image the aperture 1 : 1 onto the input transparency in plane LG. The transform of this input is formed at plane FH, where it is interfered with the reference beam to record a Fourier hologram.

A photodetector with an annular mask is placed in plane FH to record the intensity of each beam in a selected spatial frequency band. This enables the beam balance ratio, i.e., the intensity ratio between the reference and signal beams, to be measured. By controlling the voltage to the EOS, the beam balance ratio can be adjusted. When the signal beam is blocked (by activating the shutter S in the signal beam path), the reconstructed image is formed by the inverse transform lens L_6 and microscope lens (OBJ) on the television camera. With the reference beam blocked (by activating the shutter S in that leg of the system) and a second image placed in the LG, the correlation of this new input and the prior one (recorded on the FH) is imaged onto the camera.

The digital and electronic support for this system consists of the manual controls for the filter synthesis and camera readout system and meters for measuring beam intensities. A PDP-8E minicomputer with a three card interface allows control of the shutters (S) and modulator (EOS) in the system.

The type of holographic filter formed at FH in Fig. 5.7 is commonly referred to as a *Vander Lugt* filter after its inventor [5.20]. The accuracy required in the positioning of this filter is quite severe. The phase distortion produced by a displacement Δv_x of the filter is

$$|\phi| \leq \pi \Delta x \Delta v_x , \qquad (5.9)$$

where Δx is the input aperture size. For a phase error less than $\pi/4$, the tolerable displacement of the filter is

$$\Delta x_v < \lambda F/4\Delta x , \qquad (5.10)$$

where the spatial coordinate "x_v" in the frequency plane is related to the spatial frequency v_x by $x_v = v_x \lambda F$.

For a system with a 35 mm input, 600 mm focal length transform lens and 633 nm He–Ne wavelength, the allowable positioning error for the filter is $\Delta x_v = 2.7$ μm. A hybrid system [5.40, 41] has been fabricated in which a 10×10 matrix of 100 holographic filters can be automatically produced on a $2'' \times 2''$ Kodak SO-120-02 holographic plate. The positioning accuracy (2.7 μm) of each filter is achieved by precise x and y motorized translation stages and accurate stepping motors. The positioning of each filter and the electronic shutter used for exposure are controlled by a PDP-11/40 minicomputer. The exposed plate is permanently fixed on a metal frame which undergoes development and fixation with the plate.

When the coordinates of one of the 100 filters are typed on the computer console, the filter stage is automatically moved to the designated filter (within 2.5 μm) during synthesis or correlation. The 35 mm film input is also under computer control and is advanced by a film transport. These latter positioning systems are major elements needed in commercial automated optical systems.

The major aspect of any optical pattern recognition application is the synthesis of the Fourier transform hologram matched filters. In filter synthesis, considerable attention must be given to the selection of the optimum parameters for the FTHs. These parameters include exposure, the ratio between reference and signal beam intensities (K-ratio), and the spatial frequency band in which the K-ratio is set [5.39]. To facilitate this analysis, the wedge/ring detector (Sect. 5.2) has been used [5.39]. When positioned at the transform plane where the reference and image beams are interfered, the ring elements in the detector array provide a simultaneous measurement of the intensity in 32 spatial frequency bands in the object. By blocking the object and then the reference beam, the K-ratio in all 32 spatial frequency bands can be determined. This greatly facilitates the otherwise lengthy analysis of the effects of the various parameters and yields more reproducible results.

5.4.2 Digitally Controlled Fourier Plane Filter

When a simple spatial filter at P_2 of Fig. 5.1 rather than an MSF as in Fig. 5.7 is required, the hybrid Fourier plane filter [5.42], whose structure is shown in Fig. 5.8, can be used. This device uses a liquid crystal as the active material. The operating principles of liquid crystal devices have been adequately reviewed in Chap. 4 and are thus not repeated here. While the use of a liquid crystal as a spatial filter material is not new, the structure of this device (Fig. 5.8) and the ability to address it by a digital computer are novel and appropriate for the present discussion.

The device's format is similar to that of the wedge/ring detector discussed in Sect. 5.2. It consists of a wedge filter with forty $9°$ segments and a ring filter with 20 concentric rings in a 2.5 cm diameter area. A cross section of the device

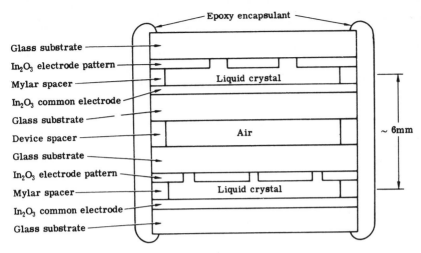

Fig. 5.8. Cross section of electroded liquid crystal Fourier plane filter [5.42]

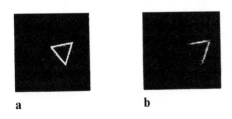

a   b

Fig. 5.9a, b. Computer controlled filtering of the input image (**a**) by removal of the zero-order term and one arm in the transform as shown in the filtered reconstruction in (**b**) [5.42]

structure is shown in Fig. 5.8. The wedge and ring filters are made separately and then oriented and cemented before mounting. The liquid crystal's dynamic scattering effect is used, but by controlling the polarization within specific scattering cones, high (10,000:1) contrast ratios and a high (50%) on-state transmission are obtained. Any of the filter segments can be separately activated. The wedge pattern permits filtering by pattern orientation, while the ring pattern enables conventional filtering based on spatial frequencies present in the image to be achieved.

The transmittance of any of the 40 possible filter segments (oppositely oriented wedges are externally connected) can be controlled by a 7 bit computer word. 30 V signals are required for operation and once a location is addressed by an ON command it remains latched until it receives an OFF signal. The low resolution of this binary filter is a limitation, but the structure has been used to remove the zero-order term in the Fourier transform (by activating the first ring of the ring filter) and one arm of a triangle (by activating the appropriate pair of wedge sections). The results of these filtering operations are shown in Fig. 5.9.

Various other forms of digitally controlled frequency plane filters are possible and have been demonstrated. The adaptive filter plane spatial light modulator (SLM) described above was electronically addressed. Alternate classes of SLMs are addressed optically, by integrated CCD arrays or by an electron beam [5.9–13]. One of the candidates for optically addressed SLMs is the PROM [5.9–13]. Such a device can be addressed sequentially point by point using various 2-D laser scanner systems. In a general filtering system, the data recorded on this frequency plane SLM (at P_2 of Fig. 5.1) can originate from a computer. The use of the PROM device as such a point by point controlled frequency plane filter has been reported [5.43].

In many instances, a complex filter function is required at P_2 of Fig. 5.1. Such a filter (e.g., a matched spatial filter) can be digitally computed and recorded sequentially on an SLM at P_2. One of the most obvious forms that such a filter can take is a computer generated hologram [5.44]. Such complex filters have successfully been recorded on the electron-beam addressed DKDP light valve [5.45]. Considerably more advanced systems of these types are possible once SLM technology has matured sufficiently.

5.4.3 Object Motion Analysis

The first hybrid optical correlation application to be considered is object motion analysis. The specific application addressed is cloud motion analysis [5.46]. A scanned photograph from the ATSIII weather satellite of the cloud pattern over Baja California and part of central Mexico is shown in Fig. 5.10a. If an FTH of the first frame in a sequence of these photographs is formed and correlated with the second frame of data, a strong correlation peak is expected if there is no motion of the cloud patterns between frames. If all of the cloud patterns moved with an equal velocity between frames, the location of the correlation peak would shift accordingly. In practice, cloud patterns move in different directions and with different velocities. To analyze this motion, various cloud pattern regions (shown encircled and numbered in Fig. 5.10a) were sequentially isolated. The location of their correlation peaks were then measured and converted to velocity vectors. The resultant plot of the velocity vectors or displacements of the indicated regions in the image are shown in Fig. 5.10b at half hour and full hour intervals.

These steps were performed by manually scanning the inputs with apertures and manually measuring the resultant displacements for the successive regions indicated. The operation thus required many hours to complete but could certainly be automated. The analysis of the optical output correlation pattern requires that the correlation peak be located and cross-correlation peaks rejected. In the half hour and hour intervals between frames, the shape of the cloud pattern can be expected to change and new FTHs may have to be synthesized [5.47] or the amplitude of the correlation peak will decrease appreciably. Considerable information on air currents and weather trends can be determined from the output data in Fig. 5.10b. The two cloud masses (22 and 32) are seen to move with the same velocity. Cloud pattern 10 is a stationary cloud on the West coast of central Mexico, while cloud pattern 17 is associated with the jet stream.

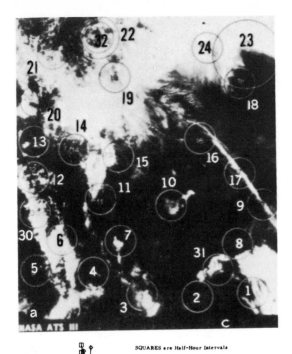

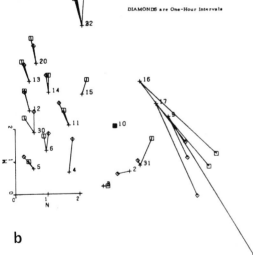

Fig. 5.10a, b. Cloud motion analysis by optical correlation [5.46]. (a) ATS III weather satellite image with overlay of various sections. (b) Displacement or velocity vectors for the various cloud sections in (a) at half hour (squares) and hour (diamonds) intervals obtained by optical correlation

5.4.4 Photogrammetric Compilation

When contour maps or ground elevation data are required from high-resolution and wide area imagery, hybrid optical techniques are clearly superior to all-digital methods. To obtain such data, two stereoscopic aerial photographs of the same ground area are first taken with the camera at different locations and

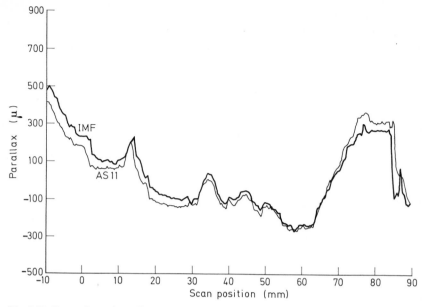

Fig. 5.11. Comparison of parallax measured on a hybrid image matched filter (IMF) correlator with parallax measured from the AS-11B automatic stereocompilation system [5.48]

aspects. Parallax differences between these photographs will be proportional point by point to the ground elevation. In most systems, these data are obtained by scanning both photographs with a small spot and cross correlating the resulting video signals. Slow manual stereo plotters and stereo compilators requiring precise equipment and skilled operators were originally used. These were later replaced by automatic systems such as the AS11B.

One optical processor for parallax extraction [5.48] using optical correlation techniques offers speed and accuracy advantages. In this scheme, an FTH is made from one of the transparencies $h(x, y)$ of a stereo pair. The second stereo transparency $f(x, y)$ is placed in the input plane P_1 in Fig. 5.1 and the FTH in plane P_2. With a small area A_f of the input illuminated, the pattern at P_3 corresponds to the cross correlation of $f(x, y)$ and $h(x, y)$. The area A_f will only correlate with the corresponding area A_h of $h(x, y)$. The location of the peak in P_3 depends on the relative positions of A_f and A_h. As these small areas change in position due to parallax, the correlation peak is correspondingly displaced. This displacement will be proportional to the parallax of the local regions. By sequentially illuminating portions of $f(x, y)$ and detecting the positions of the correlation peaks, the parallax over the entire input can be measured.

More extensive details on the basis for this method of parallax extraction and descriptions of other optical techniques for contour generation are presented in [Ref. 5.6, Chap. 5]. To implement the method described above, a telecentric laser scanner (folded telescope and two mirror galvonometer with axes at right angles) produces a beam that always appears to arise from the same point as the signal beam used during synthesis of the FTH. This beam is used to scan the input transparency under control of a D/A converter coupled to a minicomputer. The location of the output correlation peak (detected by a vidicon camera mounted on a translation stage) is then measured by translating the camera until the correlation peak falls on a set of electronic crosshairs on the camera display. Its position is then measured using linear potentiometers on the camera's translation stage. This procedure avoids problems associated with the calibration of the video scan.

Parallax data measured from 1:16,000 scale stereo imagery from the Canadian Photogrammetric test range using the optical correlator and the AS11B system are shown in Fig. 5.11. A 100 mm scan in 0.5 mm steps with a 1 mm diameter laser beam was made. The overall precision in the location of the correlation peak was 8 μm. Drift and hysteresis in the scanner and deflector, the different x position of the two scans and orientational errors in positioning the transparencies contribute to the errors between the two parallax measurement systems. After removing these errors by a least mean square error routine, the rms differences in parallax were found to be 30 and 24 μm for the holographic (IMF) and AS11B system. These results are in good agreement considering the low sophistication of the holographic system.

5.4.5 Water Pollution Monitoring

The third hybrid optical correlator to be discussed is used in water pollution monitoring [5.40, 41]. The object of this processor is the identification and counting of biological specimens called diatoms (water algae). Various features of this problem make the use of optical correlation techniques attractive. Diatoms are rigid silicon dioxide skeletons, quite symmetric about one and in some cases two axes. They possess a fairly consistent uniformity among a given species and vary in size by less than 15%. Optical discrimination is possible because of the detailed structure of the diatom.

The input used is a 35 mm positive film transparency made by phase contrast microscopy techniques on prepared water sample slides. A matched filter bank of 100 FTHs of typical diatoms is positioned in the filter plane. Averaged FTHs are used to account for diatom size variations and input depth of focus problems. In this system, one averaged FTH can be used to detect the presence of several different diatom species. Some uncertainty, therefore, exists in the output. This may be detrimental in character or pattern recognition. However, in water pollution monitoring the diatom distribution of different averaged FTHs, not that of the same average FTH, is the important feature.

With one input slide in position, the set of FTHs are positioned sequentially in the filter plane. The output plane contains correlation peaks of light. This pattern is scanned by a vidicon, digitized to 6 bits and stored in core and on magnetic tape for further study. This process is then repeated for each FTH with

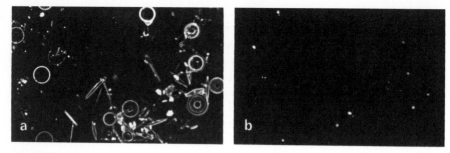

Fig. 5.12a, b. Pattern recognition of a diatom species [5.40]. (a) Input image; (b) optical correlation plane

the input fixed. A typical input sample of diatoms and the optical output are shown in Fig. 5.12. Nine correlation peaks whose locations correspond to the positions of the large circularly shaped diatoms in the input are apparent. Problems associated with size variations, orientations and depth of focus of the diatom inputs are presently under consideration.

5.4.6 Hybrid Frequency Plane Correlator (FPC) System

The next hybrid optical correlator system will be discussed in detail here and in Sect. 5.5 to provide the engineering details of one system, and for the optical/digital interface discussion to follow in Sect. 5.5. The basic structure of the hybrid processor is shown in Fig. 5.1. Both the Fourier transform and correlation planes of the system are interfaced to a PDP-11/15 minicomputer through vidicons and a special purpose interface.

Many of the applications of this hybrid processor lie in the area of signal processing and are discussed in a companion volume in this series [5.6]. To provide an appreciation for the required interface operations, a brief summary of several signal and image processing operations is provided below. The main signal processing application considered has been radar signal processing. This is a viable optical processing application because of the high data rate, the need for real-time processing and the similarity of the operations required to those inherently available in an optical processor.

a) Signal Processing

Data from various types of radars have been optically processed on this hybrid system using a real-time electron beam addressed DKDP spatial light modulator [Ref. 5.6, Chap. 8] and [5.49–51]. In each instance, the received radar data were first heterodyned and then used to modulate the beam current of the write gun in the DKDP system in proper synchronization with the deflection of the electron beam. For each type of radar system, the input data were recorded on the DKDP target crystal in a different format. The interpretation of the output data format is also different for each type of radar system. In each case, the input data format is controlled by a PDP-11/15 minicomputer and the output plane pattern is analyzed by the same minicomputer through an optical/digital interface (Sect. 5.5).

For simplicity, only the case of a linear phased array optical correlator is considered to demonstrate the concept of input data format control and the output plane analysis required. Since detailed descriptions of phased array radar signal processing exists elsewhere, the present theoretical discussion will be brief. The heterodyned signal received at the nth element of a linear phased array is

$$V_n(t) = p(t) \cos f_{if}(t - n\tau_1 - \tau_2), \tag{5.11}$$

where τ_1 and τ_2 are the delays corresponding to the target's azimuth angle θ and range R. These signals are recorded on N lines on the DKDP target crystal at plane P_1 of Fig. 5.1 by appropriately modulating the write gun's beam current in synchronization with the scan rate of the electron beam.

A cylindrical/spherical lens system L_2 images the P_1 pattern vertically and Fourier transforms it horizontally, producing the Fourier transform of the signal on each line at P_2 of Fig. 5.1. The phase of the signal on each line at P_1 will be proportional to τ_2 or R with an incremental phase proportional to τ_1 or θ between adjacent lines.

A multichannel 1-D MSF of the transmitted reference coded waveform is stored at P_2. The 2-D transform formed by L_3 produces the autocorrelation of the transmitted coded waveform at P_3. However, the correlation peak at P_3 is deflected vertically by a distance proportional to τ_1 and horizontally proportional to τ_2. The output plane pattern at P_3 can then be described by

$$p \circledast p * \delta(x_2 - K_1 R) * \delta(y_2 - K_2 \sin\theta). \tag{5.12}$$

The location of the output correlation peak is thus proportional to the target's azimuth angle θ and fine range R.

For a pulse burst radar in which the relative velocity rather than the azimuth θ is of interest, the returns from successive pulses received by the antenna can be recorded on successive lines in the input plane P_1 of Fig. 5.1. A cylindrical/spherical lens system is again used for L_2 and a multichannel MSF of the coded waveform is recorded at P_2. The P_3 output pattern has a correlation peak whose coordinates are now proportional to the target's fine range and fine velocity [5.52].

Planar phased array radar data has also been optically processed on the DKDP device [5.49]. First the $N \times N$ planar array is thinned by choosing N array elements pseudo-randomly (e.g., using the Durstenfield algorithm) such that there is one element in any row or column. The position information of these N elements is stored in PDP 11/15 core and used to control the starting location of the radar signals on the light valve. The starting location for the data recorded on each line of the DKDP crystal is shifted by an amount proportional to the horizontal array element selected in that row. If a 2-D Fourier transform rather than a correlation is performed on this input data, the first-order diffraction term in the optical transform plane will lie on the vertical axis for a boresight target, and be displaced from the horizontal axis by an amount proportional to the $i-f$ frequency of heterodyning. As the target moves from boresight, the horizontal and vertical displacements in the location of the first-order diffraction term will be proportional to the target's azimuth and elevation angles.

In all of these radar signal processors, the location of the correlation peak (or the first-order diffraction term in the Fourier transform systems) is proportional to the desired target data. In each frame, the optical/digital interface analyzing the output plane must detect these "important" peaks of light, determine their coordinates and convert them to target angle, range or velocity data. Multiple targets with various cross sections arise in practice, so that the actual detection problem is more complex and involves the tracking and sorting of various targets as well as simply detecting peaks of light. A simple thresholding operation is not always adequate for the detection of radar targets, rather both thresholding and integration with time (or distance) are needed to facilitate decisions on the presence or absence of true peaks of light (radar targets). These points will be elaborated on in Sect. 5.4.7b.

In practice, the returned radar data must be velocity and range gated and synchronized with the line scan rate of the DKDP light valve. To enhance the separation, shape and detectability of the terms in the transform plane, successive horizontal lines in the conventional television raster scan are not always used. This positioning is under control of the PDP-11/15 and varies with the radar system used. When fully operatonal in a multifunctional radar environment, data from the processing of primary radars are used as initial values in the processing of secondary radar data.

b) Image Processing

A wide variety of image correlation and pattern recognition operations have been implemented on this hybrid FPC system [5.53]. In the analysis of aerial imagery for the purpose of locating the position of certain objects or to correct the flight path of airborne vehicles, matched filters of the objects to be located or areas along the flight path are formed and stored at plane P_2 in Fig. 5.1. Transparencies of on-line imagery from a real-time sensor are then placed in the input plane and correlated with the stored filters. In airborne applications, the input imagery is often produced from an on-board sensor or radar.

A representative example of this processing [5.54] is shown in Fig. 5.13. The input aerial image in Fig. 5.13a was recorded on a photo-DKDP SLM at P_1 and correlated with an MSF of the reference scene region in Fig. 5.13b. The resultant optical output correlation plane pattern is shown in Fig. 5.13c. As shown, the presence of a correlation peak indicates that the reference is present in the input scene, whereas the location of the correlation peak indicates where the reference is in the input sensor's field of view. This latter data is of use as input to the missile's guidance system for flight path correction, etc.

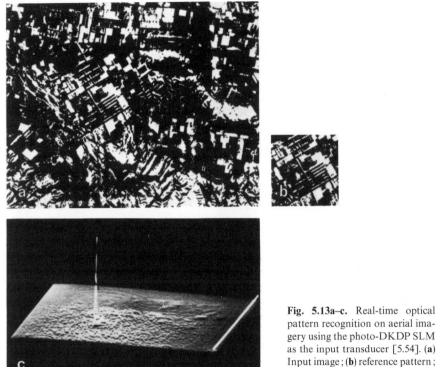

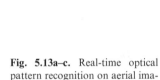

Fig. 5.13a–c. Real-time optical pattern recognition on aerial imagery using the photo-DKDP SLM as the input transducer [5.54]. **(a)** Input image; **(b)** reference pattern; **(c)** output correlation pattern

5.4.7 Optical Word Recognition

A most attractive application of hybrid optical pattern recognition is the recognition of key words in a large data base [5.55]. This application provides an excellent example of the parallel processing features of an optical pattern recognition system and the ability of such a system to detect the presence of and locations of multiple occurrences of the key object. In Fig. 5.14, we show an example of this using the real-time photo-DKDP SLM as the input transducer. The input text pattern shown in Fig. 5.14a was recorded on the SLM at P_1 of Fig. 5.1 and an MSF of the key word "PROFESSOR" was placed at P_2 on film. The output optical correlation plane pattern is shown in Fig. 5.14b. It consists of four peaks of light, whose locations correspond to the four positions of the reference word in the input paragraph. More advanced aspects of this optical word recognition (OWR) application are described below and in [5.55].

a) Normalization and Weighted MSF Systems

The OWR application described in Fig. 5.14 provides an excellent example of three other hybrid processors. Two of these are discussed in here and the third in

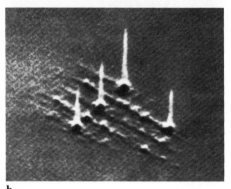

PROFESSOR
ASSOCIATE PROFESSOR
PROFESSOR AND HEAD
SECRETARY TO HEAD
PROFESSOR

a

b

Fig. 5.14a, b. Real-time optical character recognition of text [5.54]. **(a)** Input; **(b)** optical output plane showing the correlation of the input paragraph with the key word PROFESSOR

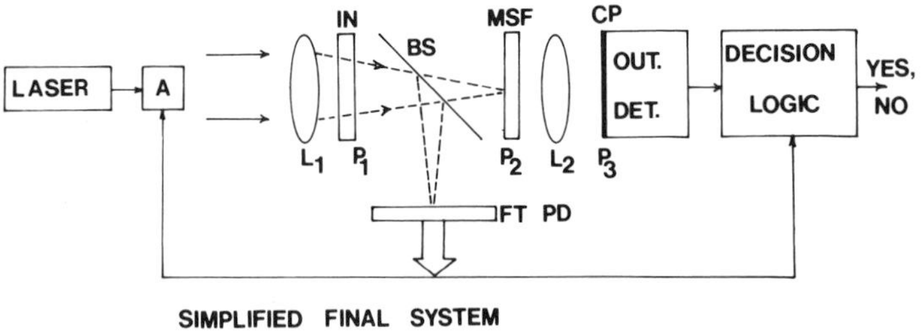

SIMPLIFIED FINAL SYSTEM

Fig. 5.15. Schematic diagram of an opto-electronic feedback processor for optical word recognition [5.55]

Sect. 5.4.7b. The data base used for the OWR analysis program was large and was recorded on microfilm. From an extensive analysis of this data base [5.55], three classes of input data were isolated: print, font and teletype text inputs. These were characterized by different dominant carrier spatial frequencies, due to the different stroke widths of the characters in each input class, and by other characteristic differences between the input and reference word (font, character spacings, and exposure differences, respectively).

The hybrid processor for this application is shown schematically in Fig. 5.15. The Fourier transform of the key word and the input text pattern are sampled at the Fourier plane with photodetectors (FTPD) and used for input and output feedback as described below. From extensive experiments, teletype input data has been found to exhibit considerable exposure differences, due to the noise and poor contrast of such data. Successful detection of the resultant optical correlation plane patterns requires normalization of the input data or the output correlation plane patterns. This is realized by sampling the dc value of the Fourier transform of the input text (by an on-axis photodiode at P_2). This data is a measure of the average input exposure. It is fed back to the input to control the laser beam intensity by attenuator (A) and hence the output correlation plane P_3 exposure. Such data normalization is essential to optimize output correlation plane detection.

This OWR case study provides yet another hybrid processor feature. The three classes of microfilm data are characterized by different character stroke widths and hence by different carrier modulation spatial frequencies. With three photodiodes appropriately placed at plane FTPD, the class of input text data can be determined [5.55]. These photodiode outputs are also of use during weighted MSF synthesis and as measurements of the information content of the input key word [5.55]. These three photodiode outputs are fed back to the output detection analysis and decision logic where they are used to set various detection threshold parameters (Sect. 5.4.7b) and to select the appropriate output correlation plane detector output to analyze (determined by the input data class, when multiple MSFs are present at P_2).

b) Output Detection Criteria

The output correlation plane detection and analysis system can also be altered to optimize correct correlation plane detection. Three output correlation plane detection techniques have been considered and experimentally analyzed for the OWR application using a microfilm data base and a sample input paragraph with four occurences of the key word. The three correlation plane detection criteria are described below with reference to the experimental data of Fig. 5.16. These data were obtained from a TV camera at P_3 of Fig. 5.15, whose output was A/D converted and fed to a digital post processor that simulated the three alternate detection schemes.

The three output correlation plane detection schemes with reference to the data of Fig. 5.16 follow.

1) Threshold Detection. In this correlation plane analysis technique, a correlation is said to occur if the output exceeds a preset threshold level V_T anywhere in the output plane. Use of this criteria for a microfilm input paragraph with four occurences of the key word is shown in Fig. 5.16a for different threshold levels. As seen, the correct number of output correlation occurences results for V_T levels from 64 to 88. This corresponds to an output SNR range of $1.5:1$.

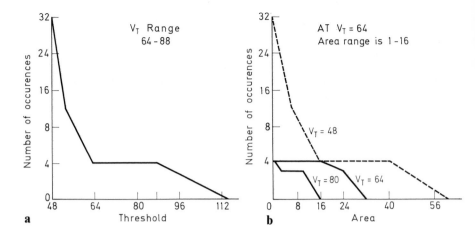

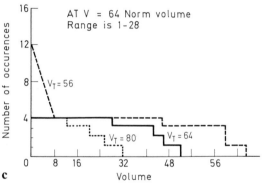

Fig. 5.16a–c. Effectiveness of (a) threshold detection, (b) area detection, and (c) volume detection in a word recognition optical correlator [5.55]

2) Area Detection. A second detection criteria uses the area under the correlation peak at a given V_T level as the output detection criterion. The results of such an analysis on the same output correlation plane pattern are shown in Fig. 5.16b. From these data we see that if $V_T = 64$ is selected, then a larger output correlation plane area range of 16:1 results, implying that better correlation plane detection results if area rather than threshold detection is used.

3) Volume Detection. The third output detection criteria is to compute the volume under the correlation peak above a given V_T level. The results of such analyses are shown in Fig. 5.16c, from which we see that with $V_T = 64$ a still better output SNR of 28:1 results if this detection criterion is used.

The above detection plane analysis techniques can be realized in hardware by video comparator, integrator, and peak detector circuitry in the output plane analysis system. Attention to the role of the output correlation plane detector in pattern recognition thus appears to be yet another tradeoff in an overall hybrid pattern recognition system.

5.4.8 Hybrid Joint Transform Correlators

Various optical correlator architectures exist that do not require formation of an MSF of the reference object. In the joint transform correlator [5.19], the two objects to be correlated f and g are placed side by side in the input plane with a center to center separation $2b$. In 1-D, we describe this input function as

$$t_1(x) = f(x+b) + g(x-b). \tag{5.13}$$

The joint transform of (5.13) is recorded on an intensity sensitive transform plane medium, whose transmittance (after exposure) is proportional to

$$I_2(v_x) = \mathscr{F}|\{t_1(x)\}|^2 = |F|^2 + |G|^2 + 2|F|\,|G|\cos[4\pi v_x b + \phi(v_x)], \tag{5.14}$$

where $\phi(v_x) = \mathrm{Arg}\,G - \mathrm{Arg}\,F$. Illumination of P_2 with a plane wave and formation of its Fourier transform yields an output plane pattern

$$u_3(x) = f \circledast f + g \circledast g + f \circledast g * \delta(x_2 + 2b) + g \circledast f * \delta(x_2 - 2b). \tag{5.15}$$

Proper choice of b insures spatial separation of the correlation from the other output terms. This correlation architecture is especially attractive for use with optically addressed transform plane SLMs (which are usually operated in reflection).

Rau [5.19] suggested a hybrid system in which the joint transform pattern was detected on a vidicon TV camera and the correlation of input characters was obtained from the output of a spectrum analyzer fed with the camera's video output. Macovski and Ramsey [5.56] transformed the coherent superposition of the two images after frequency shifting the light illuminating one image and produced an equivalent video signal. A temporal ac term in their transform plane intensity consisted of a carrier (at the shift frequency) modulated by the complex transform product. Detection of the ac time distribution in this system requires a nonintegrating detector, however.

The schematic diagram of a hybrid coherence measure or equalization correlator [5.57] is shown in Fig. 5.17. The two functions to be correlated are placed side by side in the input plane; their joint transform is formed on a photodetector and fed to a post-electronic processor. To best understand the operation of this system, we consider the anatomy of the joint transform pattern.

We assume approximately equal power spectra $|F| \simeq |G|$. If $f = g$, the I_2 pattern in (5.14) is

$$I_2(v_x) = 2|G|^2(1 + \cos 4\pi v_x b). \tag{5.16}$$

If f is displaced with respect to g, e.g., the input separation is $2.5b$, then (5.14) becomes

$$I_2(v_x) = 2|G|^2(1 + \cos 5\pi v_x b). \tag{5.17}$$

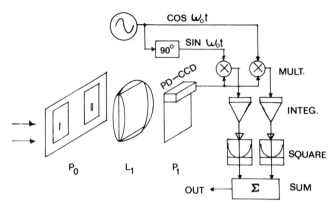

Fig. 5.17. Schematic diagram of a coherence measure hybrid correlator

Comparing (5.16) and (5.17), we see that a shift in the location of g with respect to f is equivalent to the addition of a linear phase term $\pi v_x b$ to the pattern in (5.16) and that its slope or derivative with respect to v_x is proportional to the shift in f. As f and g become different, the joint transform pattern approaches (5.14). Comparison to (5.14) shows that as f and g become different, a random spatial phase modulation is present on the fringe pattern in (5.16) or (5.17). If the strengths of f and g differ, e.g., $f = ag$, then

$$I_2(v_x) = |G|^2(1 + a + 2a\cos 4\pi v_x b) \tag{5.18}$$

or the ac modulation is reduced.

Consider first the case when the electronic post processor performs the Discrete Fourier Transform (DFT) of I_2. From (5.16) and (5.17) we see that the DFT of $I_2(v_x)$ will have a peak in v_x proportional to the location of g with respect to f. Now consider the case when $f \neq g$. From (5.14) and (5.16), we see that the coherence or carrier phase variance of $I_2(v_x)$ is a measure of the similarity of f and g. We can easily measure this coherence with a phase locked loop (PLL) fed by $I_2(v_x)$. The ac content or variance of the control voltage of the voltage controlled oscillator in the PLL directly provides the desired coherence measure. Thus, with a PLL output analysis system, we can automatically tell if a correlation is present and on which line it occurs. The DFT of this signal line tells where on the line the correlation occurs. This *coherence measure correlator* [5.57] is most appropriate for multichannel 1-D correlations as in the Tercom cruise missile guidance system, and in ambiguity surface range vs Doppler computations in radar or sonar signal processing.

In a more sophisticated version of this hybrid system, we consider the practical case when g has been degraded by a linear space-invariant impulse response h, so that the received signal is now $f = g * h$. This will degrade the correlation output $f \circledast g$. However, if h is known, a priori or can be estimated, we can compensate for it by varying the phase modulator in the DFT post processor

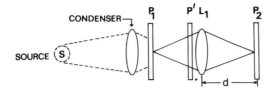

Fig. 5.18. Schematic diagram of a hybrid noncoherent correlator

by arg H. This effectively removes the degrading effects of h from the signal. We term this system an *equalization correlator* [5.57]. Successful experimental demonstrations of all properties of these two hybrid correlators have been obtained on signals, text, and aerial imagery [5.57].

5.4.9 Hybrid Incoherent Correlators

Our entire prior discussion has concentrated on coherent optical systems, however, noncoherent systems represent yet another class of optical processors that has been improved by recent application of hybrid techniques [5.58–61]. In the classic noncoherent optical correlator, plane P_1 with intensity transmittance f is imaged through a pupil plane P' with transmittance P' onto plane P_2 (Fig. 5.18). The intensity distribution at P_2 is then $I_2 = f * |h|^2$, where $h = \mathcal{F}\{P'\}$ and $\mathcal{H} = \mathcal{F}\{|h|^2\} = P' \circledast P'$. The limitations of this system are that h^2 must be real and nonnegative. In this section, we discuss several hybrid methods by which bipolar correlations can be realized using incoherent optical processors.

One of the most attractive hybrid techniques for noncoherent processing is to encode the two functions f and g with a spatial frequency carrier, correlate these encoded functions, and then electronically filter or decode the resultant correlation to obtain the bipolar correlation. In the general description of this technique, f and g are encoded by the spatial frequency carrier u_c as

$$f_e = (1/2)|f|[1 + \cos(2\pi u_c x_1 + \arg f)], \tag{5.19a}$$

$$g_e = (1/2)|g|[1 + \cos(2\pi u_c x_1 + \arg g)]. \tag{5.19b}$$

The basic incoherent correlation with f_e in P_1 and a Fresnel hologram of g_e at P' yields at P_2 (for the nonzero terms)

$$I_2 = f_e \circledast g_e = (1/4)|f| \circledast |g| + (1/4)|f \circledast g| \cos[2\pi u_c x_2 + \arg(f \circledast g)]. \tag{5.20}$$

The undesired first term in (5.20) can be removed by an electronic band-pass post-filter after scanning the I_2 pattern at P_2 to convert it to a time waveform. The magnitude and phase of the desired correlation $f \circledast g$ can then be extracted by envelope and phase electronic post processing.

Various versions of this basic hybrid correlator are possible and have been experimentally demonstrated. In one system, neither f nor g is encoded. Rather, a sinusoidal grating with intensity transmittance $(1/2)(1 + \cos 2\pi u_c x_2)$ is placed at P_2. P_2 is then imaged onto a new output plane P_3 through a second pupil plane P'' in a system similar to that of Fig. 5.18. The P_2 grating shifts the spectrum of the correlation to $+u_c$. The P'' pupil function has zero response at $x' = x_c \propto u_c$, thus high-pass filtering the correlation (removing dc). Post electronic filtering and decoding is used as before to extract $f \circledast g$ from the encoded correlation.

An alternate formulation involves encoded images separated into dc and ac components, with phase modulation zero or π when g_{ac} is positive and negative, respectively. In the final version of this hybrid system, we insure that the dc levels of f and g are 0.5 and use square-wave carrier modulation to encode f and g. Only a simple exclusive-OR operator is then needed to encode the imagery.

5.4.10 Other Advanced Hybrid Processors

Two other recently described hybrid optical/digital processors deserve mention, as they promise to have a major impact on the future of optical processing. The first system is a vector/matrix multiplier. A vertical LED input array (whose outputs represent the elements A_n of a vector) is imaged vertically and expanded horizontally onto a mask (with transmittance B_{mn}, corresponding to the elements of the matrix B). The output from B is imaged horizontally and focussed vertically onto a horizontal linear photodiode output array (whose output C_n is the vector matrix product $C = AB$). With the proper mask at B, the output is the DFT of the input [5.62].

In a second version of this system [5.63], the mask is replaced by one with transmittance $(I - B)$, the vector C is added to the system's output $A(I - B)$, and the resultant output is returned to the LED inputs. If we denote the system's input at cycle i by A_i, the final output at cycle i (and hence the system's input for cycle $i+1$) is

$$A_{i+1} = A_i(I - B) + C. \tag{5.21}$$

When $A_i = A_{i+1}$ (within some value E), the resultant A is the solution of the vector/matrix equation $C = AB$ or

$$A = CB^{-1}. \tag{5.22}$$

These and similar general vector/matrix processors are most readily fabricated and appear to be of immense use in many diverse applications.

Another most attractive new hybrid pattern recognition processor can be realized using a modified form of the optical system of Fig. 5.1 in which L_2 images P_1 onto P_2. We denote the amplitude transmittances of P_1 and P_2 by $f(x, y)$ and $g(x, y)$ respectively. The Fourier transform lens L_3 produces an on-axis output at P_3 described by

$$I_3 = \iint f(x, y)g(x, y)dxdy. \tag{5.23}$$

In this novel hybrid pattern recognition system [5.64], the outputs at P_3 can be made to be the moments

$$m_{pg} = \int\int f(x, y) x^p y^p \, dx \, dy \qquad (5.24)$$

of the input object $f(x, y)$ by proper choice of the mask $g(x, y)$ at P_2. When the $g(x, y)$ mask is absent, $I_3 = m_{00}$. When $g(x, y) = x, y$, or xy, the P_3 outputs are m_{10}, m_{01}, m_{11}, respectively. When the properly encoded mask is used for $g(x, y)$, all moments m_{pq} of the mask can be generated in parallel at different spatial locations in P_3.

The digital post processor is then used to compute the absolute invariant moments ϕ_n from these m_{pq}. These moments have been shown [5.65] to be invariant to any geometrical errors in $f(x, y)$. This resultant hybrid processor represents a new noncorrelation and non-MSF approach to pattern recognition. The low dynamic range moments m_{pq} are all generated in parallel by the optical system and require only a fixed mask at P_2 which needs to be fabricated off-line only once. The large dynamic range moments ϕ_n are computed in a special purpose hardware digital processor to the necessary accuracy.

5.5 Optical Digital Interface

In the previous sections, various versions and topologies for a hybrid optical/ digital processor have been discussed. A common element in all of these systems is the actual interface between the optical and digital sections of the system. In the case of the wedge/ring detector, the detector itself is the interface and the entire digital analysis is performed in software. In the other systems, a vidicon or photodiode array is used as the detector; the video output is analyzed, usually digitized, and further processed by hardware or software.

5.5.1 General Considerations

In the interfaces to be discussed, the optical output pattern is analyzed by both hardware and software. Data reduction is a major feature of these interfaces and decisions are often made directly on the raw video data prior to its digital analysis. This type of interface design is consistent with the overall philosophy of a hybrid processor in which the data is processed optically rather than digitally because of its large space and time bandwidth product. One design goal for the interface is thus to reduce the space-time bandwidth product of the data presented to the digital processor. If this is *not* accomplished, the throughput of the system will be greatly limited and the data rates for the digital and optical portions of the system will not be compatible.

The optical output format is quite conducive to these functions. In terms of the resolution and gray scale specifications of the detector system, these remarks imply that little will be gained by digitizing the entire output plane to a resolution of 100 lines/mm (or even 1000×1000 points) with a 64 bit gray scale level. This would only require the storage and subsequent manipulation and analysis of millions of bits per frame. These were the very operations that were to be avoided by optically processing the data.

A data rate of 30 frames/s is a fairly well accepted norm for real-time processing. It also corresponds to that of television and thus, with no actual loss of generality, this interface discussion will be restricted to television-based camera detection systems. If some generalities can be tolerated, one can view an optical diffraction or correlation pattern as a collection of peaks of light. The amplitude and location of the "important" peaks of light are what must be determined. The question of what constitutes an "important" peak of light is determined by the analysis and decision logic in the interface. To facilitate this analysis, the optical plane being digitally analyzed is divided into a matrix of cells. Rather than fabricating a matrix of detector elements, this partitioning is performed electronically on the video signal from the camera. This approach allows the resolution to be altered by program control as the application warrants but requires good geometric registration in the camera detector.

An estimate of the required resolution can be obtained from an analysis of several of the hybrid processor system applications discussed. In the case of linear or planar phased array radar data processing, the number of resolution cells in the radar's search space is simply the number of elements in the radar array [Ref. 5.6, Chap. 8] and [5.49–51]. Commercially available camera systems can thus easily accomodate array sizes of 1000×1000 elements or more. However, such radar arrays are not economically feasible at present and a 100×100 element array and subsequent 100×100 cell resolution on the camera appear adequate. Similar remarks apply to the resolution of a pulsed radar system. If 100 pulsed radar returns are processed, the fine resolution of the radar is one one-hundredth of the width of the coarse velocity or range bin. In the real pulsed data processed optically on the hybrid DKDP system, a fine range resolution of 1 m and a fine velocity resolution of 1 m/s was obtained by processing only 100 pulsed radar returns.

For the case of image correlation, an interface resolution of 100×100 would imply that objects as small as 1 % of the width of the input image could be detected. This appears to be adequate for most applications; image sectioning (magnifying portions of the image) can always be used if finer resolution detail is required.

In the optical processing of wideband signal data in a folded spectrum (2-D Fourier transform of a raster scanned input pattern), the interface resolution must equal the space-bandwidth product of the input (often 1000×1000 or more). However, in most of these cases only changing or difference signals are actually desired. If time is available for a coarse spectrum search followed by a fine resolution spectrum analysis, then a lower interface resolution may be acceptable.

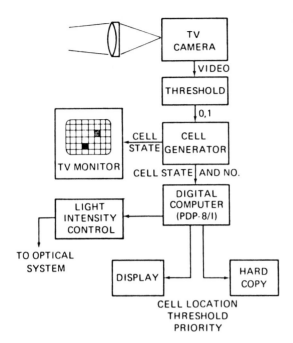

Fig. 5.19. Block diagram of an optical/digital interface [5.46]

The number of gray levels required in the interface depends greatly on the particular application and the presence of cross correlations. For most aerial image analysis, the presence of cross correlations do not appear to be a major problem. Rather, the detection of the correlation in the presence of unavoidable system distortions and noise is of concern. By properly selecting the parameters associated with the FTH synthesis, the peak correlation and cross-correlation intensities can be adjusted to satisfy the requirements of a given application and camera detector.

While the data analysis and display aspects of the interface are emphasized in this discussion, the digital system can clearly also be used to control and position optical components (often using the results of the digital analysis of the optical data).

5.5.2 Block Diagram

The block diagram of one version of an optical/digital interface [5.46] devised at Harris is shown in Fig. 5.19. The interface itself is built on three cards which plug into the omnibus of the PDP-8/I minicomputer used. The video signal is thresholded and then arranged by the cell generator. The system produces an $N \times M$ array of binary words in the computer memory. The value of each word corresponds to the maximum (not the average) light intensity in each of the $N \times M$ cells into which the optical output is formed. The concept of cell

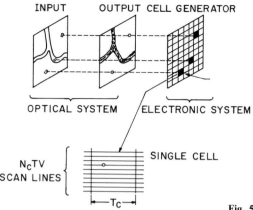

CELL GENERATOR FUNCTION

Fig. 5.20. Example used to demonstrate the concept of cell generation in an optical interface [5.66]

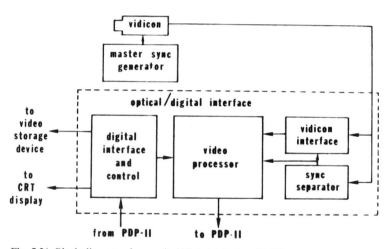

Fig. 5.21. Block diagram of an optical/digital interface [5.53]

generation [5.66] is shown in Fig. 5.20. If the video level anywhere within a cell (shown as N_c video scan lines of time scan length T_c) exceeds the threshold level, the binary value "1" is assigned to that cell. The cell generator for the interface in Fig. 5.19 consists of high-speed, double-buffered memory composed of shift registers and other logic circuits. The state of each cell is held in the flip-flop in the corresponding shift register. The normal resolution of this specific interface is a 32×32 array of cells. The software control for this system is written in a conversational on-line language FOCAL and requires very little memory space.

A block diagram of the optical/digital interface used in one of the hybrid processors at Carnegie-Mellon University is shown in Fig. 5.21 [5.53, 67]. The digital section of this hybrid processor consists of a PDP-11/15 minicomputer with 20 K of core, a highspeed DR-11-B interface which provides direct memory access to the PDP-11/15, an X–Y CRT display and raster scan monitor, teletype and paper tape reader. An isometric display, level slicing and false color coding are presently being added, together with an interface consisting of an array of 32 microprocessors and 4 K buffer memory. The present discussion concerns only the existing and operational interface system.

The interface itself has three sections: vidicon interface, video processor and digital interface and control section. This interface can be used anytime that optical data must be digitally analyzed or displayed. Closed circuit television cameras are normally used as the detectors and connecting link between the optical and digital portions of the system. Higher resolution and extended dynamic range devices or charge coupled device arrays could be used in place of the cameras if the application warrants. In the initial version of this interface, the contents of the correlation or output planes are continuously scanned at 30 frames/s by a 525 line, raster scanned, 2:1 interlaced vidicon with AGC disabled. The synchronization signals for the vidicon are kept in phase with the input data rate by the master sync generator shown in Fig. 5.21.

The vidicon interface section of the processor is used to convert the output video signal from the vidicon into a bit pattern. This is similar to the thresholder in Fig. 5.19. The video processor section of the interface is used to arrange and compress this binary pattern for storage, display, or analysis. This constitutes the cell generator referred to in Fig. 5.19. Various commands from the PDP-11/15 are used to control the vidicon interface, video processor, storage and display systems, etc. These commands originate in the digital interface and control section of the interface. All timing for the interface is derived from the horizontal and vertical blanking signals. A master sync generator in the interface is used to provide these signals; however, composite video signals (video plus synchronization signals) can also be used with a sync separator circuit (used to extract only the sync signals from the composite video plus sync signals) included in the system.

In this initial version of the interface, the resolution can be varied from 8×8 to 96×240 under program control. This variable resolution extends the flexibility of the interface and thus far has proven adequate. The threshold level for each cell is also variable under program control to any of thirty-two levels. In most cases, ten threshold levels have provided adequate level separation. The horizontal and vertical resolution as well as the threshold level are specified at the start of each frame. Once a start signal from the PDP-11/15 is received, all operations on one frame of data are automatic and start at the beginning of each field of video data.

A PDP-11/15 control word determines the horizontal resolution. For a horizontal resolution of ten, the 53 µs line scan time would be divided into ten 5.3 µs segments. The threshold unit in the interface compares the video signal within each element to a threshold level V_T. If the video signal anywhere within an element exceeds V_T, that element is assigned the binary value "1", otherwise

"0". For the case of a 10 bit horizontal resolution, the sampled video signal will be represented by a 10 bit word. This binary data is buffered in the interface and transferred, in blocks of up to sixteen bits, to the video processor section of the interface. With 64 bit horizontal resolution, four data transfers of 16 bits each are needed.

These successive horizontal lines are assembled into a digital image by the video processor section of the interface. The vertical programmed resolution is used to control this section of the system. The interface performs this digital image arrangement for each frame of data. At the full 240 vertical line resolution of the interface, no horizontal field lines are combined; while at a 120 line vertical resolution, every other line is combined, etc. In a 60×60 element digitized image, four successive horizontal scan lines and 0.88 μs (53/60) of vertical scan time are combined into one picture element.

In many cases a simple threshold detection is adequate to determine the binary value of a given cell. In other cases (e.g., radar target detection and text correlation), a combination of thresholding and integration is needed to properly distinguish the state of a given resolution cell. In the case of photogrammetric processing, a peak detection circuit is useful to determine parallax error and contour data. These additions have been incorporated into this hybrid interface. The entire process of digitizing and thresholding each scan line, forming the digital picture elements and transferring this data to the PDP-11/15 must be completed in the field time (15 ms) of commercial television. For speed considerations and to minimize the data transferred to the PDP-11/15, the amount of data transferred is minimized by the variable programable interface resolution to be no more than the specific application warrants. It is limited only by the data transfer rate of the DR11-B interface to the PDP-11/15 and the memory capacity of the minicomputer.

The vidicon interface section consists of 16 IC DIP packages on one double-height board. The programmable clock consists of an 8 bit synchronous up/down counter. The frequency control word from the PDP-11/15 is parallel loaded into this counter register. The threshold circuit is a simple differential comparator fed by an 8 bit D/A converter connected to the threshold control word from the PDP-11/15. The output of this comparator is transferred to a shift register by the thresholder. A new picture element is loaded into the shift register every T_C seconds (the horizontal time resolution of one picture element). The divide-by-sixteen counter determines the end of each 16 bits (or the end of a video line) and transfers the shift register data to a 16 bit buffer. Upon generation of the "end" signal by the counter, the data word is ready to be transferred to the video processor.

The design of the video processor and digital interface section of this interface was determined by the modular computer elements (known as register transfer modules, RTMs [5.68]) from which the interface was constructed. Data and control modules comprise an RTM system. A common 21 pin bus with 16 data lines is used for interconnection. Each module operates on one or more 16 bit words or 8 bit data bytes. Control inputs on each module enable various logical

Table 5.1. Basic optical/digital interface operations

Thresholding	Change detection
Integration	Data tracking
Peak detection	Time history
Addition	Dynamic memory allocation
Subtraction	Mensuration:
Pattern generation	location of points
Line generation	distance between points
Thinning	number of points in frame
Filling in	number of points in cluster
Analog data	number of points in pattern

operations to be performed. The main RTM module in any system is the general purpose arithmetic unit, which can perform eleven arithmetic operations on two stored data words. The bus into which all modules plug is the data path for all transfers from the control part of the RTM system. Once the controller is specified, so is the system. This is one of the advantages of the use of RTMs in a special purpose system design.

5.5.3 Fundamental Operations

In the hybrid systems and applications discussed in Sects. 5.2–4, the need for various interface operations has been established. While the specific operations required of the digital subsystem change somewhat with the application and are added to as more applications are pursued, five fundamental operations can be enumerated:

    Data detection
    Key feature extraction
    Measurement
    Data manipulation
    Display.

In the interface, these operations have been incorporated in modular form to maintain flexibility in the overall system. Each of these fundamental operations consists of various self-contained hardware and/or software routines. Any routine can be used in combination with any other one and in any of the five fundamental operations.

There is clearly some overlap in the four fundamental operations above. For example, proper data detection often requires key feature extraction, while data display usually requires manipulation of the original data. A brief description of certain hardware and software sub-operations used to implement the five fundamental operations is included in this section. Frequent reference will be made to various applications in which these operations are desirable.

The basic operations possible in the present interface are listed in Table 5.1. Thresholding of the video data to determine the state of a resolution cell has been previously discussed. Since the light energy in a correlation peak rather than its peak SNR is often a better basis on which to determine the state of a cell (this is especially true in radar signal processing), this detection feature has been included in the system by adding an integrating filter and counter register in series with the threshold comparator. This integrating detector can be bypassed when the application warrants. Another aid in detection (especially in extracting parallax data from stereo imagery) is the use of a peak sample and hold circuit in this detection path.

In the digital analysis of the transform plane in radar signal processing, it is necessary to remove the central dc term to facilitate detection of the first-order diffraction terms. This involves the subtraction of frames of data or the generation of a pattern (the dc term) and its subsequent subtraction from the entire transform pattern. These operations must be performed prior to the digital analysis in these cases [5.51]. Subtraction of frames of data is also useful in the analysis of wideband folded spectrum data [Ref. 5.6, Chap. 8], especially when the application is change detection. In this case, only those signals which have changed between frames are retained for further analysis. A related use is data tracking in which the time history of a certain radar target (in an azimuth, range, etc., vs time display) is desired. Subtraction and the results of prior analysis are used to locate the target in each frame, while the addition of frames of data is needed to form the composite tracking and time history display. The line generation feature, whereby lines can be constructed between various points on the display, is useful in displaying such output data.

The tracking of radar targets, especially in a multitarget environment, usually requires access to the actual analog level in specific resolution cells (this is proportional to the target's cross section). An A/D converter in the video line, synchronized to the resolution cell generation and detection system, allows transmission of this data to the analysis section of the interface. Similar features are necessary to determine topographic data from imagery in photogrammetric applications.

Actual numerical data from the optical pattern is also necessary. In correlation as well as signal processing applications, the cell resolution and detection systems usually yield clusters of points. The number of points in these clusters must be reduced to determine the coordinates of the center of the cluster (or peak detection coordinate data). The results of one such analysis [5.51] in which optically processed output data for targets at different angles were computed by the interface and compared to the theoretical values with excellent results. In this case, clusters of points were obtained for each target. These clusters were then thinned, the central coordinates determined, digitally converted to target azimuth angles and plotted against the actual values.

The ability to determine the number of cells of binary value "1" for given resolution and detection parameters and the number of points per cluster as a function of threshold level and integration time have proven to be useful pedogogical operations in image analysis. The typical computer print-out in Fig. 5.22 shows the number of points in several clusters, the central coordinates of each cluster and the distance (in resolution cell units) between successive clusters.

RUN
RESOLUTION   120,96

			DISTANCE FROM	
LOCATION		SIZE	PREVIOUS	
47	76	19		
15	82	22	32.55764	
52	99	23	40.71855	
14	42	25	68.50547	
53	54	23	40.80441	
89	19	22	50.20956	

Fig. 5.22. Computer print-out of coordinates of clusters, number of points per cluster and distance from prior point obtained from the optical/digital interface

5.5.4 Software Operation System

Control of the external hardware and software routines for efficient real-time operation of the interface is provided by an extensible software operating system. The high-level computer language used provides:

1) intermediate and deferred program execution modes;
2) arithmetic capability consistent with image and data processing requirements;
3) control features for various peripherals in real time;
4) common image processing features as system procedures;
5) convenient program editing features;
6) extensibility in device control and image processing procedures.

The Digital Equipment Corporation single user Basic System was chosen because of the above features and its low storage requirements. This system was modified to allow added user supplied external routines to be used. The operator works only with the Basic operating system and does not independently execute the software routines nor control the hardware functions. The Basic system can store operator generated programs and use them in deferred mode or by a single operator command. Data and parameters from the operator must be passed through the Basic system and control parameters transferred by Basic variables. These variables are algebraic symbols representing a number. The value assigned to a variable is stored and not changed until so specified in a new statement. Certain external routines control the interface hardware by initiating data flags, calculating control words, and transferring control to the hardware. The eleven external routines used by the system are listed in Table 5.2. Examples of several of these operations and brief descriptions of them follow.

Since Basic only allows the use of one external routine, the interpreter was modified to include an external CALL statement (as in Fortran). When CALL is

Table 5.2. External routines used to implement software operations

IMAGE	THIN
ADD	COORS
LOAD	VIDEO
STORE	LIST
SEARCH	EDIT
GEN	

decoded by the interpreter, control jumps to a dispatch routine. This is the link between the Basic interpreter and the user-supplied machine language syntax routines. It controls the Basic variable parameters and decodes the external routines. The last two routines in Table 5.2 (LIST and EDIT) are used to debug the system.

The LIST routine is used to print the contents of memory on the teletype. It can also be used to print the contents of certain registers and to keep a list of mnemonics and their locations on file. The EDIT routine enables the contents of a memory location to be changed.

The IMAGE routine controls the allocation of image storage and maintains a file of attributes for each image. To efficiently utilize memory space, image storage is allocated dynamically. Image storage is in contiguous core, and no more core than is actually in use is ever reserved.

The ADD routine is obviously used to add or subtract two frames. Since both images are binary, image addition is the logical inclusive OR of the 16 bit words describing the images. In subtraction, each bit in the destination image that corresponds to a set bit in the source image is cleared; this corresponds to the in-place logical AND of the destination image with the complement of the source image.

The LOAD and STORE routines are used to read data from and store data on paper tape. The SEARCH routine enables the coordinates of all points within a search window to be listed on the teletype. The extent of the search window can be varied by the user. This operation is useful in calculating target angles in phased array radar data processing, and in the storage of binary images for later display or analysis. The GEN routine allows the operator or computer to generate any image frame. The THIN routine reduces the bit pattern of a binary-digitized image to a skeleton or single point by a combination of point search and iterative techniques. All picture elements in a cluster are located by a recursive technique once any one element has been located.

The VIDEO routine calculates the control words for the interface, initiates the image digitization process and provides the cell resolution, threshold, integration, and peak detection data. This comprises the interface and control section in Fig. 5.21.

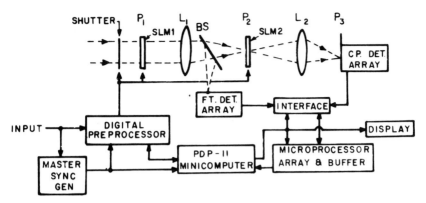

Fig. 5.23. Schematic diagram of a microprocessor-based hybrid optical/digital processor [5.69]

5.5.5 Microprocessor Based Hybrid System

The schematic block diagram of the most recent Carnegie-Mellon University hybrid processor [5.69] is shown in Fig. 5.23. As seen, the top portion of this system is a frequency plane correlator, whereas the lower portion is the digital system. Its two major unique features are 1) the use of a microprocessor array and buffer memory to provide improved flexibility in the operations possible, and 2) the use of a digital preprocessor. These two aspects are thus emphasized in the following discussion.

The digital preprocessor consists of a 2 M-bit store which can be loaded at up to video rates with a total content of a $512 \times 512 \times 8$ bit frame which can be randomly accessed in 30 ms. This component allows the optical processor and its output detectors to operate at a fixed rate of 30 frames per sec (thus greatly simplifying interfacing synchronization) and yet enables the system to operate on input data at other rates (e.g., biomedical, sonar, etc.). The ability to randomly access the contents of this memory is of immense use: in space-variant processing (when various input coordinate transformations preprocessing is required); in format control for signal processing (e.g., array radar, wideband folded spectrum, etc.); in performing digital preprocessing operations (such as edge enhancement, etc.).

The microprocessor system controls the various data detection schemes possible. It consists of an array of Intel 3000 series microprocessor elements arranged as a 16 bit 250 ns cycle time processor with $1 \text{ K} \times 32$ bits of alterable instruction and control storage. This storage can be written into only from the PDP-11. To the microprocessor, it is a read only memory (ROM). The buffer memory is a 2-port, 400 ns, $4 \text{ K} \times 16$ bit store. It is used by the microprocessor to add and subtract frames, compile point time history data, etc. It is usually loaded from the digitized video output and its contents subsequently fed to the PDP-11 for further processing. This maintains the high speed of the microprocessor

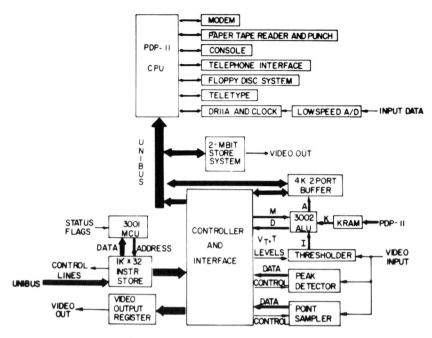

Fig. 5.24. Block diagram of the microprocessor section of the hybrid optical/digital processor of Fig. 5.23 [5.69]

output section at TV frame rates. Only when data is transferred to the PDP-11 over the UNIBUS (400 ns settling time) does the data transfer rate drop to 1.5–2 μs per 16 bit word. The variable cell resolution, data reduction operations such as retention of only the location and amplitude of the key peaks, use of the 4 K buffer memory and inclusion of interframe operations within the microprocessor section, all help to minimize data flow to the PDP-11 and maintain system speed as well as reducing memory space requirements.

A block diagram of the microprocessor section of the interface is shown in Fig. 5.24. The KRAM store is a separate memory containing constants. It is under PDP-11 control and is used in fixed pattern and mask generation and for storage of V_T and T values as noted earlier. The 3002 ALUs (arithmetic logic units) perform the basic mathematical operations of any ALU. The 3001 MCU (microcontrol unit) serves as an instruction look ahead, provides control inputs for the instruction store and checks and sets status flags. The interface and controller consists of decoders and multiplexers to route data under control of flags and lines set by the microprocessor system.

Three main input buses exist. The I bus is the most used. It is fed from the FT and CP detector arrays. The remaining buses are used mainly for data transfer. The M bus is fed from a 4 input multiplexer whose bus inputs are from the A/D, the 4 K buffer, the data lines, etc. Two output buses exist. The A bus contains the 4 K memory address. The D bus is the most used. It feeds all external registers and the 4 K buffer. (The designation is selected by an 8 line decoder fed from 3 bits of the instruction store word. Five of the 32 instruction and control store word bits provide the flag signals needed for the various output devices.)

The operations possible in this microprocessor-based hybrid system are all achieved by various digital methods. The merit of this present system lies in the fact that the operations chosen can be automatically changed by the microprocessor instruction and control and that the microprocessor portion of the system has a full data throughput of TV rates.

5.6 Summary and Conclusion

The design philosophy for a hybrid processor has been presented through examples of several systems and by brief discussions of various applications. The diffraction pattern sampling system (Sect. 5.2) is the most developed and hardened but is also the simplest and least powerful. The optical preprocessor system (Sect. 5.3) uses extensive digital analysis of optical light distributions (usually Fourier transform planes or texture variance images). Considerable advancements have also been made in the use of the digital system to control and correct real-time input devices, and for automated synthesis, viewing, and positioning of holographic spatial filters. The most powerful optical configuration is the correlator. The examples of hybrid correlators for various applications presented in Sect. 5.4 fully demonstrate the vast potential that such systems offer.

This survey of existing hybrid processors provided an understanding of the specifications, operations and analysis required in the optical/digital interface. The interface used in the hybrid system at Carnegie-Mellon University was discussed in considerable detail (Sect. 5.5) with operations, hardware, and software included. A second-generation version of this interface (using an array of microprocessors with a buffer memory in the interface) increases the flexibility, processing, analysis, and display capabilities of the digital system with no loss in speed and with minimal increase in memory storage requirements.

The advantages of a hybrid processor are apparent from the system descriptions and applications presented. The actual processing (or preprocessing) of the input data is performed in the optical section of the processor while the analysis, control, decision making and display aspects of the data processing are handled in the digital section. The specific operations performed in each section depend on the application, but the basic idea in all cases is to use the parallel processing features of an optical system and the programability offered by a digital system and to properly format and reduce the space time bandwidth of the data. This approach results in a system with the best advantages of both optical and digital processing, a more powerful and flexible overall system, and a more cost-effective processor.

Acknowledgments. The assistance of many corporations and individuals who provided data for this chapter is gratefully acknowledged. Special thanks are due to the Department of Defense, Yale

University, Radiation Inc. Division of Harris Corp., Ampex, and Recognition Systems Inc. The author wishes to thank the present and past graduate students in his Information Processing Group at Carnegie-Mellon University and the Office of Naval Research, National Aeronautics and Space Administration, Air Force Office of Scientific Research, and National Science Foundation for financial support of his work included in this chapter.

References

5.1   Dig. Intern. Opt. Comput. Conf., (1974), Zurich, IEEE Cat. No. 74CHO862-3C
5.2   Proc. Soc. Photo. and Instr. Engrs., Vol. 52, San Diego, CA (Aug. 1974)
5.3   Dig. Intern. Opt. Comput. Conf., (1975), Wash. D.C., IEEE Cat. No. 75CHO941-5C
5.4   Proc. Elec. Opt. Sys. Des. Conf., (1975), Anaheim
5.5   B.J.Thompson: Proc. IEEE **65**, 62 (1977)
5.6   D.Casasent (ed.): *Optical Data Processing*: Applications in Topics in Applied Physics, Vol. 23 (Springer, Berlin, Heidelberg, New York 1978)
5.7   J.Bordogna, S.Keneman, J.Amodei: RCA Rev. **33**, 227 (1972)
5.8   S.Lipson: Recyclable Incoherent-to-Coherent Image Converters, in *Advances in Holography*, ed. by N.Farhat (M.Dekker, New York 1976)
5.9   D.Casasent: Proc. IEEE **65**, 143 (1977)
5.10   D.Chen, Z.D.Zook: Proc. IEEE **63**, 1207 (1975)
5.11   Opt. Eng. (Special Issue) **17** (July 1978)
5.12   D.Casasent: "Coherent Light Valves", in *Applied Optics and Optical Engineering*, Vol. VI, ed. by R.Kingslake and B.J.Thompson (M.Dekker, New York 1979)
5.13   Proc. Soc. Photo. and Instr. Engrs. Vol. 83, San Diego, CA (Aug. 1976)
5.14   H.M.Smith (ed.): *Holographic Recording Materials*, Topics in Applied Physics, Vol. 20 (Springer, Berlin, Heidelberg, New York 1977)
5.15   T.S.Huang, H.L.Kasnitz: Proc. Soc. Photo. and Instru. Engrs., Seminar on Computerized Imaging Techniques (1967)
5.15a   T.S.Huang (ed.): *Picture Processing and Digital Filtering*, 2nd ed., Topics in Applied Physics, Vol. 6 (Springer, Berlin, Heidelberg, New York 1979)
5.16   D.Casasent: Laser Focus, 30 (Sept. 1971)
5.17   D.Casasent: Proc. Opt. Comput. Symp. (April 1972) Darien, Conn., IEEE Cat. No. 72CHO687-4C
5.18   D.Casasent: IEEE Trans. C-**22**, 852 (1974)
5.19   J.Rao: J. Opt. Soc. Am. **56**, 1490 (1966); **57**, 798 (1967)
5.20   A.Vander Lugt: IEEE Trans. IT-**6**, 386 (1960)
5.21   H.Kasdan, D.Mead: Proc. Elec. Opt. Sys. Des. Conf., Anaheim, CA (1975)
5.22   G.Lendaris, G.Stanley: Proc. IEEE **58**, 198 (1970)
5.23   F.T.Ulaby, J.McNaughton: Photogram. Eng. 1019 (Aug. 1975)
5.24   G.Lukes: Proc. Soc. Photo. and Instr. Engrs. **45**, 265 (1974)
5.25   P.Kruger, W.B.Thompson, A.F.Turner: IEEE Trans. SMC-**4**, 40 (1974)
5.26   E.L.Hall, R.P.Kruger, A.F.Turner: Opt. Eng. **13**, 250 (1974)
5.27   S.C.Suffin, P.A.Cancilia, H.L.Kasdan: Am. J. Pathol., 1976
5.28   R.Kopp, J.Lisa, J.Mendelsohn, B.Pernick, H.Stone, R.Wohlers: J. Histochem. and Cytochem. **22**, 598 (1974)
5.29   G.C.Salzman, J.M.Crowell, C.A.Goad, K.M.Hansen, R.D.Hibert, P.M.La Bauve, J.C.Martin, M.L.Ingram, P.F.Mullaney: Clinical Chem. **21**, 1297 (1975)
5.30   H.L.Kasdan: Dig. Intern. Opt. Comput. Conf. (1975) Washington D.C., p. 19, IEEE Cat. No. 75CHO941-5C
5.31   H.L.Kasdan, N.Jensen: Proc. Conf. on Comput. Graphics Pat. Recog. and Data Struct. (May 1975) p. 50, IEEE Cat. No. 75CHO981-1C
5.32   N.George, H.Kasdan: Proc. Elec. Opt. Sys. Des. Conf., Anaheim, CA (1975) p. 494
5.33   H.Stark: IEEE Trans. C-**24**, 340 (1975)
5.34   H.Stark: Opt. Engr. **13**, 243 (1974)

5.35 H.Stark: Proc. Intern. Opt. Comput. Conf., Zurich (1974) IEEE Cat. No. 74CHO862-3C
5.36 H.Stark, E.Garcia: Appl. Opt. **13**, 648 (1974)
5.37 D.Lee, H.Stark: J. Opt. Soc. Am. **65**, 191 (1975)
5.38 Harris Electro Optics Center: "Spatial Filter Techniques for Word/Character Recognition",
      Final Tech. Rept. RADA-TR-73-47 (March 1973)
5.39 D.Casasent, A.Furman: Appl. Opt. **17**, 1652 and 1662 (1978)
5.40 S.Almeida, J.Kim-Tzong Eu: Proc. Elec. Opt. Sys. Des. Conf. Anaheim, CA (1975) p. 268
5.41 S.Almeida, J.Kim-Tzong Eu: Appl. Opt. **15**, 510 (1976)
5.42 R.Aldrich, F.Krol, W.Simmons: IEEE Trans. ED-**20**, 1015 (1973)
5.43 R.A.Spraque, P.Nisenson: SPIE **83**, 51 (1976)
5.44 A.Lohmann, D.Paris: Appl. Opt. **7**, 651 (1968)
5.45 G.Goetz: Appl. Phys. Lett. **17**, 63 (1970)
5.46 A.Vander Lugt: Proc. IEEE **62**, 1300 (1974)
5.47 F.B.Rotz, M.O.Greer: Dig. Conf. Laser Eng. and Applic., Washington D.C. (June 1971) p. 65
5.48 R.B.Rotz: Dig. Internat. Opt. Comput. Conf., Washington D.C. (1975) p. 162, IEEE Cat. No.
      75CHO941-5C
5.49 D.Casasent, F.Casasayas: IEEE Trans. ASE-**11**, 65 (1975)
5.50 D.Casasent, F.Casasayas: Appl. Opt. **14**, 1364 (1975)
5.51 D.Casasent: Dig. Internat. Opt. Comput. Conf., Washington D.C. (1975) p. 5, IEEE Cat. No.
      75CHO941-5C
5.52 D.Casasent, E.Klimas: Appl. Opt. **17**, 2058 (1978)
5.53 D.Casasent, W.Sterling: IEEE Trans. C-**24**, 348 (1975)
5.54 D.Casasent, T.Luu: Appl. Opt. **18**, 3307 (1979)
5.55 D.Casasent, F.Caimi, J.Hinds: Opt. Eng. **18**, 716 (1980)
5.56 A.Macovski, S.Ramsey: Opt. Commun. **4**, 319 (1972)
5.57 D.Casasent, A.Furman: Appl. Opt. **17**, 3418 (1978)
5.58 A.Lohmann: Appl. Opt. **16**, 261 (1977)
5.59 W.Rhodes: Appl. Opt. **16**, 265 (1977)
5.60 W.Stoner: Appl. Opt. **16**, 1451 (1977); **17**, 2454 (1978)
5.61 A.Furman, D.Casasent: Appl. Opt. **18**, 660 (1979)
5.62 J.W.Goodman, A.R.Dias, L.M.Woody: Opt. Lett. **2**, 1 (1978)
5.63 D.Psaltis, D.Casasent, M.Carlotto: Opt. Lett. **4**, 348 (1979)
5.64 D.Casasent, D.Psaltis: Opt. Lett. **5**, 395 (1980)
5.65 M.Hu: IRE Trans. IT-**8**, 179 (1962)
5.66 F.B.Rotz: Proc. Elec. Opt. Sys. Des. Conf., Anaheim, CA (1975)
5.67 D.Casasent, P.Rapp: IEEE Trans. C-**27**, 732 (1980)
5.68 C.G.Bell, J.L.Eggert, J.Grason, P.Williams: IEEE Trans. C-**21**, 495 (1972)
5.69 D.Casasent, J.Hackwelder, P.DiLeonardo: Proc. Soc. Photo and Instr. Engrs. **117**, 26 (1977)

6. Linear Space-Variant Optical Data Processing

J.W. Goodman

With 11 Figures

The data-processing capabilities of coherent and incoherent optical analog systems are, for the most part, restricted to the class of *linear* operations. Exceptions to this rule exist, but such nonlinear methods generally require some form of pre-distortion of the input data, or alternatively a nonlinear recording medium somewhere in the optical system.

Within the class of linear operations it is customary to distinguish two subclasses: *space-invariant* operations and *space-variant* operations. These types of operations will be defined more precisely in the section to follow, but for the moment it suffices to describe them in simple terms: a space-invariant operation affects all points within the input field identically, while a space-variant operation can affect different input points differently.

The most common operations for which optical analog data processors are used are convolution, correlation, and Fourier analysis. All three are linear operations, but the first two are space invariant while the last is space variant. Fourier analysis is easily performed with coherent optical systems, due to the inherent Fourier transforming properties of a positive lens, and they represent a rather special case because of this property. Aside from this one operation, the vast majority of optical processors are limited to space-invariant operations.

A wide variety of linear, space-variant operations exists which are of considerable importance in the field of data processing, but which do not fall into the category of easily achieved operations. Nonetheless, interest in space-variant *optical* data processing is fairly recent. A paper of considerable historic importance was published by *Lohmann* and *Paris* in 1965 [6.1]. *Cutrona* recognized at an early date that coherent optical system configurations could be found that would perform space-variant linear operations on one-dimensional inputs [6.2]. Interest in space-variant *digital* image processing was sparked by the works of *Robbins* and *Huang* [6.3], and *Sawchuk* [6.4]. Some early patents concerned with optical space-variant transformations are referred to in a recent review article by *Monahan* et al. [6.5]. In addition, *Stephans* and *Rogers* [6.6] have described methods for Fourier transformation with incoherent light, a space-variant operation. Other more recent references will appear in later discussions.

Our purpose in this chapter is to review many of the basic properties of space-variant linear operations, and to discuss a variety of means by which they can be performed optically. No attempt will be made to summarize all the applications of this type of operation, for such a summary would be a formidable task indeed. However, in the course of our discussions we will mention some of these applications, and hopefully the reader will be able to pursue the subject matter further with the help of the references given.

6.1 Linear, Space-Variant Operations

In order to understand the nature of space-variant operations more thoroughly, it is helpful to first dwell at least briefly on the mathematical formalism used to describe them. A familiarity with this formalism is, in fact, essential to an understanding of the various optical approaches that can be taken to realize such operations.

6.1.1 Input-Output Relationships

A *system* may be regarded in the abstract sense as a mapping of a set of possible input functions into a set of output functions. We may represent a system as an operator $\mathscr{S}\{\cdot\}$ which transforms each possible input function $f(x, y)$ into a unique output function $g(x, y)$,

$$g(x, y) = \mathscr{S}\{f(x, y)\}. \tag{6.1}$$

We call a system *linear* if it obeys two basic properties:

Additivity: $\mathscr{S}\left\{\sum_k f_k(x, y)\right\} = \sum_k g_k(x, y),$

Homogeneity: $\mathscr{S}\{bf(x, y)\} = bg(x, y).$ $\tag{6.2}$

Here $g_k(x, y)$ represents the response that would be obtained if $f_k(x, y)$ *alone* were present at the input, while b is any constant.

An input function $f(x, y)$ can be decomposed into a set of impulses or Dirac delta functions through the well-known "sifting" formula

$$f(x, y) = \iint\limits_{-\infty}^{\infty} f(\xi, \eta)\delta(x - \xi, y - \eta)d\xi d\eta, \tag{6.3}$$

where $\delta(x - \xi, y - \eta)$ is a unit-volume impulse located at $(x = \xi, y = \eta)$. The conditions of additivity and homogeneity imply that for a linear system, the response $g(x, y)$ can be expressed in terms of the responses to all possible input impulse functions

$$g(x, y) = \mathscr{S}\{f(x, y)\} = \iint\limits_{-\infty}^{\infty} f(\xi, \eta)\mathscr{S}\{\delta(x - \xi, y - \eta)\}d\xi d\eta. \tag{6.4}$$

We call $\mathscr{S}\{\delta(x-\xi, y-\eta)\}$ the *impulse response* of the system, noting that it represents the response at (x, y) to a unit impulse at (ξ, η).

The impulse response may be represented in several different forms [6.7, 8]. For our purposes we utilize two different forms, each of which has its own realm of usefulness:

$$\mathscr{S}\{\delta(x-\xi, y-\eta)\} = h_1(x, y; \xi, \eta), \tag{6.5a}$$

$$\mathscr{S}\{\delta(x-\xi, y-\eta)\} = h_2(x-\xi, y-\eta; \xi, \eta). \tag{6.5b}$$

h_1 and h_2 are simply two different symbols for the same quantity. The form h_1 emphasizes that the impulse response depends on both where the impulse was applied [(ξ, η)] and where we observe the output [(x, y)]. The form h_2 uses a different emphasis, noting that this response depends on where the impulse was applied and where, *with respect to that point*, we observe the output [$(x-\xi, y-\eta)$]. Using these two forms of the impulse response, we obtain the input-output relationships

$$g(x, y) = \int\!\!\int_{-\infty}^{\infty} h_1(x, y; \xi, \eta) f(\xi, \eta) d\xi d\eta, \tag{6.6a}$$

$$g(x, y) = \int\!\!\int_{-\infty}^{\infty} h_2(x-\xi, y-\eta; \xi, \eta) f(\xi, \eta) d\xi d\eta. \tag{6.6b}$$

Such relationships are known as *superposition integrals*.

It is now possible to distinguish between space-invariant and space-variant linear operations. If the impulse response (h_1 or h_2) depends only on the difference of coordinates $(x-\xi, y-\eta)$, then the system is called *space-invariant*. In this case

$$\left.\begin{matrix} h_1(x, y; \xi, \eta) \\ h_2(x-\xi, y-\eta; \xi, \eta) \end{matrix}\right\} = h(x-\xi, y-\eta). \tag{6.7}$$

Stated in physical terms, for such systems the impulse response depends only on where we observe with respect to the position of the applied impulse, not on the absolute position of the impulse.

For the case of a space-invariant linear system, the input-output relationship takes the more familiar form of a *convolution integral*

$$g(x, y) = \int\!\!\int_{-\infty}^{\infty} h(x-\xi, y-\eta) f(\xi, \eta) d\xi d\eta \tag{6.8}$$

which is represented in shorthand notation by $g = h * f$. For comparison purposes, we note that the cross correlation integral takes the closely related form

$$g(x, y) = \iint\limits_{-\infty}^{\infty} h^*(\xi - x, \eta - y) f(\xi, \eta) d\xi d\eta \tag{6.9}$$

which is equivalent to the convolution of f with $h^*(-x, -y)$, i.e., h reflected about the origin and conjugated. In this sense, then, the cross-correlation integral is, like the convolution integral, a space-invariant operation.

Our interest in this chapter is with operations that must be expressed in the form (6.6a) or (6.6b), and can *not* be expressed as a simple convolution.

6.1.2 Frequency Domain Representations

In this section we derive various frequency-domain representations of both the output $g(x, y)$ and its Fourier transform

$$\hat{g}(v_X, v_Y) = \iint\limits_{-\infty}^{\infty} g(x, y) e^{-j2\pi(v_X x + v_Y y)} dx dy \tag{6.10}$$

for a linear space-variant system. We consider both of the representations of (6.6a and b). The results we obtain will later be used to find optical implementations of space-variant operations, and will also be useful in formulating discrete versions of the continuous operations described by (6.6).

We begin with the representation

$$g(x, y) = \iint\limits_{-\infty}^{\infty} h_1(x, y; \xi, \eta) f(\xi, \eta) d\xi d\eta \tag{6.6a}$$

and attempt to express it in terms of frequency-domain quantities. Let $f(\xi, \eta)$ be replaced by its Fourier integral representation

$$f(\xi, \eta) = \iint\limits_{-\infty}^{\infty} \hat{f}(v_3, v_4) e^{j2\pi(v_3 \xi + v_4 \eta)} dv_3 dv_4 . \tag{6.11}$$

Substituting this expression in (6.6a), and integrating first with respect to (ξ, η), we obtain

$$g(x, y) = \iint\limits_{-\infty}^{\infty} \hat{h}_1(x, y; -v_3, -v_4) \hat{f}(v_3, v_4) dv_3 dv_4 , \tag{6.12}$$

where

$$\hat{h}_1(x, y; -v_3, -v_4) = \iint\limits_{-\infty}^{\infty} h_1(x, y; \xi, \eta) e^{j2\pi(v_3 \xi + v_4 \eta)} d\xi d\eta . \tag{6.13}$$

The representation (6.12) is a partial frequency-domain representation, for the kernel $\hat{h}_1$ remains a function of both the output variables (x, y) and the frequency variables (v_3, v_4).

To obtain a representation of g with a kernel that depends only on frequency variables, we return to (6.6a) and replace h_1 by a four-dimensional mixed-sign Fourier integral representation

$$h_1(x, y; \xi, \eta) = \int\!\!\int\!\!\int\!\!\int_{-\infty}^{\infty} \hat{\hat{h}}_1(v_1, v_2; -v_3, -v_4)$$

$$\cdot \exp[j2\pi(v_1 x + v_2 y - v_3 \xi - v_4 \eta)] \, dv_1 dv_2 dv_3 dv_4, \tag{6.14}$$

where

$$\hat{\hat{h}}_1(v_1, v_2; -v_3, -v_4) \triangleq \int\!\!\int\!\!\int\!\!\int_{-\infty}^{\infty} h_1(x, y; \xi, \eta)$$

$$\cdot \exp[-j2\pi(v_1 x + v_2 y - v_3 \xi - v_4 \eta)] \, dx dy d\xi d\eta. \tag{6.15}$$

Substituting (6.14) in (6.6a) and performing the integrals with respect to ξ and η we obtain

$$g(x, y) = \int\!\!\int\!\!\int\!\!\int_{-\infty}^{\infty} \hat{\hat{h}}_1(v_1, v_2; -v_3, -v_4) \hat{f}(v_3, v_4)$$

$$\cdot \exp[j2\pi(v_1 x + v_2 y)] \, dv_1 dv_2 dv_3 dv_4. \tag{6.16}$$

Thus we have found a full four-dimensional frequency domain representation for the output.

Similar forms for $g(x, y)$ can be found starting from (6.6b),

$$g(x, y) = \int\!\!\int_{-\infty}^{\infty} h_2(x - \xi, y - \eta; \xi, \eta) f(\xi, \eta) d\xi d\eta. \tag{6.6b}$$

In this case we represent h_2 as a four-dimensional mixed-sign Fourier integral,

$$h_2(x - \xi, y - \eta; \xi, \eta) = \int\!\!\int\!\!\int\!\!\int_{-\infty}^{\infty} \hat{\hat{h}}_2(v_1, v_2; -v_3, -v_4)$$

$$\cdot \exp\{j2\pi[v_1(x - \xi) + v_2(y - \eta)$$

$$- v_3 \xi - v_4 \eta]\} \, dv_1 dv_2 dv_3 dv_4. \tag{6.17}$$

Substitution of this expression in (6.6b) and integration with respect to (ξ, η) yields

$$g(x, y) = \int\!\!\int\!\!\int\!\!\int_{-\infty}^{\infty} \hat{\hat{h}}_2(v_1, v_2; -v_3, -v_4) \hat{f}(v_1 + v_3, v_2 + v_4)$$

$$\cdot \exp[j2\pi(v_1 x + v_2 y)] \, dv_1 dv_2 dv_3 dv_4. \tag{6.18}$$

From this expression we see directly that the 2-D Fourier spectrum of the output $g(x, y)$ is given by

$$\hat{g}(v_1, v_2) = \int\int_{-\infty}^{\infty} \hat{h}_2(v_1, v_2; -v_3, -v_4) \hat{f}(v_1 + v_3, v_2 + v_4) dv_3 dv_4 . \tag{6.19}$$

If the system of interest happens to be space *invariant*, then

$$h_2(x - \xi, y - \eta; \xi, \eta) = h(x - \xi, y - \eta) \cdot 1 \tag{6.20}$$

and the 4-D transform $\hat{h}_2$ takes the form

$$\hat{h}_2(v_1, v_2; -v_3, -v_4) = \hat{h}(v_1, v_2)\delta(v_3)\delta(v_4), \tag{6.21}$$

where $\hat{h}$ is the 2-D transform of h. Substitution of (6.21) in (6.19) then yields

$$\hat{g}(v_1, v_2) = \hat{h}(v_1, v_2)\hat{f}(v_1, v_2) \tag{6.22}$$

which is the familiar multiplicative transform relation that applies for space-invariant systems.

Table 6.1 summarizes the various frequency-domain representations of a space-variant impulse response.

Table 6.1

$$\hat{h}_1(x, y; -v_3, -v_4) = \int\int_{-\infty}^{\infty} h_1(x, y; \xi, \eta)$$
$$\cdot \exp[j2\pi(v_3\xi + v_4\eta)] d\xi d\eta$$

$$\hat{h}_1(v_1, v_2; -v_3, -v_4) = \int\int\int\int_{-\infty}^{\infty} h_1(x, y; \xi, \eta)$$
$$\cdot \exp[-j2\pi(v_1 x + v_2 y - v_3\xi - v_4\eta)] dx dy d\xi d\eta$$

$$\hat{h}_2(v_1, v_2; -v_3, -v_4) = \int\int\int\int_{-\infty}^{\infty} h_2(\alpha, \beta; \xi, \eta)$$
$$\cdot \exp[-j2\pi(v_1\alpha + v_2\beta - v_3\xi - v_4\eta)] d\alpha d\beta d\xi d\eta$$

6.1.3 Discrete Representation of Linear Space-Variant Operations[1]

Our purpose in this section is to derive a discrete version of the superposition integral relating the input and output of a space-variant linear system. Certain optical approaches to realizing space-variant processors are based on the

---

1 For a related discussion, see [6.38].

results of this discrete approach. We begin with a one-dimensional analysis, and later consider the two-dimensional case. Attention will be restricted to superposition integrals with kernels in the form of h_2.

We suppose at the start that the one-dimensional input $f(\xi)$ contains no frequency components outside the interval $(-B_i, B_i)$ in the frequency domain. Thus $\hat{f}(v_3) = 0$ when $v_3 > B_i$. We refer to B_i as the *input* bandwidth. By the Whittaker Shannon sampling theorem [6.9], the input $f(\xi)$ can be represented in terms of its samples taken at intervals of $1/2B_i$,

$$f(\xi) = \sum_{m=-\infty}^{\infty} f\left(\frac{m}{2B_i}\right) \operatorname{sinc}\left[2B_i\left(\xi - \frac{m}{2B_i}\right)\right]. \tag{6.23}$$

The bandwidth of the output $g(x)$ can be found by examining the expression (6.19) for the spectrum $\hat{g}$ of the output. In the one-dimensional case this equation becomes

$$\hat{g}(v_1) = \int_{-\infty}^{\infty} \hat{h}(v_1; -v_3)\hat{f}(v_1 + v_3)dv_3. \tag{6.24}$$

Now let the width of $\hat{h}$ along the v_3 axis be limited to the interval $(-B_v, +B_v)$; we call the bandwidth B_v the *variational bandwidth* (cf. [6.10]), for its reciprocal is a measure of the distance over the input line for which the system is space invariant. The output bandwidth can be determined by examining (6.24). $\hat{g}(v_1)$ must go to zero when $\hat{f}(v_1 + v_3) = 0$, and the highest value of v_1 corresponding to a non-zero value of $\hat{f}$ occurs when $v_3 = -B_v$. At this point $v_1 - B_v = B_i$, or $v_1 = B_v + B_i$. Of course, $\hat{h}(v_1; -v_3)$ could go to zero at a lower value of v_1. Thus, an expression for the output bandwidth is given by the inequality

$$B_o \leq B_v + B_i. \tag{6.25}$$

$g(x)$ can be represented exactly as

$$g(x) = \sum_{k=-\infty}^{\infty} g\left(\frac{k}{2B_o}\right) \operatorname{sinc}\left[2B_o\left(x - \frac{k}{2B_o}\right)\right]. \tag{6.26}$$

The impulse response $h_2(x - \xi; \xi)$ is really a function of only two variables, $h_2(z; \xi)$, where $z \triangleq x - \xi$. The bandwidth of the impulse response $h_2(z, \xi)$ can be determined by observing when its two-dimensional Fourier transform $\hat{h}(v_1; -v_3)$ goes to zero. The width of $\hat{h}$ in the v_3 dimension [corresponding to the bandwidth of $h(z, \xi)$ in the ξ direction] has been given previously as B_v. The width of $\hat{h}$ in the v_1 direction can only be said to be greater than or equal to the width of $\hat{g}(v_1)$ as seen from (6.24). If we call this bandwidth B_H, we have $B_H \geq B_o$.

In principle we could sample the impulse response every $1/2B_H$ for its first variable and every $1/2B_v$ for its second variable, and we could sample the output every $1/2B_o$ and the input every $1/2B_i$. However, to avoid an awkward interpolation step, which would destroy the simple matrix-vector form we desire, it is found to be necessary to sample both the second variable of h_2 and f at the *same* rate. Similarily, it is necessary to sample the first variable of h_2 and g at the same rate. Since there is no harm done by sampling a function faster than the limiting Nyquist rate, and we know that $B_H \geq B_o$, we can simply sample both the first variable of h_2 as well as the output g at the rate $2B_H$. Similarly, we sample the second variable of h_2 and f at the same rate, that rate being the larger of the two rates. Assuming that $B_i \geq B_v$, we choose the rate $2B_i$, in which case the impulse response can be expressed as

$$h_2(x-\xi;\xi) = \sum_{k=-\infty}^{\infty} \sum_{l=-\infty}^{\infty} h_2\left(\frac{k}{2B_H} - \frac{l}{2B_i}; \frac{l}{2B_i}\right)$$

$$\cdot \mathrm{sinc}\left[2B_H\left(x - \frac{k}{2B_H}\right)\right] \mathrm{sinc}\left[2B_i\left(\xi - \frac{l}{2B_i}\right)\right]. \tag{6.27}$$

Substituting (6.23 and 27) in the 1-D superposition integral (6.6b), we obtain

$$g(x) = \sum_{k=-\infty}^{\infty} \sum_{l=-\infty}^{\infty} \sum_{m=-\infty}^{\infty} h_2\left(\frac{k}{2B_H} - \frac{l}{2B_i}; \frac{l}{2B_i}\right) f\left(\frac{m}{2B_i}\right)$$

$$\cdot \mathrm{sinc}\left[2B_H\left(x - \frac{k}{2B_H}\right)\right]$$

$$\cdot \int_{-\infty}^{\infty} \mathrm{sinc}\left[2B_i\left(\xi - \frac{m}{2B_i}\right)\right] \mathrm{sinc}\left[2B_i\left(\xi - \frac{l}{2B_i}\right)\right] d\xi. \tag{6.28}$$

The integral has value $\dfrac{1}{2B_i}$ for $m = l$, and zero otherwise. The expression for $g(x)$ thus becomes

$$g(x) = \sum_{k=-\infty}^{\infty} \sum_{l=-\infty}^{\infty} h_2\left(\frac{k}{2B_H} - \frac{l}{2B_i}; \frac{l}{2B_i}\right) \left[\frac{1}{2B_i} f\left(\frac{l}{2B_i}\right)\right]$$

$$\cdot \mathrm{sinc}\left[2B_H\left(x - \frac{k}{2B_H}\right)\right]. \tag{6.29}$$

Comparison of (6.26 and 29) [where the output in (6.26) has been oversampled at the rate $2B_H$] shows that

$$g\left(\frac{k}{2B_H}\right) = \sum_{l=-\infty}^{\infty} h_2\left(\frac{k}{2B_H} - \frac{l}{2B_i}; \frac{l}{2B_i}\right) \cdot \frac{1}{2B_i} f\left(\frac{l}{2B_i}\right). \tag{6.30}$$

Finally, defining

$$g_k = g\left(\frac{k}{2B_H}\right)$$

$$h_{kl} = h_2\left(\frac{k}{2B_H} - \frac{l}{2B_i}; \frac{l}{2B_i}\right)$$

$$f_l = \frac{1}{2B_i} f\left(\frac{l}{2B_i}\right), \tag{6.31}$$

we have the simple relationship between sample values

$$g_k = \sum_{l=-\infty}^{\infty} h_{kl} f_l. \tag{6.32}$$

If we apply, at the input of the filter, a function which is nonzero only on an interval[2] of length L_i, then

$$g_k = \sum_{l=0}^{M-1} h_{kl} f_l, \tag{6.33}$$

where $M = 2B_i L_i$. The resulting output will in general be nonzero on a somewhat longer interval (length L_o), and hence the number of nonzero output samples is $N = 2L_o B_H$. The relationship (6.33) can now be expressed as a matrix product

$$\begin{bmatrix} g_0 \\ g_1 \\ \vdots \\ g_{N-1} \end{bmatrix} = \begin{bmatrix} h_{00} & h_{01} & \cdots & h_{0,M-1} \\ h_{10} & h_{11} & \cdots & h_{1,M-1} \\ \vdots & & & \\ h_{N-1,0} & h_{N-1,1} & & h_{N-1,M-1} \end{bmatrix} \begin{bmatrix} f_0 \\ f_1 \\ \vdots \\ f_{M-1} \end{bmatrix}. \tag{6.34}$$

In a more abbreviated form we write this matrix equation as

$$g = [H] f, \tag{6.35}$$

where g is of length N, f of length M, and $[H]$ is an $M \times N$ matrix.

We conclude that a one-dimensional superposition integral can, to a good approximation, be equivalently expressed as a matrix-vector product. A similar result is possible in the two-dimensional case. While the strictly analogous representation would be to write the output g as a 2-D matrix, the input f as a 2-D matrix, and the impulse response as a 4-D tensor, more commonly the

---

2 Such a function cannot be strictly bandlimited, but it can be approximately so.

input and output samples are scanned column by column to compose lexico-graphically ordered column vectors, the input vector being of length M^2 and the output vector of length N^2. The system matrix becomes a block matrix of size $M^2 \times N^2$. For details the reader may wish to consult *Andrews* and *Hunt* [6.11]. Again, the simple matrix-vector product of (6.35) serves as an adequate description of the filtering operation.

One final matrix representation is also of some use. If the impulse response has rectangular separability, i.e.,

$$h_2(x-\xi, y-\eta; \xi, \eta) = h_{2X}(x-\xi; \xi) h_{2Y}(y-\eta; \eta), \tag{6.36}$$

then the $M^2 \times N^2$ block system matrix can be expressed as the Kronecker product of two $M \times N$ matrices

$$[H] = [H]_X \otimes [H]_Y. \tag{6.37}$$

In this case the stacked matrix-vector equation (6.35) can be reduced to

$$[G] = [H]_X [F] [H]_Y^t, \tag{6.38}$$

where $[F]$ is an $M \times M$ matrix of samples of the 2-D input, and $[G]$ is an $N \times N$ matrix of samples of the 2-D output. For further details, the reader may again consult reference [6.11]. We conclude that in the case of a separable 2-D linear system, a fully two-dimensional matrix representation is possible in which the output 2-D matrix is a product of three 2-D matrices, including the 2-D input matrix.

6.2 Optical Implementations of Space-Variant Linear Operations

We consider now a wide variety of optical methods for realizing space-variant linear operations. Solutions based on ray optics are discussed first. Next, a class of techniques based on holographic multiplexing of filters is described. Thirdly, a variety of methods capable of performing *one-dimensional* linear space-variant operations are considered. Fourth, a method known as coordinate transformation processing is described. Lastly, we consider several methods for realizing space-variant operations which are based on a discrete or matrix formulation of the filtering problem.

6.2.1 Implementations Based on Simultaneous Control of the Position and Direction of Rays

To implement a two-dimensional linear space-variant operation, we must somehow realize an impulse response that depends on *four* independent variables. Herein lies the chief difficulty encountered in realizing such operations optically, for we usually have only two spatial variables (the transverse coordinates) available, and certainly we can never use more than three spatial variables. Of course the possibility of using wavelength, polarization, or time as an independent variable is always present, but usually systems based on these variables become either very complex or very slow when inputs of large space-bandwidth product are to be processed.

If we restrict ourselves to the realm of geometrical or ray optics, the number of independent variables available to us can be expanded[3]. A ray passing through a plane at any particular coordinates (ξ, η) is also characterized by its two direction cosines (u, v), which can be controlled quite independently of ξ and η. Of course, when diffraction is taken into account, an uncertainty relationship exists between the variables (ξ, η) and (u, v), and it becomes impossible to control both position and direction of a ray with arbitrary precision. However, for the moment we neglect this limitation.

Several examples of space-variant optical systems that rest on independent control of position and direction can be found. Perhaps the simplest example is the coherent optical method for performing coordinate transformations demonstrated by *Bryngdahl* [6.12, 13]. Suppose that we wish to subject an image to an invertible coordinate transformation described by

$$x = Z_1(\xi, \eta) \qquad \xi = z_1(x, y)$$

$$y = Z_2(\xi, \eta) \qquad \eta = z_2(x, y). \tag{6.39}$$

Such a transformation may be regarded as a linear, space-variant filtering operation with impulse response

$$h_1(x, y; \xi, \eta) = |\mathcal{J}| \delta[z_1(x, y) - \xi, z_2(x, y) - \eta], \tag{6.40}$$

where $|\mathcal{J}|$ is the Jacobian of the transformation, given by

$$|\mathcal{J}| = \left\| \begin{array}{cc} \dfrac{\partial \xi}{\partial x} & \dfrac{\partial \xi}{\partial y} \\[2mm] \dfrac{\partial \eta}{\partial x} & \dfrac{\partial \eta}{\partial y} \end{array} \right\|. \tag{6.41}$$

(The symbol $\| \cdot \|$ indicates the magnitude of the determinant.) From a purely geometrical point of view, a method such as shown in Fig. 6.1a could in principle be used to realize this type of operation. A thin glass plate is manufactured with a surface profile such that, due to refraction, the ray

---

3 I am indebted to Professor G. L. Rogers, whose comments in a lecture provided much of the insight on which this discussion is based.

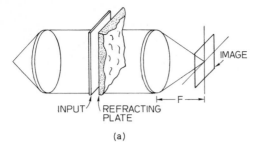

(a)

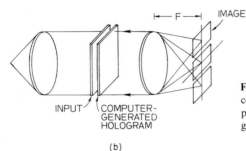

(b)

Fig. 6.1a, b. Optical methods for realizing a coordinate transformation with (**a**) a glass plate of variable thickness; (**b**) a computer generated hologram

emerging at coordinates (ξ, η) is traveling with direction cosines

$$u = \frac{Z_1(\xi, \eta)}{F}$$

$$v = \frac{Z_2(\xi, \eta)}{F},$$

(6.42)

where F is the focal length of the lens that follows. The lens maps all rays traveling with direction cosines (u, v) into a unique location (x, y) in the rear focal plane, with

$$x = Fu$$
$$y = Fv.$$

(6.43)

Thus the ray passing through point (ξ, η) arrives at focal plane coordinates $u = Z_1(\xi, \eta)$, $v = Z_2(\xi, \eta)$.

For most interesting geometrical transformations, the glass plate of Fig. 6.1a would be impossible to realize due to difficulties of manufacturing. *Bryngdahl* [6.12, 13] realized geometrical transformations by replacing the glass plate with a computer-generated hologram, as shown in Fig. 6.1b. This hologram is a generalized grating in which the grating frequency varies as a function of the coordinates (ξ, η). In fact the local spatial frequencies of the

Fig. 6.2. Experimental demonstrations of coordinate transformations generated by gratings with an error function frequency variation [6.12]. (Courtesy of *O. Bryngdahl*)

grating at location (ξ, η) are chosen to be

$$v_\xi = \frac{Z_1(\xi, \eta)}{\lambda F}$$

$$v_\eta = \frac{Z_2(\xi, \eta)}{\lambda F},$$

(6.44)

where λ is the wavelength of the light. The hologram will then impart, to the first diffraction order of the transmitted light, direction cosines given by (6.42), as desired[4].

Examples of typical experimental results obtained by *Bryngdahl* are shown in Fig. 6.2. In these cases the coordinate transformation was generated by gratings having an error function frequency variation.

---

4 It is somewhat ironic that this system, which depends on geometrical optics as its basic foundation, also depends on diffraction to control ray directions.

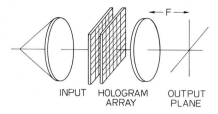

INPUT   HOLOGRAM   OUTPUT
          ARRAY   PLANE

Fig. 6.3. Generation of a fully general space-variant system using a computer-generated hologram array

An extension of *Bryngdahl*'s technique can readily be imagined such that fully general impulse responses $h_1(u, v; \xi, \eta)$ can be realized, rather than simply coordinate transformations. The input may be considered to consist of an array of small squares, or pixels, each with a particular value of amplitude transmittance. The hologram is constructed such that, behind the pixel with central coordinates (ξ, η) lies a small hologram which generates, in the focal plane of the lens, an amplitude distribution $h_1(u, v; \xi, \eta)$. Thus each input pixel is followed by a small hologram which generates precisely the right impulse response for that pixel, as illustrated in Fig. 6.3. This technique is closely related to, but conceptually simpler than, that described by *Gibin* and *Tverdokhleb* [6.14].

The limitations of this holographic approach to generation of a variable impulse response are found to be quite severe when diffraction is taken into account. Suppose that we have $N \times N$ pixels in the input, and we wish to create an image which likewise has $N \times N$ pixels. In the most general case, a hologram residing behind any single input pixel must be capable of sending light to any of the $N \times N$ output pixels. To achieve $N \times N$ resolvable directions of diffraction, each hologram must be capable of containing $N \times N$ fringe periods. Hence, to map an $N \times N$ input into an $N \times N$ output requires a total hologram with $N^2 \times N^2$ resolvable elements. The optical system following this hologram must have a space-bandwidth product of at least $N^2 \times N^2$. Thus if the optical system is limited to, e.g., 3000×3000 resolvable spots, the input and output must be limited to about 55×55 pixels. The fundamental reason for this limitation lies in the uncertainty relationship linking position and direction.

6.2.2 Implementations Based on Holographic Multiplexing Techniques

In order to realize a space-variant coherent optical processor, we wish to create a situation in which different input points evoke different impulse responses at the output. One approach to realizing such a situation is to multiplex a number of holographic filters on a single recording medium, and in addition to code the recorded filters in such a way that they are selectively activated by different points or groups of points in the input field.

The use of thick holograms for recording the filters provides one mechanism for achieving this type of multiplexing (see *Deen* et al. [6.15]). The optical processing system is identical with a conventional coherent optical processor, except that a thick holographic filter is used in the frequency domain, as shown

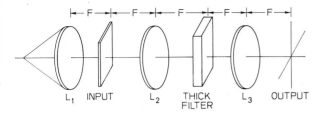

Fig. 6.4. Space-variant processes using a thick holographic filter

in Fig. 6.4. The filter contains a multitude of superimposed holograms, each recorded with a reference beam having a different angle of incidence. An input point located at (ξ, η) illuminates the filter with a plane wave having an angle of incidence uniquely determined by (ξ, η). Ideally, the Bragg effect is depended upon to assure that this incident plane wave will evoke an image response $h_1(x, y; \xi, \eta)$ generated by only one of the stored holograms. A different input point, located at (ξ', η'), selectively activates another of the stored holograms to evoke a response $h_1(x, y; \xi', \eta')$ appropriate for that point. The Bragg extinction phenomenon is depended upon to assure that there is no significant "cross-talk" between the holograms.

The thick hologram approach does not suffer from the same limitations on space-bandwidth product suffered by the systems described in Sect. 6.2.1. However, it has its own particular set of limitations, in this case rather fundamental limitations associated with the phenomenon of Bragg reflection. The most serious of these limitations arises from the fact that there is not one unique angle that will evoke a response from a given hologram stored in a thick filter, but rather, an entire cone of angles, all of which satisfy the Bragg condition (*Krile* et al. [6.16]). Hence it is fundamentally impossible, even with a filter of unlimited thickness, to assure that every different input point evokes a different impulse response. Therefore, the technique is limited to rather dilute input fields, or equivalently to relatively small numbers of input points. Further difficulties arise in attempting to multiplex a large number of holograms on one recording medium due to limitations of the medium's dynamic range. Finally, the finite thickness of the medium effectively "blurs" the selectivity of the Bragg cones, prohibiting the selective recall of extremely large numbers of different impulse responses.

An alternative approach to multiplexing holograms is to use phase-coded reference beams [6.16]. In this case, each filter impulse response is recorded using a reference beam at a different angle, but in addition each reference beam is distorted by passage through a unique portion of a diffuser. The input to be filtered is followed by the same diffuser used in the holographic recording process. To the extent that a single reference beam incident on the recording medium has an autocorrelation function consisting of a sharp narrow spike, while two different incident reference beams have a cross-correlation function that is nearly zero everywhere, each input point source will evoke an output which consists primarily of the required impulse response. Thus the different stored impulse responses are selectively evoked by different input points.

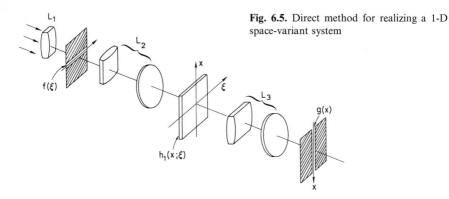

Fig. 6.5. Direct method for realizing a 1-D space-variant system

The limitations associated with this technique have not yet been fully explored, but two factors place important limits on the number of different impulse responses that can be simultaneously stored. One limit is posed by the fact that the cross-correlation function of two diffusers is never zero, but rather is a broad diffuse function of relatively small amplitude. This cross correlation contributes a background which becomes more and more severe as the number of stored holograms and the number of input points increase. The rise of this background can be partially suppressed by use of a thick recording medium. A second limit is associated with the finite dynamic range of the recording medium, which prevents an arbitrarily large number of holograms from being multiplexed.

6.2.3 Techniques for Performing One-Dimensional Linear Space-Variant Operations

A variety of optical techniques are known for performing linear space-variant filtering of one-dimensional inputs. The fact that the space-variant impulse response depends only on two variables, rather than four, makes the one-dimensional case considerably easier to attack than the full two-dimensional case.

The earliest description of a system capable of performing space-variant linear operations on 1-D inputs is that of *Cutrona* [6.2]. More recent work by *Rhodes* and *Florence* [6.17] stimulated further thinking along related lines. Most recently, two different groups have simultaneously published extensions of Cutrona's ideas [6.18, 19].

A direct method for realizing the one-dimensional superposition integral

$$g(x) = \int\limits_{-\infty}^{\infty} h_1(x; \xi) f(\xi) d\xi \tag{6.45}$$

is shown in Fig. 6.5. The input function $f(\xi)$ is introduced as an amplitude transmittance variation within a horizontal slit in plane P_1. The cylindrical lens

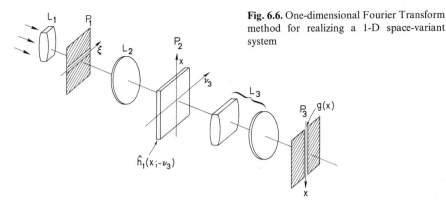

Fig. 6.6. One-dimensional Fourier Transform method for realizing a 1-D space-variant system

L_1 is used simply to supply an intense line of illumination. The spherical-cylindrical optics L_2 image the input function horizontally, but spread the input vertically. Hence a single point source in the input slit illuminates plane P_2 with a vertical line of light. A mask with amplitude transmittance $h_1(x; \xi)$ is placed in plane P_2. The light leaving the mask passes next through the spherical-cylindrical optics L_3, which image in the vertical direction and Fourier transform in the horizontal direction. The output slit in plane P_3 selects the zero-frequency component of that transform, thus performing the integral in the ξ-direction required by (6.45).

As described in the literature, this system is usually assumed to utilize coherent light. However, incoherent light can also be used, in which case the words "amplitude transmittance" above should be changed to "intensity transmittance". An incoherent version of this system will be described more fully in Sect. 6.2.5.

A second method for performing the 1-D superposition integral is illustrated in Fig. 6.6. This system uses one spherical lens and one spherical-cylindrical lens combination (in addition to the cylindrical illumination lens). The filter in this case is holographic, so coherent light must be used. As an aid to understanding this system, we note that $g(x)$ can be expressed as the 1-D analog of (6.12).

$$g(x) = \int_{-\infty}^{\infty} \hat{h}_1(x; -v_3)\hat{f}(v_3)dv_3, \tag{6.46}$$

in which $\hat{f}(v_3)$ is the 1-D Fourier transform of the input $f(\xi)$, and $\hat{h}_1(x; v_3)$ is the 1-D transform of $h_1(x; \xi)$ with respect to ξ. Again lens L_1 simply provides a line of illumination. The input $f(\xi)$ is again entered in plane P_1. Lens L_2 performs a two-dimensional Fourier transform of the input, producing the 1-D transform $\hat{f}(v_3)$, spread uniformly in the vertical direction, in plane P_2. A mask with amplitude transmittance $\hat{h}_1(x; -v_3)$ is placed in P_2. The spherical-cylindrical lens combination L_3 images vertically but Fourier transforms horizontally. A vertical slit in plane P_3, centered in one of the first-order diffraction output components, provides the output function $g(x)$.

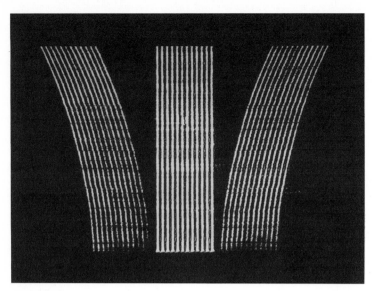

Fig. 6.7. Example of the output of the system of Fig. 6.6 (output slit removed)

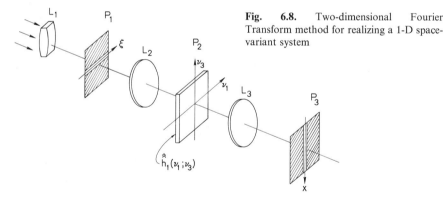

Fig. 6.8. Two-dimensional Fourier Transform method for realizing a 1-D space-variant system

An example of the output of such a system is shown in Fig. 6.7 (without the vertical output slit). The input is in this case a Ronchi ruling (20 lines/cm), and the filter is designed to perform a simple coordinate distortion

$$g(x) = f(\sqrt{x}). \tag{6.47}$$

The output consists of a zero-order image of the input (spread vertically) and two first-order images. A vertical output slit properly placed in one of the two first-order images will indeed produce the desired output.

A third and final method for performing 1-D filtering is shown in Fig. 6.8. This method is based on a 1-D version of (6.16),

$$g(x) = \iint\limits_{-\infty}^{\infty} \hat{h}_1(v_1; -v_3)\hat{f}(v_3) e^{j2\pi v_1 x} dv_1 dv_3, \tag{6.48}$$

Fig. 6.9. General form of a coordinate transformation processor

where $\hat{h}_1(v_1 ; v_3)$ is the 2-D Fourier transform of $h_1(x ; \xi)$. Lens L_1 again supplies a line of illumination, and the input $f(\xi)$ is entered in the slit of plane P_1. Spherical lens L_2 Fourier transforms in two dimensions, producing $\hat{f}(v_3)$ spread out vertically in plane P_2. A holographic mask with effective amplitude transmittance $\hat{h}(v_1 ; -v_3)$ is placed in this plane. Spherical lens L_3 performs a 2-D Fourier transformation, and the output slit, properly located in the plane P_3, yields the desired output $g(x)$.

For further discussion of the methods described in this section, the reader is referred to [6.19]. Related publications may also be of interest [6.20, 21].

6.2.4 Coordinate Transformation Processing

A certain class of linear space-variant filtering operations has a remarkable property: if the input coordinate system is subjected to a properly chosen coordinate transformation, and if the output coordinate system is also properly transformed, the space-variant operation becomes a space-invariant operation. Figure 6.9 illustrates the sequence of operations involved. The use of this type of filtering, which is known as "coordinate transformation processing", was pioneered in the digital domain by *Robbins* and *Huang* [6.3] and *Sawchuk* [6.4]. Most extensive use of this type of filtering in the optical domain has been made by *Casasent* and *Psaltis* [6.22, 23].

We illustrate this type of operation with a straightforward one-dimensional example (see *Bracewell* [6.24]). The Abel transform is defined by

$$g(x) = 2 \int_x^\infty \frac{f(\xi)\xi d\xi}{\sqrt{\xi^2 - x^2}}. \tag{6.49}$$

Clearly from the integral definition, this is a linear operation. However, it is also quite clear that the impulse response

$$h(x ; \xi) = \frac{2\xi}{\sqrt{\xi^2 - x^2}} U(\xi - x) \tag{6.50}$$

[where $U(\xi)$ is a unit step function] depends on both ξ and x independently. Therefore this operation is space variant.

The Abel transform can be converted to a space-invariant form if the following transformations of input and output coordinates are made:

$$\bar{\bar{\xi}} = \xi^2$$
$$\bar{x} = x^2 .$$

(6.51)

Defining

$$G(\bar{x}) \triangleq g(\sqrt{\bar{x}})$$
$$F(\bar{\bar{\xi}}) \triangleq f(\sqrt{\bar{\bar{\xi}}})$$

(6.52)

we obtain

$$G(\bar{x}) = \int_{-\infty}^{\infty} \frac{U(\bar{\bar{\xi}} - \bar{x})}{\sqrt{\bar{\bar{\xi}} - \bar{x}}} F(\bar{\bar{\xi}}) d\bar{\bar{\xi}} ,$$

(6.53)

which *does* represent a space-invariant operation.

A closely related simplification results if coordinate transformations are applied to the complex Mellin transform [6.24] defined by

$$g(x) = \int_{0}^{\infty} \xi^{jx-1} f(\xi) d\xi .$$

(6.54)

In this case we let $\xi = \exp(-\bar{\xi})$ and define

$$F(\bar{\xi}) \triangleq f(e^{-\bar{\xi}}).$$

(6.55)

With this change of variables (6.54) becomes

$$g(x) = \int_{-\infty}^{\infty} e^{-jx\bar{\xi}} F(\bar{\xi}) d\bar{\xi}$$

(6.56)

which is recognized as a simple Fourier transform. While we have not produced a space-invariant operation, nonetheless we have reduced the Mellin transform to an operation readily performed with a coherent optical data processing system.

Many potentially useful applications exist for coordinate transformation processing, including restoration of motion blurred, distorted imagery [6.4, 25], restoration of images degraded by certain aberrations [6.3, 26], and scale-invariant pattern recognition [6.22, 23].

A price in terms of input space-bandwidth product is invariably paid in coordinate transformation processing. The smallest resolution element of the input data after coordinate distortion must be no smaller than a single resolution element of the optical processor that follows. On the other hand, many resolution elements of the input data will be enlarged by the coordinate distortion, such that they occupy several resolution elements of an optical processor. Hence the space-bandwidth product of the original input data must be smaller than the space-bandwidth product of the optical processor, the exact amount by which it is smaller depending on the particular coordinate transformation to be performed [6.27].

At present the most practical method for achieving the required coordinate transformations at the processor input is by means of a nonlinear scanning device which inputs the data onto some form of integrating coherent optical light valve. A fully optical and parallel method for performing coordinate transformations on input data of high space-bandwidth products would be extremely welcome, but at this time no suitable methods are known.

6.2.5 Matrix Multiplication Implementations

We have seen in Sect. 6.1.3 that linear, space-invariant filtering operations can be implemented in discrete form. The input $f(\xi)$ is represented by a vector f consisting of samples of $f(\xi)$, the impulse response is represented by a matrix $[H]$, and the output by a vector g. Thus, to perform a linear space-variant operation, it suffices to be able to multiply a vector by a matrix. This result is valid in both the one-dimensional and the two-dimensional cases. We now consider various optical methods for implementing such matrix operations.

Multiplication of two matrices by coherent optical techniques was proposed by *Heinz* et al. [6.28] in 1970, and was demonstrated experimentally by *Jablonowski* et al. [6.29] in 1972 for the simple case of 2×2 matrices. The matrix product

$$[G] = [H][F] \tag{6.57}$$

is equivalent to a multitude of matrix-vector operations

$$g = [H]f, \tag{6.58}$$

the various vectors f being the column vectors of $[F]$ and the vectors g being the column vectors of $[G]$. The method of matrix multiplication used in these references was based upon the idea that the inner product of two vectors (e.g., any row of $[H]$ with any column of $[F]$) is equivalent to a cross-correlation operation evaluated for zero relative shift of the two functions being correlated. The matrix $[F]$ was input to a conventional coherent optical processor as a transparency containing a properly arranged set of spots of various densities. A Fourier hologram of another array of spots, representing the matrix $[H]$, was recorded and this hologram was used in the frequency plane of the processor. A properly selected set of spots at the output represented the output matrix $[G]$.

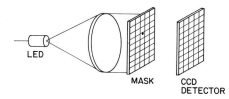

Fig. 6.10. Serial matrix-vector multiplication using a CCD detector [6.35]

Some problems with accuracy were encountered and were attributed to nonlinearities of the recording media. This method is unsuited for multiplying matrices containing negative or complex-valued elements, since only the squared modulus of the elements of [G] can be detected at the output unless rather complex interferometric measurement techniques are used. This latter problem is common to all coherent optical approaches to matrix multiplication. The reader may also wish to consult a recent paper by *Tamura* and *Wyant* [6.30], which describes an alternative coherent method for matrix-multiplication, a method very similar to one described in Sect. 6.2.3 for performing continuous operations.

Schneider and *Fink* [6.31] have described an incoherent method in which several spatially separated sources illuminate, from different angles, a transparency containing the elements of one matrix. This matrix is thereby projected onto another mask containing the elements of a second matrix. By using a cylindrical lens and a segmented spherical lens, the proper products of elements are added to produce elements of the final matrix in the output plane. Diffraction effects pose a severe limitation in this system. As the complexity of the matrices increases (i.e., the space-bandwidth product grows), "cross-talk" begins to limit the accuracy of the system. At first glance it would appear that any incoherent system such as this would be limited to matrices with real and non-negative elements. This turns out not to be the case, as we shall describe later.

Krivenkov et al. [6.32] have described an incoherent method for multiplying three 2-D matrices together. As implied by (6.38), such a system can be used to perform 2-D space-variant operations, provided the impulse response is separable in rectangular coordinates. Such is the case for the 2-D discrete Fourier transform, the 2-D Hadamard transform, and many other interesting transformations.

A very important and promising method for multiplying a matrix by a vector using incoherent light has been described by *Bocker* [6.33] and by *Bromley* [6.34]. An improved version of the system was later described by *Monahan* et al. [6.35]. This method is illustrated in Fig. 6.10. The values of the components of the vector *f* (assumed real and non-negative) are entered serially as brightness values of the light emitted by a single light-emitting diode (LED) source. This light diverges and falls upon a matrix mask. For the moment, the elements of the matrix [H] are assumed to be real and non-negative, and are represented by the areas of transparent openings in an otherwise opaque mask. The light transmitted by the mask falls upon a two-dimensional charge coupled device (CCD) light sensor, which is used in a rather unconventional integrating mode. Assume that the charges are transferred in the horizontal direction, i.e., into the paper in Fig. 6.10. The intensity associated with the first pulse of light emitted by the LED is made proportional to f_0, the first element of the input vector. The charge deposited in the kth detector element in the first vertical column of elements is then

$$Q_k^{(1)} = h_{k0} f_0 . \tag{6.59}$$

These charges are clocked one column further to the right, and a second pulse of light, this time proportional to f_1, is sent. The total charge accumulated in the kth element of the second vertical column is

$$Q_k^{(2)} = h_{k0} f_0 + h_{k1} f_1 . \tag{6.60}$$

After M pulses and M charge transfers, the total charge accumulated in the kth element of the last column of detector elements is

$$Q_k^{(M)} = \sum_{l=0}^{M-1} h_{kl} f_l , \tag{6.61}$$

which is precisely the kth element of the desired output vector g. All N components of the output vector are, in principle, available in parallel, but only after M cycles of the clock.

This matrix multiplier is capable of multiplying 500×500 matrices with real and non-negative elements by a 500 length vector with real and non-negative elements with data rates (output samples per second) of up to 10 MHz. If complex matrices and complex vectors are of interest, then (as we shall see shortly) either the matrix and vector must be shrunk, or multiple passes through the system must be performed, with different matrices on each pass. The limitation of the speed of the device lies in the CCD detector. While multiplications are performed optically, additions are performed electronically, with the speed being limited as a result.

A related but faster system has been proposed by *Goodman* et al. [6.21] and is illustrated in Fig. 6.11. In this case the vector f is input in parallel with perhaps 100 high-speed light-emitting diodes. The matrix mask is similar to that used in the previous system. At the output is a row of 100 avalanche photodiodes. Each LED illuminates a vertical column of the matrix mask and each detector senses the integral of the light transmitted through a row of the mask. Thus both multiplications and additions are performed optically. With commercially available LED's and detectors, data throughputs up to 10^{10} samples per second appear possible when the elements of both the vector and the matrix are non-negative and real. Its chief disadvantage is the limited length of the input vector, a limitation that arises simply from the difficulties associated with

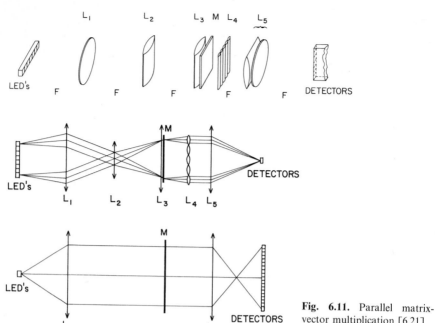

Fig. 6.11. Parallel matrix-vector multiplication [6.21]

realizing the parallel electronics. This limitation may be alleviated as more fully integrated arrays of LED's and photodetectors become available.

Finally, we describe a method by which complex matrices can be multiplied by complex vectors with incoherent systems. Let the input vector be decomposed as follows[5]:

$$f = f^{(0)} + f^{(1)} e^{j\frac{2\pi}{3}} + f^{(2)} e^{j\frac{4\pi}{3}}, \tag{6.62}$$

where $f^{(0)}$, $f^{(1)}$, and $f^{(2)}$ each contain only real and non-negative elements. Similarly, let the matrix $[H]$ be decomposed as

$$[H] = [H]^{(0)} + [H]^{(1)} e^{j\frac{2\pi}{3}} + [H]^{(2)} e^{j\frac{4\pi}{3}}. \tag{6.63}$$

If the output vector is decomposed this way too, then the three desired components of the output vector can be found by the larger matrix-vector product

$$\begin{bmatrix} g^{(0)} \\ g^{(1)} \\ g^{(2)} \end{bmatrix} = \begin{bmatrix} [H]^{(0)} & [H]^{(2)} & [H]^{(1)} \\ [H]^{(1)} & [H]^{(0)} & [H]^{(2)} \\ [H]^{(2)} & [H]^{(1)} & [H]^{(0)} \end{bmatrix} \begin{bmatrix} f^{(0)} \\ f^{(1)} \\ f^{(2)} \end{bmatrix}. \tag{6.64}$$

---

5 A decomposition into biased real and imaginary parts can also be made, but if this is done, provision must be made for evaluating and cancelling unwanted bias terms, and the output dynamic range is decreased somewhat. See [6.39].

Thus to perform the product of a complex matrix times a complex vector requires a loss by a factor of three in the length of the vector processed. This method has been successfully used to perform discrete Fourier transforms of complex input data [6.21]. Further discussions of methods for performing complex operations with incoherent light can be found in [6.36, 37].

6.3 Concluding Remarks

Relatively little attention has been paid in the past to methods for performing space-variant linear filtering operations with optical systems. There is no doubt that such operations are important in practice. It seems likely that, with the increasing attention such problems are now receiving, significant further progress will be made. A variety of known methods have been summarized here, but no doubt many other different approaches will evolve in the future. Hopefully, the above review will help in clarifying the relationship between various present and future methods, and perhaps may even stir further interest in this area.

Note in Print. Two other reviews of the subject of space-variant optical processing have been written since the original submission of this chapter [6.40]. The interested reader may wish to consult these additional reviews of the subject.

Acknowledgement. The support of the National Science Foundation in the area of space-variant optical data processing is gratefully acknowledged.

References

6.1 A.W.Lohmann, D.P.Paris: J. Opt. Soc. Am. **55**, 1007 (1965)
6.2 L.J.Cutrona: "Recent Developments in Coherent Optical Technology", in *Optical and Electro-Optical Information Processing*, ed. by J.T.Tippett, D.A.Berkowitz, L.C.Clapp, C.J.Koester, A.Vanderburgh, Jr. (MIT Press, Cambridge 1965) Chap. 6
6.3 G.M.Robbins, T.S.Huang: Proc. IEEE **60**, 862 (1972)
6.4 A.A.Sawchuk: Proc. IEEE **60**, 854 (1972)
6.5 M.A.Monahan, K.Bromley, R.P.Bocker: Proc. IEEE **65**, 121 (1977)
6.6 N.W.F.Stephans, G.L.Rogers: Phys. Ed. **9**, 331 (1974)
6.7 T.Kailath: "Channel Characterization: Time-Variant Dispersive Channels", in *Lectures on Communication System Theory*, ed. by E.J.Baghdady (McGraw-Hill, New York 1961) Chap. 6
6.8 R.J.Marks II, J.F.Walkup, M.O.Hagler: Appl. Opt. **15**, 2289 (1976)
6.9 J.W.Goodman: *Introduction to Fourier Optics* (McGraw-Hill, New York 1967)
6.10 R.J.Marks II, J.F.Walkup, M.O.Hagler: J. Opt. Soc. Am. **66**, 918 (1976)
6.11 H.C.Andrews, B.R.Hunt: *Digital Image Restoration* (Prentice-Hall, Englewood Cliffs, N.J. 1977)
6.12 O.Bryngdahl: Opt. Commun. **10**, 164 (1974)
6.13 O.Bryngdahl: J. Opt. Soc. Am. **64**, 1092 (1974)
6.14 I.S.Gibin, P.E.Tverdokhleb: "Information Processing in Optical Systems of Holographic Memory Devices", in *Optical Information Processing*, ed. by Y.E.Nesterikhin, G.W.Stroke, W.E.Kock (Plenum Press, New York 1976)

6.15 L.M.Deen, J.F.Walkup, M.O.Hagler: Appl. Opt. **14**, 2438 (1975)
6.16 T.F.Krile, R.J.Marks II, J.F.Walkup, M.O.Hagler: Appl. Opt. **16**, 3131 (1977)
6.17 W.T.Rhodes, J.M.Florence: Appl. Opt. **15**, 3073 (1976)
6.18 R.J.Marks II, J.F.Walkup, M.O.Hagler, T.F.Krile: Appl. Opt. **16**, 739 (1977)
6.19 J.W.Goodman, P.Kellman, E.W.Hansen: Appl. Opt. **16**, 733 (1977)
6.20 P.Kellman, J.W.Goodman: Appl. Opt. **16**, 2609 (1977)
6.21 J.W.Goodman, A.Dias, L.M.Woody: Opt. Lett. **2**, 1 (1978)
6.22 D.Casasent, D.Psaltis: Proc. IEEE **65**, 77 (1977)
6.23 D.Casasent, D.Psaltis: Opt. Eng. **15**, 258 (1976)
6.24 R.N.Bracewell: *The Fourier Transform and its Applications* (McGraw-Hill, New York 1965)
6.25 D.Casasent, A.Furman: Appl. Opt. **16**, 1955 (1977)
6.26 A.A.Sawchuk, M.J.Peyrovian: J. Opt. Soc. Am. **65**, 712 (1975)
6.27 D.Casasent, D.Psaltis: Opt. Commun. **23**, 209 (1977)
6.28 R.A.Heinz, J.O.Artman, S.H.Lee: Appl. Opt. **9**, 2161 (1970)
6.29 D.P.Jablonowski, R.A.Heinz, J.O.Artman: Appl. Opt. **11**, 174 (1972)
6.30 P.N.Tamura, J.C.Wyant: "Matrix Multiplication Using Coherent Optical Techniques", in
      Optical Information Processing, Vol. 83, Proc. S.P.I.E., ed. by D.Casasent, A.A.Sawchuk
      (S.P.I.E., Redondo Beach, CA 1977)
6.31 W.Schneider, W.Fink: Opt. Acta **22**, 879 (1975)
6.32 B.E.Krivenkov, S.V.Mikhlyaev, P.E.Tverdokhleb, Y.V.Chugui: "Non-coherent Optical
      System for Processing of Images and Signals", in *Optical Information Processing*, ed. by
      Y.E.Nesterikhin, G.W.Stroke, W.E.Kock (Plenum Press, New York 1976) pp. 203–217
6.33 R.P.Bocker: Appl. Opt. **13**, 1670 (1974)
6.34 K.Bromley: Opt. Acta **21**, 35 (1974)
6.35 M.A.Monahan, R.P.Bocker, K.Bromley, A.C.H.Louie, R.D.Martin, R.G.Shepard: "The
      Use of Charge Coupled Devices in Electro-Optical Processing", Proc. 1975 Intern. Conf. on
      the Application of Charge-Coupled Devices (Naval Electronics Laboratory Center, San
      Diego, Ca., Oct. 1975)
6.36 A.W.Lohmann: Appl. Opt. **16**, 261 (1977)
6.37 J.W.Goodman, L.M.Woody: Appl. Opt. **16**, 2611 (1977)
6.38 R.J.Marks II, J.F.Walkup, M.O.Hagler: IEEE Trans. on Circuits and Systems CAS-**25**, 228
      (1978)
6.39 J.W.Goodman, A.R.Dias, L.M.Woody, J.Erickson: "Some New Methods for Processing
      Electronic Image Data Using Incoherent Light", Proc. of ICO-11, Madrid, Spain (1978) p. 139
6.40 J.F.Walkup: Opt. Eng. **19**, 339 (1980);
      W.T.Rhodes: "Space-Variant Optical Systems and Processing", in Applications of the
      Optical Fourier Transform, ed. by F.Stark (Academic, New York 1981) Chap. 8

7. Nonlinear Optical Processing

S. H. Lee

With 52 Figures

In the first six chapters of this volume we have primarily been concerned with linear processing, although it has been mentioned that nonlinear processing can be performed using hybrid (optical analog/electronic digital) systems by carrying out the nonlinear portions of processing with electronics [7.1, 2]. In this chapter we discuss nonlinear optical processing. Examples of nonlinear operations which have been successfully demonstrated with optics include logarithm, exponentiation, intensity level slicing, thresholding, analog-to-digital conversion, logics and bistability. Logarithm, exponentiation, intensity level slicing and thresholding are useful image processing operations, whereas thresholding, analog-to-digital conversion, logic and bistability are important toward developing a futuristic digital optical processor. (Digital optical processing holds the promise of real-time parallel processing inherent with optics, while maintaining accuracy and flexibility inherent with digital systems.) To implement these nonlinear operations, several schemes, devices and systems are presently available. They are the half-tone screen process, theta modulation, nonlinear optical devices and systems, with and without feedback. Their principles of operation are discussed below.

7.1 Half-Tone Screen Process

When a slowly varying object function $g_i(x, y)$ is contact printed through a half-tone screen of transmission $T(x, y)$ onto a high contrast film, a half-tone image $g_m(x, y)$ consisting of a dot array results. The size of the dots is dependent upon both $g_i(x, y)$ and $T(x, y)$ as illustrated in Fig. 7.1 [7.3, 4]. This is described mathematically by

$$g_m(x, y) = \text{step}[g_i(x, y) T(x, y) - I_c]$$
$$= \begin{cases} 1 & \text{if} \quad g_i(x, y) T(x, y) \geq I_c \\ 0 & \text{if} \quad g_i(x, y) T(x, y) < I_c, \end{cases} \tag{7.1}$$

where I_c is the clipping level of the film and is determined by the film and the exposure time. By properly controlling the transmission profiles of the half-tone screens, the size of the dots in the half-tone image will be non-linearly related to $g_i(x, y)$. Then, upon low-pass filtering either by eye or with an optical processor,

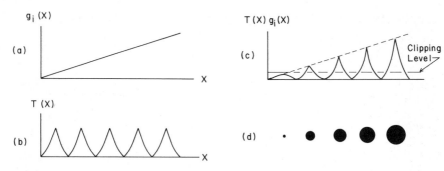

Fig. 7.1a–d. The halftone screen process [7.3]: (**a**) continuous-tone light distribution input; (**b**) transmission characteristics of the halftone screen; (**c**) light distribution falling on hard clipping film; (**d**) halftone image recorded on hard clipping film

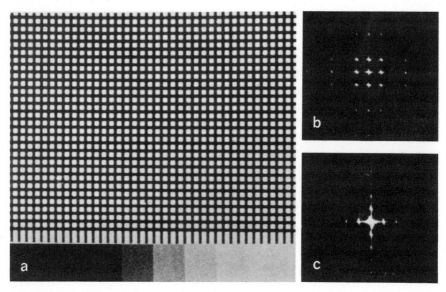

Fig. 7.2a–c. The effect of the logarithmic transformation on the Fourier spectrum. (**a**) The original pattern of two multiplied gratings perpendicular to each other. The dynamic range is from 0 to 2 in density. A step tablet at the bottom of the photo shows the grey levels; (**b**) normal spectrum of the linearly copied crossing gratings, with intermodulation; (**c**) spectrum of the logarithmically transformed crossed grating obtained using the logarithmic contact screen [7.3]

the half-tone image will yield a filtered image $g_o(x, y)$ which is nonlinearly related to $g_i(x, y)$ in a monotonic manner.

Experimentally this principle has been verified for the logarithmic transformation by modulating or coding the input image with a logarithmic contact screen, which was obtained by making a contact negative duplicate of a Kodak Gray Contact Screen (100 lines per inch, elliptical dot) on Kodak Contrast Process Ortho film [7.3]. In Fig. 7.2 an input image which is the product of two

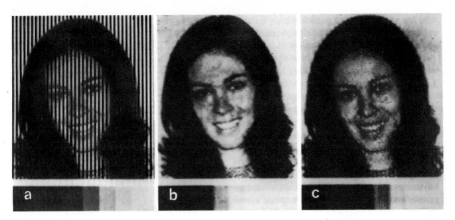

Fig. 7.3a–c. Simple logarithmic filtering. (**a**) The original pattern composed of two pictures (a girl's face and a grating) multiplicatively. The dynamic range is from 0 to 2 in density; (**b**) logarithmic filtering to remove grating; (**c**) linear filtering to remove grating [7.3, 4]

input components is converted by logarithmic transformation into the sum of the two input components. In this illustration the two input components are grey tone gratings oriented perpendicular to each other. It is noted that when the transmittance through the two crossed component gratings are recorded through the logarithmic contact screen, the resultant coded image yields a spectrum which is the superposition of the two spectra from the two component gratings, with each component grating yielding spectral contents along one spectral axis only. On the other hand, when the same transmittance through the two crossed gratings are recorded linearly without using the logarithmic screen, the resultant image yields a spectrum which is the convolution of the two spectra from the two component gratings, thus yielding the inter-modulation spectral components off-axis.

A logarithmic filtering experiment has also been performed to compare with a conventional linear filtering experiment. Figure 7.3a shows the original pattern, which is the product of a continuous tone picture (a girl's face) and a Ronchi type of grating. A simple coherent optical filtering geometry was used to filter out the spectrum of the grating by placing absorbing spots in the frequency plane at the locations of the grating harmonics. Figure 7.3b, c show the results of logarithmic filtering and linear filtering, respectively. (In both cases the same absorbing filter was used in the frequency plane.) The superior result of logarithmic filtering suggests that it should be useful in separation of multiplicative noise from signal and speckle noise reduction [7.4].

To achieve nonmonotonic nonlinear effects, higher diffraction orders from the half-tone image must be selected, instead of low-pass filtering [7.5–8]. To understand the concepts involved, consider the half-tone image to consist of many localized regions. In every localized region we have a simple rectangular grating whose grating width w is dependent on the transmittance of the original

Halftone Image

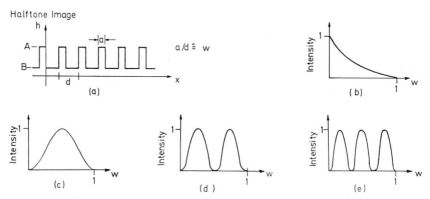

Fig. 7.4a–e. Diffraction from a rectangular grating [7.6]: (**a**) the rectangular grating; (**b**) zeroth diffraction order; (**c**) first diffraction order; (**d**) second diffraction order; (**e**) third diffraction order

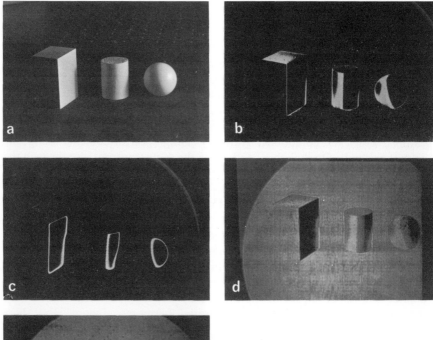

Fig. 7.5a–e. Nonmonotonic nonlinear processing with halftone screen [7.5]: (**a**) original photograph of geometrical figures to be processed; (**b**) level sliced at one setting; (**c**) level sliced at another setting; (**d**) quantified to three levels; (**e**) notch filtered (transmitting all levels except one)

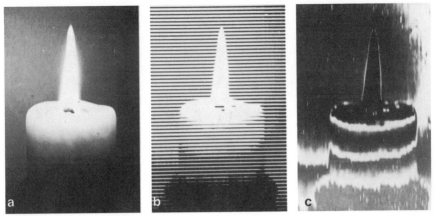

Fig. 7.6a–c. Isophotes linearly spaced in log-intensity [7.7–8] (a) original image; (b) halftone image; (c) isophotes. The three isophotes each represent a doubling of the input amplitude

Fig. 7.7a–g. Results of optical bit-plane generation compared to digital electronic bit-plane generation [7.8]. (a) Original image; (b) first bit plane optically generated; (c) second bit plane optically generated; (d) third bit plane optically generated; (e) first bit plane electronically generated; (f) second bit plane electronically generated; (g) third bit plane electronically generated

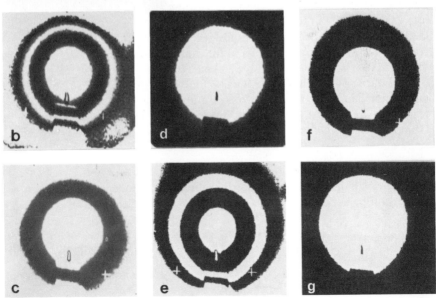

input in that region. The diffraction from the rectangular grating to higher orders will be nonmonotonically dependent on the grating width, though the dependence is monotonic at the zero-order, as shown in Fig. 7.4. Since the width of the grating in each localized area in the half-tone image is controlled by the transmittance of the original input in the same area, the diffraction to higher orders will be nonmonotonically dependent on the input also. Experimentally, the concepts have been verified for the level slicing operation (Fig. 7.5), in isophote production (Fig. 7.6) and analog-to-digital conversion (Fig. 7.7). The first diffraction order was selected for the level slicing and the third diffraction order for producing the three isophotes.

7.2 Theta Modulation Techniques

The theta modulation technique involves converting each pixel of the original object $g_i(x, y)$ into a grating cell whose grating angle θ is proportional to the amplitude transmittance of the pixel [7.9]

$$\theta(x, y) = Kg_i(x, y)$$
$$K = \pi/\text{Max } g_i. \tag{7.2}$$

Hence, the object in its theta modulated form, $g_m(x, y)$, is made up of a collection of grating cells. An example of the modulation scheme is illustrated in Fig. 7.8. When the modulated signal is illuminated with a collimated, coherent beam in a coherent optical processing system, light will be diffracted into various angles in the Fourier plane. It is interesting to note, however, that the light from all elemental gratings in $g_m(x, y)$ oriented in the same angle, which corresponds to all image elements of the same transmittance in $g_i(x, y)$, will be diffracted into one angle in the Fourier plane (Fig. 7.8c). Now, if a filter is placed in this Fourier plane whose transmission function $h(\theta)$ is a nonlinear function of the azimuth angle θ, the output image amplitude $g_o(x, y)$ will be nonlinearly related to $g_i(x, y)$. An example is given in Fig. 7.8d. The following three examples of $h(\theta)$ will further illustrate the versatility of this nonlinear processing technique.

Example 1
If the filter $h(\theta)$ has an exponential transmittance dependence on θ, the output $g_o(x, y)$ will be exponentially related to the original object $g_i(x, y)$.

Example 2
If the filter $h(\theta)$ is zero for $0 < \theta < \theta_0$ and unity for $\theta_0 < \theta < \pi$, the thresholding operation results with the elimination in the output image of any regions of amplitude transmittance below a certain value corresponding to θ_0. By varying θ_0, the threshold level can be changed.

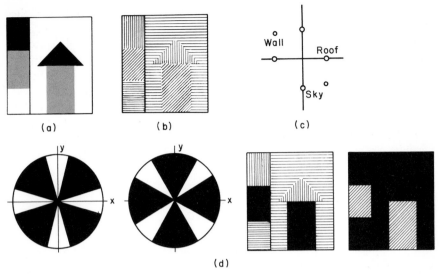

Fig. 7.8a–d. Principle of theta modulation [7.9]. (**a**) object with grey ladder; (**b**) same object in theta-modulated form; (**c**) diffraction pattern of theta-modulated object; (**d**) result of demodulation procedure for one modulated object with two different demodulation masks

Example 3

If the filter $h(\theta)$ is a slit oriented at one angle θ, the output image will be an equiamplitude or equidensity line image, i.e., in the image $g_o(x, y)$ sharp lines appear, representing the contour for one amplitude value in $g_i(x, y)$. If the filter $h(\theta)$ consists of multislits, instead of a single slit, oriented at equiangular spacings, the output image becomes a contour map of equiamplitudes.

Generally, the principles of theta modulation can be extended to include the encoding of each pixel of $g_i(x, y)$ by a corresponding grating cell whose grating frequency (instead of grating angle) is proportional to the amplitude transmittance of that pixel. In this case, the filtering will be carried out along one axis in the Fourier plane [e.g., $h(v_y)$] instead of around the azimuth angle $h(\theta)$. Moreover, it is interesting to note that once $g_i(x, y)$ is encoded by the theta or frequency-modulation techniques, the same encoded image can be processed in many ways to provide different outputs (e.g., images having different threshold levels or equidensity contours with different equidensity contour spacings). This interesting fact may be of practical importance in various image processing applications.

7.2.1 Encoding by a Scanning Interferometric Pattern System

To accomplish encoding we can design a scanning interferometric pattern system which combines the function of laser beam scanning and interferometric pattern generation (Fig. 7.9a) to produce $g_m(x, y)$ of good resolution [7.10]. The

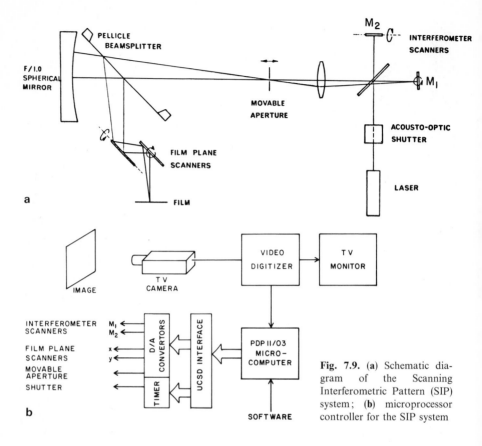

Fig. 7.9. (a) Schematic diagram of the Scanning Interferometric Pattern (SIP) system; (b) microprocessor controller for the SIP system

image $g_i(x, y)$ to be encoded is first input to a PDP 11/03 microcomputer via a video interface (Fig. 7.9b). Then, a laser beam is introduced through an acousto-optic shutter into a Michelson interferometer formed by two galvanometric mirrors and a beam splitter. Under microcomputer control, the interferometric mirrors generate an interference pattern of an orientation and frequency (corresponding to the grey level of the pixel) on an aperture, which defines the size and shape of the grating cell. An $f/1.0$ imaging system and another set of galvanometric scanners then place the grating cell at the appropriate location on the film plane. Since an entire set of fringes is recorded simultaneously, rather than scanned line by line, the writing time per cell and scanner accuracy requirements are greatly reduced.

Figure 7.10 demonstrates the complete theta modulation process. The image to be encoded (256×256 pixels $\times 20$ grey levels) is stored in the video interface for computer access (Fig. 7.10a). Under computer control, the scanning interferometric pattern system exposes a mosaic of gratings on the film, the angles and frequencies corresponding to the grey levels of the pixels they replace. A magnified portion of the encoded image is shown in Fig. 7.10b.

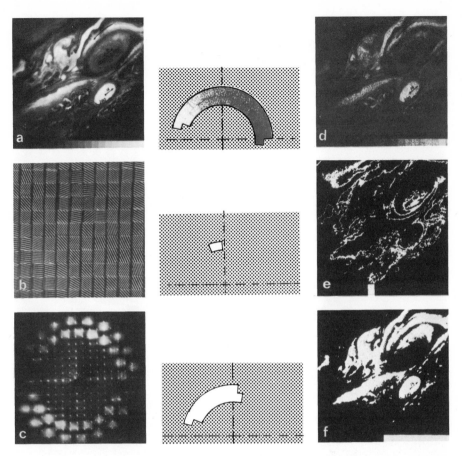

Fig. 7.10a–f. The complete theta modulation process. (**a**) The original object of 20 grey levels $g_i(x, y)$; (**b**) the central part of the coded image $g_m(x, y)$; (**c**) the Fourier transform of $g_m(x, y)$; (**d**) the contrast enhanced image obtained by exponential filtering of $g_m(x, y)$; (**e**) the result of level slicing (slicing the twelfth level); (**f**) the result of the thresholding at level 10. The central column shows the Fourier plane filters used to obtain the results of (**d**)–(**f**)

The Fourier transform of the encoded image (Fig. 7.10c) clearly shows the various grey levels. (Grey level assigment proceeds counterclockwise from the 3 o'clock position, with odd number levels on the inner circle, and even number levels on the outer.) The intensity of each spot indicates the relative frequency of occurrence of that grey level in the image. Reconstructed (demodulated) images using various Fourier plane filters are shown in Fig. 7.10d–f. In Fig. 7.10d, an exponentially incremented neutral density filter produces a contrast enhanced version of the original image. (Note the modified grey scale at the bottom of the image.) In Fig. 7.10e, the image is level sliced by transmitting a single intermediate grey level through the filter. In Fig. 7.10f, a binary mask performs

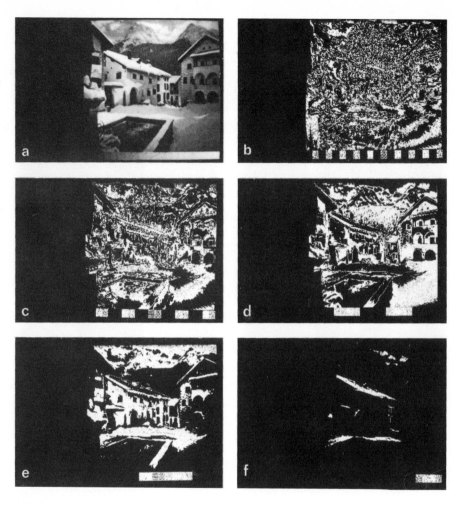

Fig. 7.11a–f. Analog-to-digital conversion. (a) Original image of 20 grey levels; (b)–(f) first to fifth bit planes. (Note the density steps at the bottom of each photograph)

thresholding, selecting the top ten grey levels. For analog-to-digital conversion, the original analog image of twenty grey levels is converted into five binary bit planes of different significance (Fig. 7.11). The nth bit plane is obtained by either blocking or transmitting each of the twenty grey levels depending on whether the nth bit of that grey level's binary equivalent is a zero or a one, respectively. For example, the least significant bit plane is obtained by blocking all the even levels (2, 4, 6...20) while transmitting all the odd levels. The most significant bit plane is obtained by blocking the first fifteen levels while transmitting levels 16 to 20.

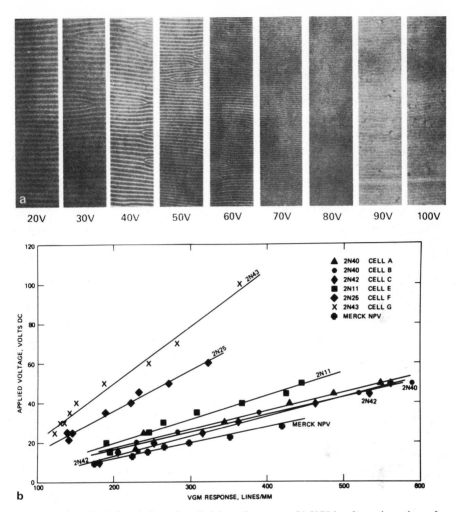

Fig. 7.12. (**a**) VGM viewed through polarizing microscope; (**b**) VGM voltage dependence for various LCs [7.12]

7.2.2 Encoding by a Variable Grating Mode Liquid Crystal Device

The results of Figs. 7.10, 7.11 were obtained from an encoded image produced by the scanning interferometric pattern system of Fig. 7.9. This system is quite fast, since one grating cell can be produced at a time rather than requiring line by line scan. However, it would be even faster to produce all grating cells simultaneously. A device which can do this is a variable grating mode (VGM) liquid crystal device [7.11, 12], although some aspects of the device's performance, such as defects in the liquid crystal cell structure, still need improvement. The primary active element of the VGM liquid crystal device is a thin

(12 μm) layer of nematic phase liquid crystal which forms periodic stripe domains in the presence of an applied static field (Fig. 7.12). The formation of the domains results in a phase grating characterized by a spatial frequency that depends on the magnitude of the voltage across the liquid crystal layer. The grating period can be optically controlled by placing a two-dimensional photoconductive layer in a series with the layer of liquid crystal to convert an optical intensity distribution into its corresponding voltage distribution.

Experimentally, the VGM liquid crystal device has been applied to demonstrate level slicing on grey tone images and logic operations on binary images. Since the principle of level slicing has just been discussed above, we shall consider only the principles of logic operations as follows: we take note first of the fact that when two binary images are used to address the input of a VGM liquid crystal device, the input intensity is limited to three levels (0, 1, 2). In response to these three intensity levels, the VGM device produces gratings of three frequencies (v_0, v_1, v_2) and contributes diffractions of the readout beam into three locations in the Fourier plane of a coherent processor (Fig. 7.13). Then, by blocking v_0 and v_1 in the Fourier plane and transmitting v_2, we obtain the AND operation at the output of the coherent processor. By blocking v_0 and transmitting v_1 and v_2, we obtain the OR operation. For the NEGATION operation, one of the binary input images must be uniformly dark and the Fourier filter allows the transmission v_0 only.

7.3 Nonlinear Devices

Nonlinear optical processing can be performed using nonlinear materials and devices. We shall describe first in this section, nonlinear devices consisting of plane parallel mirrors with object amplitude or phase modulations, or with saturable absorber between the mirrors. Then, nonlinear electro-optic devices of PROM ($Bi_{12}SiO_{20}$) or liquid crystal with or without optical feedback will be discussed.

7.3.1 Nonlinear Processing with Fabry-Perot Interferometer (FPI)

The plane parallel Fabry-Perot interferometer of Fig. 7.14 has interesting transmission characteristics which can be exploited for nonlinear processing. The transmission characteristics change depending on the kind of medium placed between the mirrors. A nonlinear operation such as contrast control has been demonstrated with amplitude modulation medium; intensity level slicing and A/D conversion have been demonstrated with phase modulation medium, and real-time image thresholding is a possibility with a saturable absorber.

a) FPI Containing Amplitude Modulation Medium

Contrast control can be achieved when the input transparency is placed between the mirrors of a FPI. Since the transmission of the FPI is described as

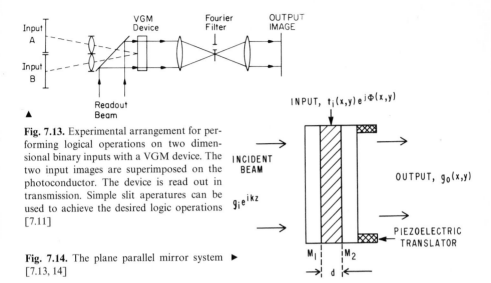

Fig. 7.13. Experimental arrangement for performing logical operations on two dimensional binary inputs with a VGM device. The two input images are superimposed on the photoconductor. The device is read out in transmission. Simple slit aperatures can be used to achieve the desired logic operations [7.11]

Fig. 7.14. The plane parallel mirror system ▶ [7.13, 14]

the coherent sum of a multiply reflected light field and the transparency placed inside the interferometer will modulate each pass, image contrast can be enhanced or reduced as the result of constructive or destructive interferences between multiple reflections depending on the mirror separation [7.13].

The output amplitudes from the coherent feedback systems of Fig. 7.14 can be easily derived:

$$\frac{g_o(x, y)}{g_i} = t_i(x, y)t_m^2[1 + r_m^2 t_i^2(x, y)e^{j\phi} + r_m^4 t_i^4(x, y)e^{j2\phi} + \ldots]$$

$$= t_i(x, y)t_m^2/[1 - r_m^2 t_i^2(x, y)e^{j\phi}],\qquad(7.3)$$

where g_i = light amplitude of input illumination, $t_i(x, y)$ = amplitude transmittance of the original image, r_m, t_m = mirror amplitude reflectance and transmittance and $e^{j\phi}$ = phase delay of light traveling between mirrors which is dependent on mirror separation. The corresponding output intensity is

$$T_c(x, y) = \frac{T_i(x, y) T_m^2}{1 + R_m^2 T_i^2(x, y) - 2R_m T_i(x, y)\cos\phi},\qquad(7.4)$$

where $T_i(x, y)$, T_m, R_m are intensity transmittances and reflectances. Therefore, by controlling the mirror separation which affects $e^{j\phi}$, the output intensities will show various contrasts (Fig. 7.15). Experimental results are illustrated in Fig. 7.16.

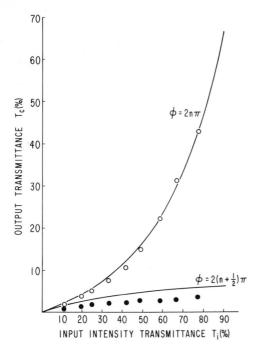

Fig. 7.15. The output vs input transfer characteristics. For the system in Fig. 7.14, $R_1 = 0.65$, $R_2 = 0.54$

b) FPI Containing Phase Modulation Medium

Intensity level slicing can be performed on images if their intensity variations are recorded as phase variations on a transparent medium located between the mirrors of a high finesse Fabry-Perot interferometer [7.14]. The transmittance, $T_p(x, y)$, of the interferometer with a phase variation $\Phi(x, y)$ recorded on the transparent medium is

$$T_p(x, y) = \frac{T_m^2}{1 + R_m^2 - 2R_m \cos[\phi + 2\Phi(x, y)]}. \tag{7.5}$$

For high finesse, large values of R_m are chosen. Then, the device acts as a narrow band filter, transmitting light only in those areas of the image where $\Phi(x, y) + (\phi/2) = n\pi$ (Fig. 7.17). If the phase variation $\Phi(x, y)$ is recorded as a monotonic function of the input intensity with a range of less than π, different values of Φ can be selected by a piezoelectric translator which controls the mirror spacing and ϕ. With $R_m = 95\%$, the full width at half maximum of T_p is about 0.1 radian, approximately 30 values of Φ or 30 gray levels of an image can be resolved within one free spectral range of interferometer scanning.

   This device can also be used to compute many different nonlinear functions of the original intensity distribution. For example, suppose the square root of the original image intensity distribution is desired and the image is recorded

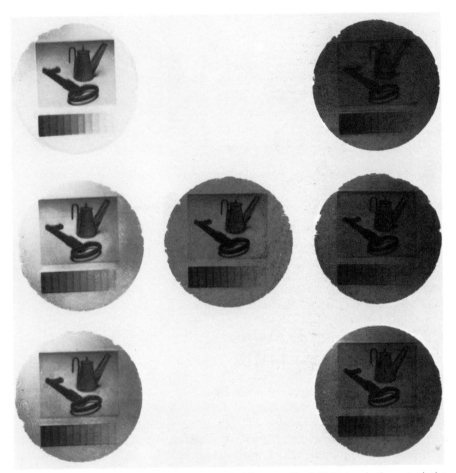

Fig. 7.16. Experimental results of contrast control with system Fig. 7.14. Picture in the center is the reference. Contrast increases in clockwise direction, starting upper right-hand corner [7.13]

such that Φ is proportional to the intensity. The output image is constructed by incrementing the mirror separation to select the various values of Φ. For each value of Φ, the intensity of the incident beam is made equal to the square root of the intensity of the original. The intensity distribution recorded in the output will thus be the square root of the intensity distribution in the input.

As another example, analog-to-digital conversion of an image can be achieved. The least significant bit plane of an eight gray level image is generated by turning the laser on when levels 1, 3, 5, and 7 are selected. It is on for levels 2, 3, 6, and 7 for recording the next most significant bit plane and it is on for levels 4, 5, 6, and 7 for recording the most significant bit plane.

Experimentally, intensity level selection was demonstrated with a bleached image recorded on a high resolution photographic plate [7.13] and with images

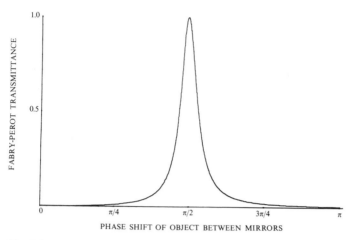

Fig. 7.17. Transmittance of high finesse Fabry-Perot interferometer. $R_m = 95\%$

recorded on photoresist and iron-doped lithium niobate [7.14]. A/D conversion was also demonstrated with photoresist (see Fig. 7.18).

c) FPI Containing Saturable Absorber

The plane parallel mirror system of Fig. 7.14 can be used for image thresholding if a good saturable absorber is placed between the mirrors (Fig. 7.19a). This device, which is also called a saturable resonator, has the interesting property that its transmission switches from a very low value to a very high value at some threshold intensity determined by the mirror reflectance and the saturable absorber, as illustrated in Fig. 7.19b [7.15–18]. If an image is projected onto the saturable resonator, light will be transmitted through it only in those areas of the image where the intensity is above the threshold level. Organic dyes are examples of saturable absorbing material. But, because of their high intensity requirements, image thresholding has yet to be experimentally demonstrated with organic dyes. With the recent development of a laser amplifier based on injection locking techniques [7.19], iodine vapor or solution and sodium vapor may offer a practical solution toward constructing a saturable resonator for image thresholding in real-time, since they promise to be 10^3 times more sensitive than the organic dyes previously studied [7.20].

7.3.2 Optical Parallel Logic Operations Performed on PROM

The principles of PROM operation have been given in Sect. 4.1.3 and are well illustrated in Fig. 4.7. The two modes of PROM operation important to logic operations are baseline subtraction and selective erasure. Baseline subtraction refers to operational dependence of the readout image contrast on the voltage

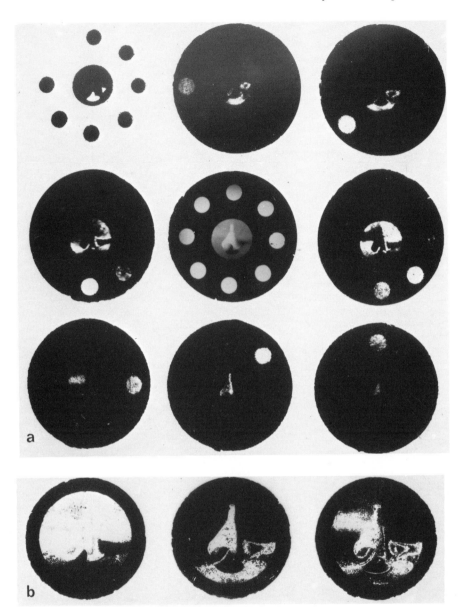

Fig. 7.18. (a) Intensity level selection with the plane parallel Fabry-Perot interferometer. The central image is grey tone input. Eight dots around the outside were given different exposures to make a grey scale. The dot in the photo at the lower right-hand corner had the highest exposure equal to the exposure of the funnel. The dot in the photo at the upper left-hand corner had zero exposure. Eight intensity levels are selected with intensity increasing from left to right and top to bottom. **(b)** Analog-to-digital conversion. A/D conversion of the grey tone image in the center of **(a)**. The most significant bit plane is on the left. The least significant bit plane is on the right. The intensity level for any point in the original can be found from these three binary images [7.14]

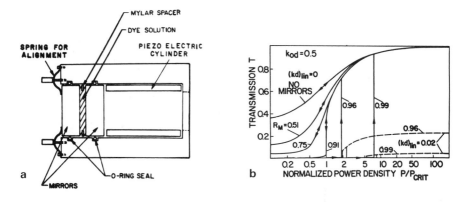

Fig. 7.19. (a) Construction of the saturable resonator. (b) Transmission of a saturable resonator versus the incident power density for different mirror reflectivities. The full curves are for the case that all losses are saturable; the dashed curves are for additional linear losses $(kd)_{lin} = 0.02$ [7.15]

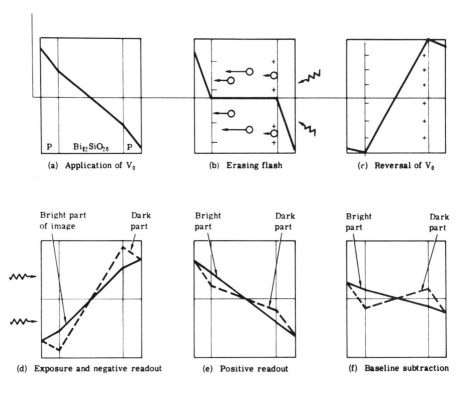

Fig. 7.20. (a)–(f) Voltage cycle used to operate PROM [7.21]. P = parylene (on the two sides of $Bi_{12}SiO_{20}$); solid lines illustrate the potential variations across the device for the bright parts of an image; dash lines illustrate the potential variations across the device for the dark parts of an image

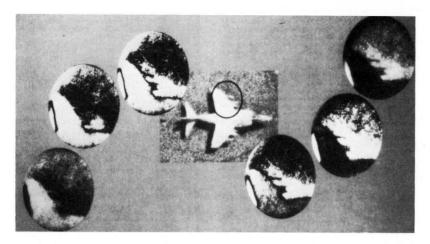

Fig. 7.21. Examples of real-time image modifications using baseline subtraction to deliberately distort the contrast of an image in order to enhance the readability of the data of interest [7.22]

applied across the device (and therefore on the voltage variation across the $Bi_{12}SiO_{20}$ crystal) during the readout portion of an operating cycle (Fig. 7.20d–f [7.21]). The series of photographs in Fig. 7.21 shows images of an airplane wing at different levels of contrast modification [7.22]. Also included is the original photograph of an airplane. The sequence of photos shows the wing in both positive and negative contrast (the first and sixth images) as well as in half and half contrast states during which most enhancement of image details occurs. Selective erasure on the PROM refers to the exposure of the PROM to a second image after baseline subtraction has been applied to the first image. Exposure to the second image changes the stored voltage pattern.

By the judicious use of selective erasure and regular baseline subtraction, a normal PROM device can be made to perform an operational NAND, NOR, OR or negative of Exclusive OR (NXOR) operations on pairs of binary patterns which sequentially illuminate the PROM. Figure 7.22 shows the experimental results for two low resolution (2×2 array) binary images. Image A consists of two horizontal strips: the upper one dark and the lower one bright. Image B consists of two vertical strips: the left one dark and the right one bright. For the OR operation, Fig. 7.22 shows a dark upper left corner, and the other three corners bright. For the NOR operation, we have the NEGATION of OR, with bright upper left corner and three dark corners elsewhere. For a NAND operation, there should be a dark lower right corner and three bright corners elsewhere. Finally, we have the two diagonal corners of upper right and lower left dark, and the other two corners bright for the NXOR operation.

PROM logic gates have the unique property that a single device can be used for any of five operations (four logic gates plus negation) with no physical

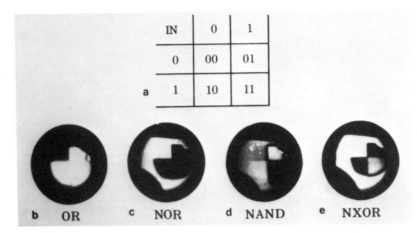

Fig. 7.22a–e. Parallel logical operations performed on a PROM. (a) Input bit patterns; (b) OR; (c) NOR; (d) NAND; (e) NXOR [7.22]

modifications. If the drive voltages and exposure shutter operation were under the guidance of a control minicomputer, relatively few logic gates could form a large computer with intermediate computing results stored in buffer memories and fedback through the same PROMs to be operated upon for different logics results. However, note that buffer memories will also need to perform a frequency conversion on the stored images because it takes blue light to write and red light to read a PROM.

7.3.3 Analog-to-Digital Conversion Performed by Birefringent Liquid Crystal Devices

The principle of operating a liquid crystal (LC) device in the birefringent mode has been described in Sect. 4.1.1. No matter whether the liquid crystal is the positive type aligned parallel to the glass substrate or the negative type aligned perpendicular to it, all LC devices operating in the birefringent mode behave much alike. A sinusoidal variation of intensity transmittance with applied voltage is observed when a LC device is placed between cross polarizers. The applied voltage is a function of the illumination of writing light, when a photoconductive film is placed at the surface of the liquid crystal cell (Fig. 4.1). Thus, the overall relationship between the intensity transmittance (for the reading beam) and the incident intensity of the writing light at any point on the device is given ideally by the sinusoidal curve shown with dashes in Fig. 7.23. Depending on the overall voltage applied across the device, the dashed curves can be driven through more than one sinusoidal cycle. The solid-line curves in Fig. 7.23 show the nonlinear transfer characteristic needed to produce the bit planes of the three bit reflected binary or Gray code and their relationship to the dashed curves of sinusoidal device characteristics. When the output of Fig.

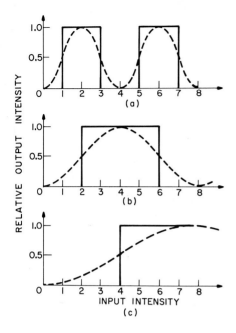

RELATIVE OUTPUT INTENSITY

(a)

(b)

INPUT INTENSITY

(c)

Fig. 7.23a–c. Nonlinear characteristic curves required for the three-bit Gray code. Solid curves are the desired characteristics for the bit plane outputs. Dotted curves are the ideal sinusoidal responses of a linear birefringent device. Parts (**a**)–(**c**) represent increasingly significant output bits [7.23]

7.23a is threshold at one half, a 1 output is produced above the threshold and a 0 output below, as shown by the curves with solid lines [7.23]. This thresholding can be done with an optical threshold device or electronically following light detection by a parallel array of sensors. The threshold output in Fig. 7.23a is the least significant bit of the three bit Gray code. The other two bits are obtained by attenuating the input intensity effectively to rescale the horizontal axis. Use of the full dynamic range (0 to 8) gives the least significant bit. Attenuating the input by a factor of 1/2 (to the range 0 to 4) gives the first cycle of the characteristic curve shown in Fig. 7.23b. The last (most significant) bit is obtained by using an attenuation of one fourth so the curves of Fig. 7.23c result. Note that any continuous input between 0 and 8 gives a unique quantized three bit output.

The system can produce these bits in parallel by placing an array of three periodically repeated attenuating strips over the writing surface of the liquid crystal device, as shown schematically in Fig. 7.24. The strips have attentuation factors of 1, 1/2, and 1/4, and the image of the strips is in registration with a parallel photodetector array with electronic thresholding in the output plane. All three bits are sensed in parallel this way. The period of the strips should be much smaller than the inverse of the maximum spatial frequency of the input picture to avoid aliasing [7.23].

To demonstrate the concept experimentally, a test target of an eight grey-level step tablet was used as input and the output in the form of three bit planes was recorded on hard-clipping film rather than a thresholding detector array (Fig. 7.25). However, a number of factors, including the optical nature of the liquid crystal and the photoconductor properties, affect the input-output

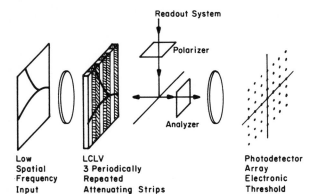

Low Spatial Frequency Input   **LCLV 3 Periodically Repeated Attenuating Strips**   **Photodetector Array Electronic Threshold**

Fig. 7.24. System for parallel A/D conversion [7.23]

Fig. 7.25. Direct analog-to-digital conversion. The eight-level analog input is shown at the top. Below is the binary-coded output in the form of three bit planes of the Gray code [7.23]

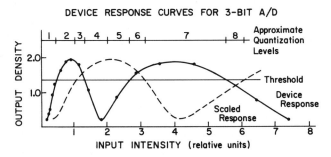

Fig. 7.26. Response curve of the liquid-crystal device used for the three bit A/D conversion. The solid curve is the measured response. The dotted curve represents the same response with a fixed attenuation of the input [7.23]

response of the LC device and produce the approximately sinusoidal characteristic (Fig. 7.26). The quasiperiodic nature of the response curves of the actual device necessitates the use of nonuniform quantization levels, which may limit A/D conversion to three bit resolution.

Another possible method to perform A/D conversion involves employing LC devices to provide the results of level slicing. When the level sliced images

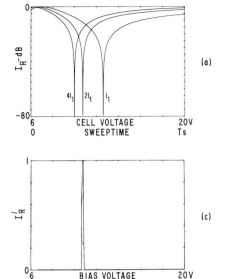

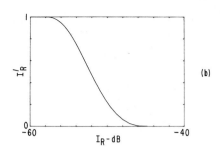

(a)

(b)

(c)

Fig. 7.27a–c. Theoretical calculations for (a) notch generator response for three separate input intensities, (b) intensity transfer function of contrast inverter, and (c) contrast inverter response to intensity notch input [7.24]

for a gray tone object are available as a function of time, they can be modulated (or weighted temporally) in a similar manner as discussed in Sect. 7.3.1 (under the section of FPI containing phase modulation medium) to yield the A/D conversion output [7.24].

Liquid crystal devices can be employed to provide the results of level slicing because a LC device, operated in the birefringent mode and placed between cross polarizers, responds to a writing light intensity I_{in} according to

$$I_R/I_1 = \sin^2 \delta(I_{in}), \tag{7.6}$$

where I_R is the light intensity reflected from the LC device due to the reading light intensity I_1 and $\delta(I_{in})$ is the phase retardation caused by I_{in}. I_R in (7.6) has a reasonably sharp dip for certain values of I_{in} and device bias voltage (Fig. 7.27a). When the device bias voltage is scanned with time, the dip in I_R will scan through different values of I_{in}. The level slicing results are obtained by allowing I_R, which is reflected from the LC device operated in the region around the dip, to illuminate a contrast inverter, which can be another LC device properly biased (Fig. 7.27b). In fact, by biasing the contrast inverter at a fixed voltage in the nonlinear region of its response, the output I'_R of the contrast inverter will exhibit a peak sharper than the dip in I_R (Fig. 7.27c). A sharpened peak helps to improve the resolution of level slicing. Now, since the I_R dip scans through I_{in} (by scanning the bias voltage across the first LC device with time), the I'_R peak will also scan through I_{in} with time, providing different level sliced images as a function of time.

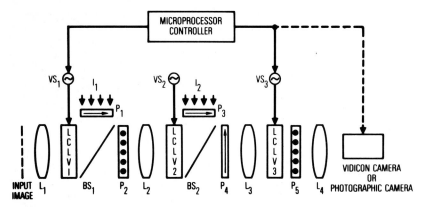

Fig. 7.28. Experimental nonlinear image processing system [7.24]

When the level sliced images are available, they can be modulated or temporally weighted by a modulator which can be a third LC device. Then, the modulator output is integrated with time by a vidicon integrator or photographic film to give the A/D conversion results. For example, the least significant bit plane of an eight gray level image is generated by turning the LC modulator on when images of 1, 3, 5, and 7 sliced levels are available for recording. The modulator is on again for recording levels 2, 3, 6, and 7 for the next most significant bit plane, and it is on for recording levels 4, 5, 6, and 7 for the most significant bit plane.

Figure 7.28 shows one possible hardware configuration for realizing the nonlinear processor. Lens L_1 images the input object onto the face of liquid crystal light valve $LCLV_1$. $LCLV_1$ provides the response of a dip as described previously. Its bias voltage is derived from a programmable voltage source VS_1 which is controlled by a microprocessor system to produce the desired sweep voltage. The read illumination I_1 for $LCLV_1$ is polarized by P_1, passes through the light valve by reflection off the beamsplitter BS_1, and then is transmitted through BS_1 to the crossed polarizer P_2. This light from P_2 which was labeled as I_R previously, is focused by lens L_2 onto the contrast inverter implemented by light valve $LCLV_2$. I_R is appropiately scaled by adjusting I_1 so that $LCLV_2$ is operated in the region of nonlinear response (for level slicing of improved resolution). The contrast inversion is completed as illumination I_2 passes through polarizer P_3, beamsplitter BS_2, the light valve $LCLV_2$, and finally out through the parallel analyzer P_4 as I'_R. The fixed bias voltage for the contrast inverter is supplied by voltage source VS_2. I'_R (i.e., the polarized light form P_4) is imaged by the system of lenses L_3 and L_4 through the temporal weighter implemented with $LCLV_3$, cross analyzer P_5, and finally onto the vidicon integrator. The bias voltage for the temporal weighter $LCLV_3$ is derived from programmable source VS_3 and is controlled in conjunction with source VS_1 by the microprocessor. The microprocessor further controls the scan time of the vidicon integrator, providing a command at the appropriate time to scan the

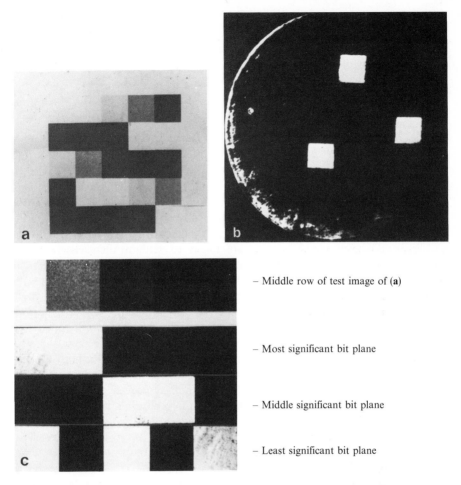

- Middle row of test image of (a)

- Most significant bit plane

- Middle significant bit plane

- Least significant bit plane

Fig. 7.29. (a) Discrete intensity test image; (b) level slice of test image; (c) A/D conversion of middle row of test image [7.23]

vidicon's photosensitive surface and sends the processed image to the display unit.

The response of the system of Fig. 7.28 to operations on variable intensity, two-dimensional input images is shown in Fig. 7.29. Figure 7.29a shows the original discrete intensity test image used as input to the system. The test pattern cells spanned a range of 23 db of input intensity with each cell being repeated a minimum of 3 times throughout the pattern. Figure 7.29b shows the integrated system response for a simple level slice operation on the test pattern.

The results of the slightly more complex operation of A/D conversion are shown in Fig. 7.29c. The third row of the input intensity test pattern of Fig. 7.29a was digitized to 3 bits, each output image bit plane requiring a separate processor operation. The microprocessor was programmed to perform the digitization over 24 db input intensity, i.e., eight distinct 3 db intensity bands spanning 24 db. The choice was somewhat arbitrary and through simple reprogramming of the microprocessor controller the digitization could be performed on the basis of absolute intensity.

7.3.4 Optical Parallel Logic Operations Performed by Twisted Nematic Liquid Crystal Devices

The electro-optic response of a twisted nematic liquid crystal (TNLC) device is different from that of a liquid crystal device operated in birefringent mode. When no voltage is applied, the nematic liquid crystal in a TNLC device is aligned parallel to the surfaces of the glass substrates and there is a 90° twist between the two boundary layers (Fig. 7.30). Then, as the voltage is increased, the intermediate liquid crystal layers begin to realign themselves in the direction of the applied field. The realignment of the intermediate LC layers will continue until the applied voltage is high enough that all of them are aligned along the field direction. After this happens, further increases in applied voltage will have no more effects on the liquid crystal alignment. Now, if we send polarized light through a TNLC device, its polarization vector will be rotated by different amounts depending on the alignment of the liquid crystal layers. The polarization rotation is 90° for zero applied voltage and 0° for high applied voltage resulting in the electro-optic response of Fig. 7.31.

The schematics of an optical parallel logic (OPAL) device which is made of twisted nematic liquid crystal and operates on two binary images, is shown in Fig. 7.32 [7.25]. The device contains a two-dimensional array of cells; each cell performs the same logic function on the two inputs it receives. Each cell is divided into two parts: one being the CdS photoconductor and the other being the patterned transparent electrode of $(In_2O_3)_{0.8}(SnO_2)_{0.2}$. The twisted nematic liquid crystal is sandwiched between this patterned transparent electrode and a continuous transparent electrode. The ijth cell receives inputs A_{ij} and B_{ij} from the ijth elements of the binary image A and B. Signal A_{ij} is absorbed by the photoconductor whereas signal B_{ij} is transmitted through the patterned transparent electrode, the twisted nematic liquid crystal and an analyzer (not shown) to provide the output C_{ij}.

In going through the liquid crystal, the polarization of the signal B_{ij} is rotated by $\pi/2$ when the voltage across the liquid crystal cell is below the threshold voltage. But, when the voltage across the liquid crystal cell is significantly larger than the threshold voltage, the twist of the liquid crystal molecules is undone as explained above, and B_{ij} passes through the liquid crystal cell with its polarization unaffected. The voltage across the liquid crystal cell is controlled by A_{ij} on the photoconductor. When the intensity of A_{ij} is zero, the photoconductor is in its dark (high impedance) state. Most of the bias voltage across the device now drops across the photoconductor and little

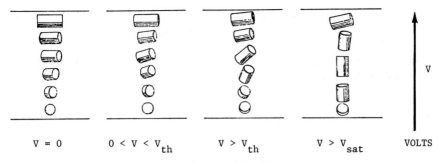

Fig. 7.30. Realignment of twisted nematic by applied voltage

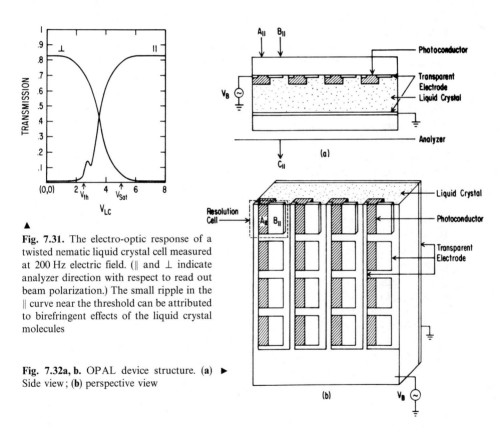

Fig. 7.31. The electro-optic response of a twisted nematic liquid crystal cell measured at 200 Hz electric field. ($\parallel$ and $\perp$ indicate analyzer direction with respect to read out beam polarization.) The small ripple in the $\parallel$ curve near the threshold can be attributed to birefringent effects of the liquid crystal molecules

Fig. 7.32a, b. OPAL device structure. (a) ▶ Side view; (b) perspective view

across the liquid crystal. But when the signal A_{ij} is maximum, it switches the photoconductor into a conducting (low impedance) state. In this case, the bias voltage is largely transferred to the liquid crystal. Thus, the interaction between the two input signals A and B is achieved cell by cell within the OPAL device (Fig. 7.33). When the output analyzer is oriented parallel to the polarization of the signal B, the output C is in logic state 1 (bright) only when both A and B are

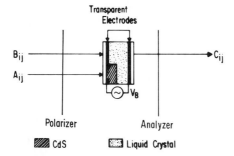

Fig. 7.33. Operation of one resolution element of an OPAL device. For analyzer ∥ to polarizer: $C_{ij} = A_{ij} \wedge B_{ij}$. For analyzer ⊥ to polarizer: $C_{ij} = \bar{A}_{ij} \wedge B_{ij}$

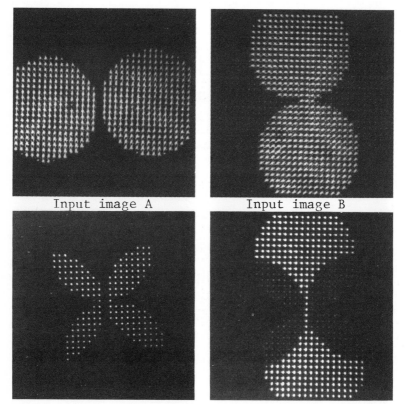

Fig. 7.34. Photographs of the inputs and the outputs of an OPAL device (consisting of 32×32 array)

in the logic state 1. This corresponds to a logic AND operation ($A \wedge B$). On the other hand, if the analyzer is oriented perpendicular to the polarization of signal B, C will be in logic state 1, if and only if A is in logic state 0 (dark) and B is in logic state 1. This corresponds to a logic operation $\bar{A} \wedge B$. Thus, the same device performs two distinct logic operations on the inputs for two orthogonal analyzer orientations (Fig. 7.34). Furthermore, the NEGATION operation can

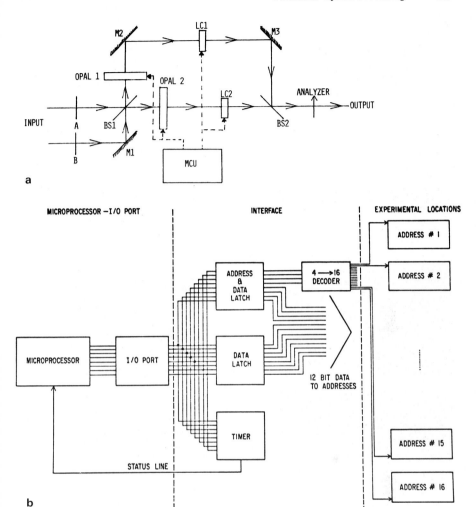

Fig. 7.35. (a) The schematic diagram of the Optical Logic Unit. M = Mirrors, BS = Beamsplitter, LC = Liquid Crystal Shutter, and MCU = Microprocessor Control Unit. —>— Optical Signal, ——→ Electronic Control Signal. **(b)** The schematic diagram of the microprocessor control unit

be considered as the special case of $\bar{A} \wedge B$, when $B = 1$, i.e., by allowing all pixels of the binary image B to be bright and orienting the analyzer perpendicular to the polarization of B, we can obtain $C = \bar{A}$.

To perform any (one of the five basic) logic operations, an optical logic unit under a microprocessor control as shown schematically in Fig. 7.35 can be designed. The optical logic unit consists primarily of two OPAL devices and two liquid crystal shutters [7.26]. The two OPAL devices are complementarily addressed by binary images A and B, i.e., A_{ij} addresses the photoconductor of

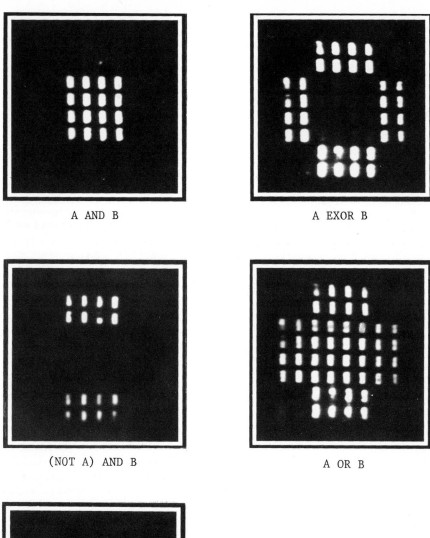

A AND B

A EXOR B

(NOT A) AND B

A OR B

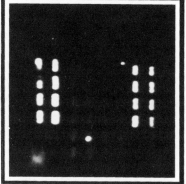

A AND (NOT B)

Fig. 7.36. Experimental results obtained with the Optical Logic Unit. (The OPAL devices used in this optic logic unit consist of an 8×8 array)

Table 7.1. Voltages applied across the four liquid crystals devices in Fig. 7.35 for performing any one of five logic operations

Voltage OPAL 1	Voltage OPAL 2	Voltage LC 1	Voltage LC 2	Output OPAL 1	Output OPAL 2	OUT OLU
V_0	V_0	V_0	V_0	$A \wedge B$	$A \wedge B$	$A \wedge B$
V_0	V_0	0	0	$\bar{A} \wedge B$	$A \wedge \bar{B}$	$A \oplus B$
0	V_0	V_0	0	0	$A \wedge \bar{B}$	$A \wedge \bar{B}$
V_0	0	0	V_0	$\bar{A} \wedge B$	0	$\bar{A} \wedge B$
V_0	0	0	0	$\bar{A} \wedge B$	A	$A \vee B$

$\wedge =$ Logic AND; $\bar{A} =$ Not A; $\oplus =$ Exclusive or; $\vee =$ OR

ijth cell of OPAL 1 and the patterned transparent electrode of ijth cell of OPAL 2. The combination of a liquid crystal shutter and the analyzer in Fig. 7.35 can be considered as a polarizer which assumes either one of two orthogonal orientations depending on the voltage applied across the shutter. Table 7.1 shows the combinations of voltage to be applied across the four liquid crystal devices in order to perform one of five logic operations. For example, when the voltage V_0 is applied across all four LC devices, the liquid crystal shutter and analyzer combination will act as an analyzer oriented parallel to the polarization of A and B. Then the output contribution from the optical branch containing OPAL 1 and LC 1 is $A \wedge B$, and that from the other branch is $B \wedge A$. As another example, when V_0 is applied across only OPAL 1, the liquid crystal shutter and analyzer combination will act as an analyzer oriented perpendicular to the polarization of A and B. Then, the output contribution from the optical branch containing OPAL 1 and LC 1 is $\bar{A} \wedge B$. From the optical branch containing OPAL 2 and LC 2, the output contribution is A because the patterned transparent electrode portion of each logic cell of OPAL 2 is addressed by A_{ij}. Without any voltage applied across OPAL 2, the twisted nematic liquid crystal rotates the polarization of A_{ij} by 90°. A_{ij} can now pass through the LC 2-analyzer combination, since no voltage is applied across LC 2 either. Finally, superposing $\bar{A} \wedge B$ on A gives $A \vee B$. Figure 7.36 shows the experimental results for the five logic operations with a horizontal bar as input A and a vertical bar as input B.

OPAL devices have also been used to implement a half-adder circuit [7.25]. A half-adder circuit in the arithmetic and logic unit of a digital computer accepts two input bits of the same significance and generates a SUM and CARRY output bit of different significance. The truth tables as well as the implementation of the half-adder circuit with AND, OR, and NEGATION gates are shown in Fig. 7.37. It can be seen from the truth table that the CARRY output is obtained by performing an AND operation on the two inputs, whereas the SUM output requires a logic operation of Exclusive OR (EXOR). This EXOR operation can be implemented with AND, OR, and

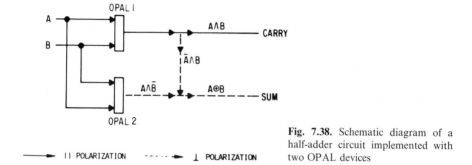

A	B	SUM	CARRY
0	0	0	0
0	1	1	0
1	0	1	0
1	1	0	1

SUM = A⊕B

CARRY = A∧B

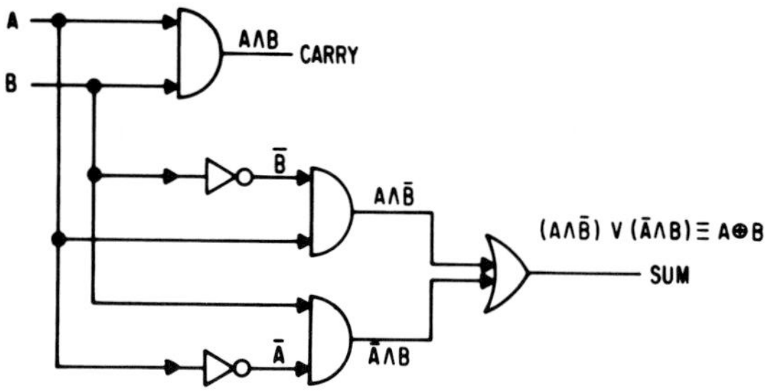

Fig. 7.37. Truth table for a half-adder circuit and its implementation with AND, OR, and NEGATION gates

Fig. 7.38. Schematic diagram of a half-adder circuit implemented with two OPAL devices

NEGATION gates using the following decomposition:

$$A \oplus B \equiv (\bar{A} \wedge B) \vee (A \wedge \bar{B}). \tag{7.7}$$

Thus, six gates are normally required to build a half-adder circuit.

To build a half-adder circuit with OPAL devices, one should take notice from Fig. 7.33 that one device can provide both $A \wedge B$ and $\bar{A} \wedge B$ outputs with two orthogonal analyzer orientations. Furthermore, the two operations encountered in the EXOR decomposition ($A \wedge \bar{B}$ and $\bar{A} \wedge B$) are mutually exclusive. Therefore, the OR operation between them can be optically performed by a simple superposition without affecting the signal level of logic state 1. Utilizing these special properties, a half-adder circuit can be constructed with two OPAL devices, instead of six. The schematic diagram of the half-adder circuit for processing binary images is shown in Fig. 7.38. OPAL device 1 (OPAL 1), in conjunction with a polarizing beamsplitter, provides the CARRY output $A \wedge B$ as well as $\bar{A} \wedge B$ (one part of the EXOR decomposition). OPAL device 2 (OPAL 2), in conjunction with a crossed analyzer, provides $A \wedge \bar{B}$ by transmitting the input signal A as the output and absorbing the input signal B in the CdS part of the cell. Combining the appropriate outputs from OPAL 1 and OPAL 2 gives the SUM output ($A \oplus B$). Figure 7.39 details the actual

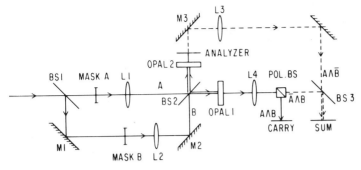

Fig. 7.39. Optical arrangement of a half-adder circuit

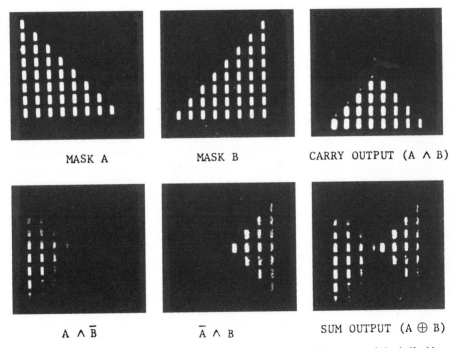

Fig. 7.40. Photographs of the inputs and the CARRY and the SUM output of the half-adder circuit. The two parts of the EXOR operation are also shown. The SUM output is a simple superposition of these two parts

optical configurations of a half-adder circuit using lenses, mirrors and be-amsplitters (polarizing and ordinary). The photographs of the experimental results obtained are displayed in Fig. 7.40. Figure 7.40a contains the two inputs and the CARRY output of the half-adder circuit. Figure 7.40b shows the photographs of the two parts of the EXOR decomposition as well as the SUM output which is a simple superposition of these two parts.

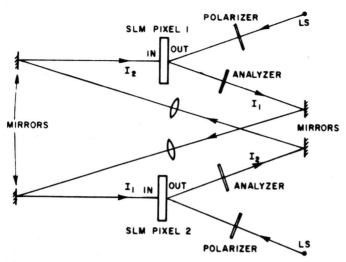

Fig. 7.41. Schematic diagram of an elementary two-dimensional optical flip-flop device [7.28]

7.3.5 Bistable Optical Spatial Devices Using Direct Optical Feedback

Both birefringent and twisted nematic liquid crystal devices have been employed to demonstrate bistable performances. Bistable devices are important components in a digital processor for carrying out memory functions (e.g., flip-flop and latching).

The optical arrangement employing birefringent liquid crystal devices of Fig. 4.1 to demonstrate an optical flip-flop is shown in Fig. 7.41. Linear polarized light of fixed intensity is reflected from the right-hand side of each LC spatial light modulator (SLM) pixel. The polarizations and hence the intensities I_1 and I_2 of the reflected beams, after they pass through the cross polarizers, are controlled by the intensities of light beams incident on the left-hand (input) side of the LC SLM pixels. The inputs and outputs of the two LC SLMs are also cross-coupled with a simple imaging arrangement [7.28].

Let the steady state relationship between I_1 and I_2 be given by

$$I_1 = \psi(I_2), \tag{7.8a}$$

where the functional form of ψ depends on the birefringent behavior of the liquid crystal (7.6), the photoconductor characteristics and the bias voltage applied across the SLM 1. Since the inputs and outputs of SLM 1 and SLM 2 are cross-coupled by optical feedback, the second relationship between I_1 and I_2 is

$$I_2 = \psi(I_1), \tag{7.8b}$$

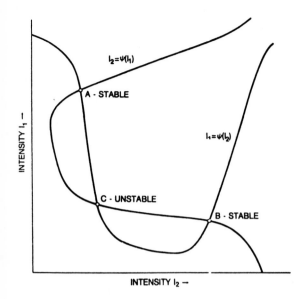

INTENSITY $I_1 \rightarrow$

$I_2 = \psi(I_1)$

A - STABLE

$I_1 = \psi(I_2)$

C - UNSTABLE

B - STABLE

INTENSITY $I_2 \rightarrow$

Fig. 7.42. Graphical solution of equilibrium states for the two-dimensional optical flip-flop [7.28]

where ψ is the response function of SLM 2 identical to the response of SLM 1 in (7.8a). The static equilibrium states of this system are the solutions to these two simultaneous equations. The solution is depicted graphically in Fig. 7.42, when the equilibrium states A, B, C are at the intersections of the two response curves generated from (7.8). To determine which of these states are stable, the time-dependent perturbation theory can be applied to give the following stability criteria:

$$\left[\frac{d\psi_1(I_1)}{dI_1}\right]\left[\frac{d\psi_2(I_2)}{dI_2}\right] \begin{array}{l} <1 \text{ stable} \\ >1 \text{ unstable}. \end{array} \qquad (7.9)$$

Applying these criteria, it is seen that states A and B are stable whereas state C is not. Switching a pixel in this system from one stable state to another can be accomplished by temporary blocking of one of the feedback paths or by addressing that pixel on one of the SLMs with a light beam. Thus, every pixel in this two-dimensional optical flip-flop can be individually set in the desired stable state.

Experimentally this system has been implemented with one SLM by dividing it into two parts and cross-coupling the inputs and outputs of these parts by a single imaging system (Fig. 7.43). The oscilloscope traces of signal I_1 and I_2 are shown in Fig. 7.44, when one pixel of the flip-flop is switched from state A (I_1 high, I_2 low) to state B (I_1 low, I_2 high). This is accomplished by temporary blocking of the feedback signal to SLM 2. The system assumes a transient state B' (both I_1 and I_2 high) and then relaxes to the stable state B when the feedback is restored.

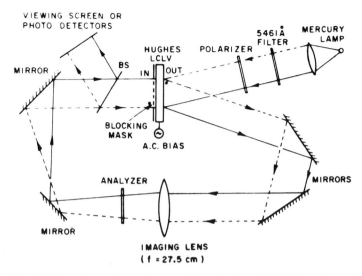

Fig. 7.43. Experimental setup for a two-dimensional optical flip-flop using one SLM. The output is imaged with unit magnification and *inversion* onto the input. This effectively divides the SLM into two parts and cross-couples their inputs and outputs (notice the dotted and the solid line crossing each other) [7.28]

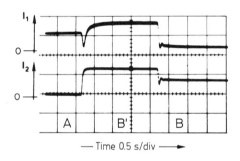

Fig. 7.44. Oscilloscope traces of pixel intensities $I_1(t)$ and $I_2(t)$ as they are being switched from state A to state B' to state B. The response time of the SLM is about 30 ms [7.28]

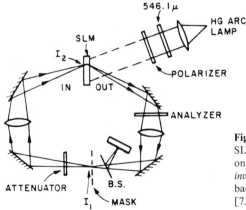

Fig. 7.45. Experimental setup for a single SLM feedback. Output of the SLM is imaged onto input with unit magnification and *no inversion*. Thus the output of a pixel is fed back to its own input through an attenuator [7.29]

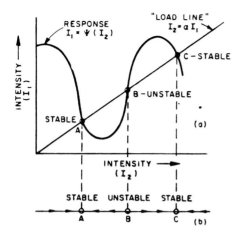

Fig. 7.46. (a) Graphical determination of equilibrium states. (b) Determination of the stability of the equilibrium states [7.29]

Similar bistable characteristics were obtained by using a single SLM and imaging the output of one pixel onto its input through an attenuator (Fig. 7.45). In this case, the two simultaneous equations to be solved for finding the equilibrium states are:

$$I_1 = \psi(I_2)$$
$$I_2 = \alpha I_1, \ 0 \leqq \alpha \leqq 1, \tag{7.10}$$

where $\psi(I)$ is the steady state response of the SLM mentioned in (7.8) and α is the attenuation constant of the feedback loop [7.29]. Figure 7.46 shows how the solution for these simultaneous equations can be obtained graphically. The stability of the three equilibrium states is determined by considerations similar to those presented for the two SLM system. Thus both of these schemes provide a two-dimensional array of optical flip-flops.

To demonstrate optical latching and thresholding, the OPAL device of Fig. 7.32 containing twisted nematic liquid crystal has been employed with optical feedback. The electro-optic response of twisted nematic liquid crystal (Fig. 7.31) has several desirable features which are helpful for obtaining a bistable performance: 1) a well-defined threshold voltage, 2) relatively sharp transition from OFF to ON state and 3) saturation. Another interesting feature associated with the device of Fig. 7.32 is the particular arrangement of the photoconductor and electrodes, which permits various sizes and shapes of patterned transparent electrode design to impedance match the photoconductor to the liquid crystal for optimizing the device performance (e.g., the area inside the hysteresis loop) or for controlling the threshold level. Figure 7.47a shows the schematic diagram of one pixel of a two-dimensional device with optical feedback.

The analysis of this system can be carried out by relating the transmission T of the liquid crystal to the voltage across it, V_{LC}, in two ways [7.27, 30]. The first relationship is an experimentally determined electro-optic response of the

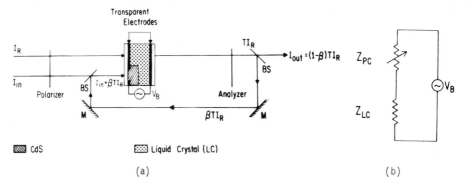

Fig. 7.47. (a) The schematic diagram of one pixel of an OPAL device with optical feedback [7.30]. I_{in} = Input Signal, I_{out} = Output Signal, I_R = Readout Signal, BS = Beam Splitter, T = Transmission of the device, β = Feedback Coefficient, V_B = Bias Voltage and M = Mirror. **(b)** The electrical equivalent circuit for a single pixel

twisted nematic liquid crystal for a parallel analyzer (Fig. 7.31). The second relationship is obtained by noting that the voltage V_{LC} depends on the light intensity on CdS. Employing the equivalent circuit of a resolution cell shown in Fig. 7.47b, one can obtain

$$V_{LC}(I_{pc}) = V_B\{Z_{LC}/[Z_{LC} + Z_{pc}(I_{pc})]\}, \tag{7.11}$$

where V_B is the bias voltage across the device, Z_{LC} is the impedance of the liquid crystal and Z_{pc} is the photoconductor impedance. One can also express Z_{pc} as:

$$Z_{pc} = Z_d/(1 + SI_{pc}), \tag{7.12}$$

where Z_d is the dark impedance and S is the photosensitivity of the photoconductor. Including the contribution from optical feedback, the light intensity on the photoconductor is:

$$I_{pc} = I_{in} + \beta T I_R, \tag{7.13}$$

where β is the feedback coefficient. Combining (7.11–13), we have

$$V_{LC}(T) = V_B\left[\frac{1 + S(I_{in} + \beta T I_R)}{1 + R + S(I_{in} + \beta T I_R)}\right], \tag{7.14a}$$

where R is the impedance ratio (Z_d/Z_{LC}). Inverting (7.14a), T can be expressed as a function of V_{LC}:

$$T(V_{LC}) = \frac{R}{S\beta I_R}\left[\left(\frac{V_{LC}}{V_B - V_{LC}}\right) - \frac{1}{R}\right] - \frac{I_{in}}{\beta I_R}. \tag{7.14b}$$

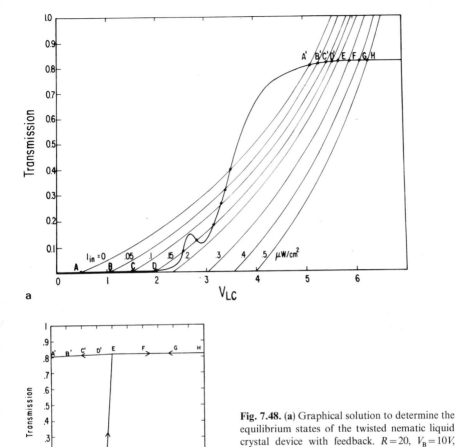

Fig. 7.48. (a) Graphical solution to determine the equilibrium states of the twisted nematic liquid crystal device with feedback. $R=20$, $V_B=10V$, $\beta=0.2$, $I_R=5\,\mu W/cm^2$. (b) The input-output characteristics obtained from the graphical solution in (a)

From this relationship we can generate a family of $T-V_{LC}$ curves, one for a different value of I_{in} (Fig. 7.48a). The equilibrium states of the feedback system for a particular value of I_{in} can be determined graphically by noting the points of intersection of the two curves corresponding to the two $T-V_{LC}$ relationships just discussed (Fig. 7.48a). The normalized input-output characteristics of the system are given in Fig. 7.48b, as I_{in} is increased from 0 to 1 and then back to 0. For zero value of I_{in}, the system can be in either a high transmission state (A') or a low transmission state (A) depending on its history. The resetting (from A' to A) can be accomplished by a momentary interruption in the feedback loop or in the bias voltage. The hysteresis characteristics shown in Fig. 7.48b make this system useful as an optical latch for binary images. The sharp transition

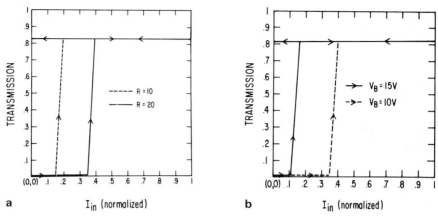

Fig. 7.49. The input-output characteristics of the twisted nematic liquid crystal device with feedback calculated for (**a**) different values of R, (**b**) different values of V_B

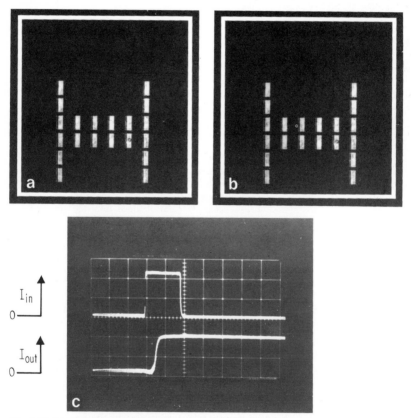

Fig. 7.50. (**a**) The input image, (**b**) the output of the OPAL device with feedback, and (**c**) the oscilloscope traces for the input and output optical signals of a single cell in the OPAL device with feedback (horizontal scale 0.2 sec/div.)

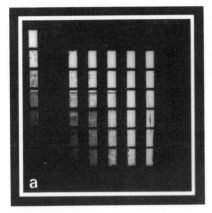 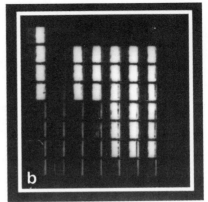

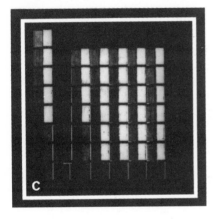

Fig. 7.51. (a) The grey scale input image; (b) the output of the OPAL device with feedback for $V_B = 15V$; (c) the output of the OPAL device with feedback for $V_B = 17V$

from OFF to ON state enables the system to also perform thresholding on a gray scale image. Different threshold levels can be obtained by changing the bias voltage V_B and the impedance ratio R (Fig. 7.49). R is controllable by changing the size and shape of the patterned transparent electrode between the CdS and the liquid crystal layers.

Experimentally, both optical latching and thresholding have been demonstrated by an 8×8 device with optical feedback. Figure 7.50a, b show the input and output of optical latching in two dimensions. Figure 7.50c shows the oscilloscope traces of the input and output optical signals of a single cell in the OPAL device with optical feedback. The output signal remains in the ON state even after the input signal is terminated, indicating latching behavior. Optical thresholding in two dimensions is shown in Fig. 7.51. The input is an image of eight grey levels and consists of the density step tablet on the left column and the image containing an inverted L shaped equidensity profile on the right. The threshold level was changed by controlling V_B.

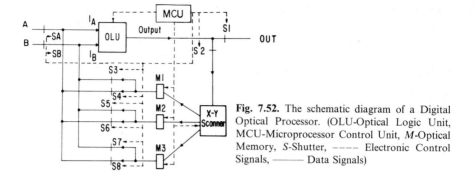

Fig. 7.52. The schematic diagram of a Digital Optical Processor. (OLU-Optical Logic Unit, MCU-Microprocessor Control Unit, *M*-Optical Memory, *S*-Shutter, ---- Electronic Control Signals, —— Data Signals)

7.4 Concluding Remarks

A number of schemes, devices and systems for nonlinear optical processing have been discussed in this chapter. They are the half-tone screen process, theta modulation technique, FPI containing various media between mirrors and electro-optic (PROM and liquid crystal) devices with or without feedback. Nonlinear optical operations useful for image processing such as logarithm, exponentiation, intensity level slicing and thresholding have been demonstrated. They certainly help to expand the horizon of optical image processing. Nonlinear optical operations important for developing a futuristic digital optical processor such as thresholding, A/D conversion, logics and bistability have also been shown to be feasible.

Figure 7.52 shows a schematic diagram of a digital optical processor which requires a minimum of hardware [7.26]. It accepts binary images converted from analog images by any one of the A/D conversion techniques described in this chapter, and processes the binary images through an Optical Logic Unit like the one shown in Fig. 35a as an example. The digital optical processor must also contain optical memory (e.g., one of those described in Sect. 7.3.5) to store the intermediate results of computation, and a control unit (Fig. 7.35b) to direct the operation of all the components mentioned above. Complex digital operations can be achieved by using the optical logic unit and optical memory repeatedly under the direction of the control unit.

At present, many researchers consider that it is highly risky to develop an all-optical digital computer because much research effort in developing fast, nonlinear optical devices (e.g., optical logic and memory devices) would be needed, besides having to solve the new problems of computer architecture so that the high degree of parallelism in processing offered by an all-optical digital system can be fully utilized. Moreover, the all-optical digital computing research faces serious competition from the all-electronic digital computing, whose processing capacity and speed have been advancing at phenomenal rates. However, some other researchers consider that the high degree of parallelism an all-optical computer has to offer will be hard to reach in the foreseeable future by an all-electronic computer. The potential development of new computer architecture which can utilize the high degree of parallelism may by itself also justify research on all-optical systems [7.31]. When the nonlinear optical device technology is further advanced, digital optical processors of higher complexity and speed may then be developed.

Note in Print. Optical image thresholding using a saturable Fabry-Perot resonator filled with a photochromic fulgide in toluene has been demonstrated since the original submission of this chapter [7.32]. The critical power density of thresholding at 5145 Å was 700 mW/cm^2.

Acknowledgement. The support of the National Science Foundation and the Airforce Office of Scientific Research in the area of nonlinear optical processing is gratefully acknowledged.

References

7.1   G. Häusler, A. Lohmann: Opt. Commun. **21**, 365 (1977)

7.2   A. Lohmann: Opt. Commun. **22**, 165 (1977)

7.3   H. Kato, J. W. Goodman: Opt. Commun. **8**, 378 (1973)

7.4   H. Kato, J. W. Goodman: Appl. Opt. **14**, 1813 (1975)

7.5   A. A. Sawchuk, S. R. Dashiell: Opt. Commun. **15**, 66 (1975)

7.6   A. W. Lohmann, T. C. Strand: Proc. Electro-Opt. System Design/Intern. Laser Conf., Anaheim **16** (1975)

7.7   T. C. Strand: Opt. Commun. **15**, 60 (1975)

7.8   T. C. Strand: Ph. D. Thesis, University of California, San Diego, CA (1976)

7.9   J. D. Armitage, A. W. Lohmann: Appl. Opt. **4**, 399 (1965)

7.10 R. Sandstrom, S. H. Lee: To be published

7.11 B. H. Soffer, D. Boswell, A. M. Lackner, P. Chavel, A. A. Sawchuk, T. C. Strand, A. R. Tanguay, Jr.: Proc. SPIE **232**, 128 (April, 1980)

7.11a A. Armand, A. A. Sawchuck, T. C. Strand, D. Boswell, B. H. Soffer: Opt. Lett. **5**, 398 (1980)

7.12 B. H. Soffer, D. Boswell, A. M. Lackner, A. R. Tanguary, Jr., T. C. Strand, A. A. Sawchuk: Proc. SPIE **218**, 81 (1980)

7.13 S. H. Lee, B. Bartholomew, J. Cederquist: Proc. SPIE **83**, 78 (1976)

7.14 B. Bartholomew, S. H. Lee: Appl. Opt. **19**, 201 (1980)

7.15 E. Spiller: J. Appl. Phys. **43**, 1673 (1972)

7.16 J. W. Austin, L. G. DeShazer: J. Opt. Soc. Am. **61**, 650A (1971)

7.17 T. Venkatesan, S. L. McCall: Appl. Phys. Lett. **30**, 282 (1977)

7.18 H. M. Gibbs, S. L. McCall, T. Venkatesan: Phys. Rev. Lett. **36**, 1135 (1976)

7.19 R. Akins, S. H. Lee: Appl. Phys. Lett. **35**, 660 (1979)

7.20 S. H. Lee: In *Optical Information Processing*, ed. by Yu. E. Nesterikin, G. W. Stroke, W. E. Kock (Plenum Press, New York 1976) p. 265

7.21 R. Sprague, P. Nisenson: Proc. SPIE **83**, 51 (1976)

7.22 B. Horwitz, F. Corbett: Opt. Eng. **17**, 353 (1978)

7.23 A. Armand, A. A. Sawchuk, T. C. Strand, D. Boswell, B. H. Soffer: Opt. Lett. **5**, 129 (1980)

7.24 J. D. Michaelson, A. A. Sawchuk: Proc. SPIE **218**, 107 (1980)

7.25 R. A. Athale, S. H. Lee: Opt. Eng. **18**, 513 (1979)

7.26 R. A. Athale, H. S. Barr, S. H. Lee, B. J. Bartholomew: Proc. SPIE **241**, 149 (1980)

7.27 R. P. Akins, R. A. Athale, S. H. Lee: Opt. Eng. **19**, 347 (1980)

7.28 U. Sengupta, U. Gerlach, S. Collins: Opt. Lett. **3**, 199 (1978)

7.29 U. H. Gerlach, S. A. Collins, U. K. Sengupta: 1979 IEEE/OSA Conf. on Laser Engineering Applications, 18.4; Opt. Eng. **19**, 452 (1980)

7.30 R. Athale, S. H. Lee: Appl. Opt. **20**, 1424 (1981)

7.31 S. H. Lee: Report of a workshop on Optical Computing Systems held at Carnegie-Mellon University and sponsored by the National Science Foundation (Sept. 14–15, 1972)

7.32 Y. Mitsuhashi: Opt. Lett. **6**, 111 (1981)

Subject Index

Topics in
Applied Physics

Founded by H. K. V. Lotsch

Volume 23

Optical Data Processing

Applications

Editor: D. Casasent

1978. 170 figures, 2 tables. XIII, 286 pages
ISBN 3-540-08453-3

Contents:
D. Casasent, H. J. Caulfield: Basic Concepts. – *B. J. Thompson:*
Optical Transforms and Coherent Processing Systems – With
Insights From Cristallography. – *P. S. Considine,*
R. A. Gonsalves: Optical Image Enhancement and Image
Restoration. – *E. N. Leith:* Synthetic Aperture Radar. –
N. Balasubramanian: Optical Processing in Photo-
grammetry. – *N. Abramson:* Nondestructive Testing and
Metrology. – *H. J. Caulfield:* Biomedical Applications of
Coherent Optics. – *D. Casasent:* Optical Signal Processing.

This is an updated summary of the present status of the
rapidly advancing field of optical data processing. It is intended for those researchers presently engaged in various aspects
of optical processing, or for those in any type of data processing, who may be contemplating the use of optical processing techniques. Anyone who desires to know what has
recently been achieved in this area – and why, will find this
book indispensable.

Volume 41

The Computer in Optical Research

Methods and Applications

Editor: B. R. Frieden

1980. 92 figures, 13 tables. XIII, 371 pages
ISBN 3-540-10119-5

Contents:
B. R. Frieden: Introduction. – *R. Barakat:* The Calculation of
Integrals Encountered in Optical Diffraction Theory. –
B. R. Frieden: Computational Methods of Probability and
Statistics. – *A. K. Rigler, R. J. Pegis:* Optimization Methods in
Optics. – *L. Mertz:* Computers and Optical Astronomy. –
W. J. Dallas: Computer-Generated Holograms.

The ever-increasing use of digital computers in optics
research has made an overview of computer-based research
methods long overdue. This volume introduces optics researchers and students to those methods which have been
developed to solve important problems in the field. The
numerous examples included from both elementary and
advanced research make it clear just how simple and effective
the algorithms really are.

Springer-Verlag
Berlin
Heidelberg
New York

Y. I. Ostrovsky, M. M. Butusov,
G. V. Ostrovskaya

Interferometry by Holography

1980. 184 figures, 4 tables. X, 330 pages
(Springer Series in Optical Sciences,
Volume 20)
ISBN 3-540-09886-0

Contents:
General Principles: Interference of Light.
Optical Interferometry. Holography. Holographic Interferometry. – Experimental Techniques: Light Sources. Hologram Recording
Materials. Setups. Experimental Aspects. –
Investigation of Transparent Phase Inhomogeneities: Features of Holographic Interferometry of Transparent Objects. Sensitivity
of Holographic Interferometry and Methods
of Changing It. Holographic Diagnostics of
Plasma. Use of Holographic Interferometry
in Gas-Dynamic Investigations. – Investigation of Displacements and Relief: The Process of Interference-Pattern Formation in
Holography. Methods of Interpreting Holographic Interferograms when Displacements
are Studied. Investigation of Surface Relief.
Flaw Detection by Holographic Interferometry. – Holographic Studies of Vibrations:
Influence of Object Displacement on the
Brightness of the Reconstructed Image. – The
Powell-Stetson Method. The Stroboholo-
graphic Method. Phase Modulation of the
Reference Beam. Determining the Phases of
Vibrations of an Object. – References. –
Subject Index.

Holography in Medicine and Biology

Proceedings of the International Workshop,
Münster, Federal Republic of Germany,
March 14–15, 1979
Editor: G. v. Bally
1979. 240 figures, 2 tables. IX, 269 pages
(Springer Series in Optical Sciences,
Volume 18)
ISBN 3-540-09793-7

Contents:
Introductory Survey. – Holography in Ortho-
pedics. – Moiré-Topography. – Holography
in Biology. – Holography in Radiology. –
Holography in Ophthalmology. – Holography
in Urology. – Holography in Dentistry. –
Holography in Otology. – Acoustical Holography. – Special Holographic Techniques. –
Index of Contributors.

B. Saleh

Photoelectron Statistics

With Applications to Spectroscopy and Optical
Communication
1978. 85 figures, 8 tables. XV, 441 pages
(Springer Series in Optical Sciences,
Volume 6)
ISBN 3-540-08295-6

Contents:
Tools from Mathematical Statistics:
Statistical Description of Random Variable
and Stochastic Processes. Point Processes. –
Theory: The Optical Field: A Stochastic Vec-
tor Field or, Classical Theory of Optical
Coherence. Photoelectron Events: A Doubly
Stochastic Poisson Process or Theory of
Photoelectron Statistics. – Applications:
Applications to Optical Communication.
Applications to Spectroscopy.

M. Young

Optics and Lasers

An Engineering Physics Approach
1977. 122 figures, 4 tables. XIV, 207 pages
(Springer Series in Optical Sciences,
Volume 5)
ISBN 3-540-08126-7

Contents:
Ray Optics. – Optical Instruments. – Light
Sources and Detectors. – Wave Optics. –
Interferometry and Related Areas. – Holo-
graphy and Fourier Optics. – Lasers. – Electro-
magnetic and Polarization Effects.

Springer-Verlag
Berlin
Heidelberg
New York